EDITOR'S PREFACE

This is the second monograph in this series devoted to the
subject of soil piping. No. 1 was a research report
devoted to intensive field work undertaken by the Institute
of Hydrology in an experimental catchment in mid-Wales.
The concern was with the hydrological significance of soil
pipes. In the present monograph the author provides an
extended research review of a range of literature on piping
from geomorphology, hydrology, soil conservation and civil
engineering. The review synthesises evidence on distribution, frequency, severity, morphometry and causal mechanisms
of piping. It provides an assessment of current knowledge
and research. It indicates that soil piping can be a major
process in soil erosion and in landsurface drainage network
development, but also that it has been very largely ignored.
Suggestions are made regarding the fundamental significance
of piping to established theories of hillslope hydrology and
geomorphology, stream response and process dominance domains.

S. Trudgill

AUTHOR'S PREFACE

When I sat down on the banks of the Burbage Brook in the
English Peak District to eat my sandwich lunch twelve years
ago almost to the day and first noticed holes in the
opposite bank which appeared to have issued water, I was
unaware of the large 'hidden' literature on these features or
indeed that I was looking at a 'soil pipe'. But I felt
strongly that I had found a feature which could be very
significant for hydrology and fluvial geomorphology. Dick
Chorley's recollection of a paper by Garald Parker started
me on the long road to this review. In the meantime, as in
so many subjects, the literature has mushroomed, so that
approximately half the references in the bibliography post-
date that fateful lunch-break.

My purpose in writing this review is to indicate the
currently accepted significance of piping to hydrologists
and geomorphologists and perhaps to indicate some ways in
which this might be extended in the near future. In so
doing, I have been forced to stray outside this immediate
field into the applied work of dam engineers and soil
conservationists which respectively accounts for a quarter
and an eighth of all the literature used in this review.
Whilst I am certain that I have not reviewed this work in
the same way as the respective specialists would have done,
I hope that I have grasped the essentials of relevance to
my central theme.

J.A.A. Jones
Aberystwyth
February 1980

iii

ACKNOWLEDGEMENTS

I have been helped immeasurably in compiling this review
not only by the very efficient and painstaking Inter-Library
Loans Service of my own College but also by the readiness
of my correspondents to discuss their observations, to send
me summaries of their own work and to lead me to literature
I was otherwise unaware of. Geo Abstracts also proved an
invaluable aid in this respect.

I must mention in particular: G.D.Aitchison, CSIRO Soil
Mechanics Section, Australia, for supplying numerous CSIRO
reports; D.J. Bowden, Colchester Institute of Higher
Education; Dr. R.B. Bryan, Department of Geography, University of Toronto; Peter Charman, New South Wales Soil Conser-
vation Service, for guiding me through much of the Service's
work; Dr. R.Cryer, Department of Geography, Sheffield
University; Dr. Z Czeppe, Geographical Institute of the
Jagiellonian University, Krakow, for a guide to East European
literature; Professor J. De Villiers, Department of Soil
Science, University of Rhodesia; Dr. D.P. Drew, Department
of Geography, University of Dublin; Graham Humphreys,
Department of Geology, Reading University; Dr.E.A. Jenne,
U.S. Geological Survey; Professor G. Kassiff, Israel
Institute of Technology; Professor M. Marker, Department of
Geography, University of Fort Hare, S. Africa; Dr. M.D.
Newson, Institute of Hydrology; Dr. E.D. Ongley, Department
of Geography, Queen's University, Ontario; G.G. Parker,
formerly U.S. Geological Survey; the late Dr. H.T.U. Smith,
Department of Geology, University of Massachusetts; Dr. M.A.
Stocking, School of Development Studies, University of East
Anglia; Dr. C.M. Wilson, Anglian Water Authority; and Dr.A.
Yair, The Hebrew University of Jerusalem, as well as present
and past colleagues in Cambridge and Aberystwyth, especially
Professor R.J. Chorley. Former students at Aberystwyth
have also helped by discussing their research findings and
making their dissertations available, namely, Bill Dovey
and David Rees (formerly MSc students in Water Resources
Technology at the Department of Civil Engineering, University of Birmingham) and Danny Lear (still at Aberystwyth).

I also offer my sincere thanks to Dr. John Dalrymple for
numerous stimulating discussion and for his useful comments
on sections 4.4 and 8.6, as well as to Irina Smith and
Janusz Orzechowski for their valuable translations from
Russian and Polish works respectively.

The following authors and publishers have kindly given
permission to reproduce or adapt illustrations:

    American Society for Testing Materials (Figures 14,
      15, 17)

Dr. T.C. Atkinson (Figure 30B)

Professor R.J. Chorley (Figure 30A)

Dr. J.B. Dalrymple (Figure 6)

M. McCaig (Figure 51)

Dr. M.D. Newson (Figures 31, 36-41)

Dr. M.A. Stocking (Figures 45, 46)

Dr. A. Yair (Figures 47, 49)

And Dr. Hugh Plommer, University College, Cambridge,
kindly gave permission for use of the frontispiece.

Quoniam usibus omnium maxime necessariae aquae videntur,
primo quae genera terrae tenues aut abundantes venas
emittant, quibus etiam signis altius depressae inveniantur,
quomodo ex fontibus vel puteis ducantur, quae nocivos aut
salubres habeant liquores, studiose scire oportet. aquae
ergo fontanae aut sponte profluunt aut saepe de puteis
abundant. quibus tales copiae non erunt, signis infra
scriptis quaerenda sunt sub terra capita aquarum et proxima
fontibus altiora puteis colligenda. ante solis itaque ortum
in locis quibus aqua quaeritur aequaliter in terra pro-
cumbatur et mento deposito per ea prospiciatur. mox videbis
in quibuscumque locis aqua lateat umores in aera supra
terram crispantes et in modum tenuis nebulae rorem spargen-
tes, quod in siccis et aridis locis fieri non potest.
quaerentibus ergo aquam diligenter erit considerandum quales
terrae sint.

*"De aquae inventione", in "De Diversis Fabricis
Architectonicae"*, M. Cetius Faventinus.

Since water seems to be what everyone most needs for all
purposes, we must study to find out first of all what kinds
of soil supply us with meagre or abundant springs, what
signs too we have that these can be struck further below
the surface, how their supplies are conveyed from fountains
or well, and which provide harmful or healthy water. Now
springs of water often gush up of their own accord; or they
can often bubble up from wells. People who do not enjoy
such plenty must use the clues that I give below to trace
springs of water below ground, and they must collect in
wells the water to be found lower down not far from springs.
So in the places where you are looking for water you must
lie level and prone on the earth before sunrise, and with
your chin let down on to the ground you must look around
you. Soon you will see in the places where water lies
concealed the wisps of vapour rising into the air and
scattering dew like fine clouds. This would be impossible
on parched, dry land. So when you are looking for water
you should consider the nature of the soils .

Translation: H. Plommer, *"Vitruvius and
Later Roman Building Manuals"*,
Cambridge University Press, 1973, p.45.

CONTENTS

LIST OF FIGURES

LIST OF TABLES

1. SIGNIFICANCE AND PRESENT STATUS OF RESEARCH

The literature on piping is now vast and theories and hypotheses on the causes and processes of pipe genesis have become legion. This is partly due to the undoubted variety of piping processes in different climatic, geomorphic and pedological environments around the world and partly it reflects the differing interests and expertise of the analysts. It is also evident that considerably more attention has been paid to the detailed causes, landforms and processes of piping than to the significance of piping in the wider context of hydrological and geomorphological systems. Indeed, piping has commonly been regarded as some form of special, localised phenomenon, a view belied by the evidence now available.

The recent interest shown in soil piping in Britain has been mainly from a hydrological viewpoint. In many ways this is largely a response to a combination of the increasing interest in hydrology, and the development of hydrogeomorphology from the more restricted fluvial geomorphology of the 1960's, with the increasing realisation of the deficiencies of the Hortonian theoretical framework, particularly in the British context.

Despite the occasional early suggestion that water may be transmitted laterally through the soil mantle to the stream (e.g. King, 1899), hydrologists in general have persisted in ignoring this aspect of the soil until very recently. Indeed, hydrological texts are still published which totally ignore it on the basis that water infiltrated into the soil is essentially lost to the storm response system (e.g. Chapman and Dunn, 1975; Viessman, Knapp, Lewis and Harbaugh, 1977). Early reference to features akin to pipes was made by Bryan (1919, 557), whose classification of springs included 'minor tubular springs' and 'hardpan springs' under the category of shallow water, which fluctuates seasonally and often in relation to rainfall events. These are usually found in 'unconsolidated sediments'. Lowdermilk (1934, 534) also referred rather vaguely to 'wet-weather springs'. Ironically, Horton himself (1936, 348) identified ephemeral outlets in the streambank, which he called 'bournes', as features which limit the rising level of the water table during storms. He said that flow from these could explain sudden increases in streamflow which are not associated with increased rainfall intensity, without resorting to the hypothesis that the whole soil profile has become saturated. However, he did not include these suggestions in his most influential summary published in 1945, in which stream hydrograph peaks clearly required overland flow to generate them. In spite of a special sub-committee of the American Geophysical Union being set up to report on shallow subsurface flow (Hursh, 1944), the combined philosophies of the unit hydrograph (Sherman, 1932) and infiltration-excess overland flow (Horton, 1945) ruled supreme, until broken in the aftermath of Hewlett's (1961) revelations (cf. Ward, 1975; Chorley, 1978).

The 'Hewlett model' of stream runoff generation, involving subsurface contributions from throughflow or shallow interflow, channel precipitation and partial contributing areas is now well established, and has led to a change of emphasis amongst engineering hydrologists seeking to estimate the runoff potential of soils away from measurement of the infiltration capacity of the soil surface to the storage and later the flow properties of the whole soil profile (Gilman and Newson, 1980). However, criticism of the model continues to concentrate particularly on the rather slow rates of subsurface seepage measured from matrix flow (e.g. Dunne and Black, 1970a and b; Freeze, 1972b, 1974). Recent suggestions regarding the hydrological function of soil pipes and localised zones of seepage are providing a counter-argument and are substantiating and extending the Hewlett model (Figs. 1-3). According to Ward (1975, 255; 1978, 44-45) pipeflow represents a dominant storm runoff process, and work at the Institute of Hydrology (Gilman and Newson, 1980) and by Jones (1975; 1978a) in mid-Wales tends to bear this out. Knapp (1979, 17, 41, 81) also alludes briefly to the role of pipeflow based on his own observations in mid-Wales, and Weyman and Weyman (1977, 14 and fig. 6) make brief reference to piping as an occasional, modified form of throughflow in their popular text; though without noting its effect on the hydrograph. Building upon the experience of Weyman (1971) and Stagg (1974) in the Mendips, Smith and Stopp (1979) have suggested that pipes are the main source of the differential response times between podzols and brown earths. The recent evidence and hypotheses concerning the hydrological role of natural piping are examined in chapter 7.

In looking at the present status of hydrogeomorphology, Gregory (1979a, 92) has noted that we now realise that the drainage network consists not simply of channel links, but an 'amalgam of several types of link, which are distinguished according to the character of channel processes'. Change within the drainage system is not simply a matter of extension and contraction (Gregory and Walling, 1968), but also a matter of change in the character of channels or elements comprising the network (Jones, 1971; Gregory, 1979a and b). Hence, identification of network components such as rills and pipes has led to studies which involve classification of the elements within a drainage net, and Gregory (1979a) points out that such classifications are potentially useful for 1) modelling of hydrological processes, 2) interpretation of recent changes and 3) greater understanding of the direction of network change in the future.

Piping clearly has geomorphological significance in terms of erosion and slope development as well as in terms of drainage channel and valley network development. Yair (1973) has pointed out that whilst creep and rainsplash have been considered the main erosion processes on upper convex hillslopes, rilling, piping and gullying are also

Fig. 1   Ephemeral pipes in a streambank in the English
          Peak District.

Fig. 2   Artificially exposed junction in soil pipe network.

Fig. 3   Waterfall in partially-collapsed soil pipe.

active and the existence of Horton's (1945) 'belt of no
erosion' must therefore be questioned. And sometime earlier,
Czeppe (1965) suggested that periglacial slope retreat may
be a function of 'less organised surface and subsurface
flow', including piping, rather than the parallel rilling
proposed by Büdel (1948). Carson and Kirkby (1972, 233)
state: 'Piping is very important for sub-surface flow and
debris removal, but its quantitative contribution cannot
yet be estimated'. Recent progress in quantifying that
contribution is described in later chapters.

     However, it is only very recently that piping has been
investigated as a significant process in landsurface evolution, with major advances in placing it in context being made
by Parker (1963) and Dalrymple, Blong and Conacher (1968).
It is still commonly accorded little space in general textbooks and there is a strong tendency to treat it as unusual.
Apart from tantalisingly brief observations by von Richthofen
(1886), Penck (1924, 323) and Passarge (1929), piping pro-

cesses were ignored by the earlier writers of geomorphological texts. More recently, Scheidegger (1961, 81; 1970)
makes only one reference to subsurface erosion, namely
'boiling' at the foot of alluvial slopes (cf section 5.2).
Pitty (1971, 209) refers briefly to piping beneath weathering
crusts and to gully extension by 'lines of concentrated flow
of sub-surface seepage' (*ibid.*, p.358). Both Leopold,
Wolman and Miller (1964, 446-447) and Morisawa (1968, 71)
make brief reference to gully extensions by collapse of
underground pipes or tunnels. Nevertheless, Leopold *et al.*
actually state that this is an important element in many
headward gully extensions and that in their experience in
semi-arid regions erosion by water flowing over the vertical
face of a gully head is 'not among the most active agents',
although they lament the lack of quantitative data on the
subject. Butzer (1978, 109-110) refers briefly to piping
in soft shales or silts, initiated by cracks, differential
permeability, rodent burrows or roots, the most severe cases
being in semi-arid areas, but Young (1972) ignores it.

Not until the work of Dalrymple, Blong and Conacher
(1968) was soil piping accorded a place in a general landscape model and not until the works of Carson and Kirkby
(1972, 232-233, 237) and Gregory and Walling (1973, 283-286,
373-374) was it accorded general roles respectively in hillslope erosion and drainage basin dynamics by geomorphological
and hydrological textbooks in English. Lustig (1970, 170)
states that it is largely under-rated as an erosional process except as a cause of failure in earth dams. According
to Gregory and Walling (1973, 286) 'pipes must contribute
significantly to water flow in the soil'.

More recently, Douglas (1977, 66-67, 102) has belied
the frequent suggestion that piping is only important in
semi-arid areas by discussing the importance of piping and
subsurface erosion in the development of slopes in 'humid
landforms', based mainly on experience in Australasia, and
in the development of gullies and drainage networks. He
points out that where gully development is by pipe collapse,
the drainage density is in fact controlled by subsurface
processes and that where pipes bring water into streamhead
hollows the pipehead acts as a balance point, similar to the
suggestion of Kirkby and Chorley (1967) for streamheads in
general, between pipehead erosion and hollow infilling in
drier weather. Both Statham (1977, 67, 135) and McCullagh
(1978, 40) also refer briefly to piping developed along
cracks or fissures and forming foci for stream channel extension.

The importance of subsurface drainage and piping to
badlands erosion has been pointed out, for example, by Parker
(1963) and Bell (1968). However, only recently has attention been focussed on the geomorphological and hydrological
processes related to piping in badland areas, and this
research is tending to revolutionise our understanding of
badlands landsurface systems. Yair, Lavee Bryan, and Adar
(1980) point out that shale badlands have commonly been
assumed to have relatively impermeable surfaces (e.g. Enge-

len, 1973), but that a high permeability is in fact suggested by the frequency with which sub-surface drainage and piping are developed in badlands. They emphasise that 'many of the quantitative data essential to understanding of badlands geomorphology are still almost completely absent', including 'clarification of the circumstances of moisture infiltration and pipe initiation'.

In continental Europe, Rathjens (1973, 169) has noted that piping has received hardly a word of notice in either French or German geomorphological texts. Yet he argues that it should be incorporated as a diagnostic feature in a system of climatic geomorphology (cf section 4.1). Franzle (1976) has since carried this suggestion a step further and incorporated piping as a major, indeed 'paramount', feature of a morphodynamic geosystem model of the tropical and sub-tropical rain forests. The most glaring exception in Europe has been in Polish literature where frequent reference to piping, mainly as an erosion process in loess landscapes, has been made from the time of Kriechbaum (1922) and Zaborski (1926) to the recent more quantitative survey by Galarowski (1976). However, there has been a marked lack of theoretical consideration in the Polish literature beyond treating piping as a special process.

Indeed, the view that piping is a 'special', localised process continues to be very prevalent, from suggestions that it is limited by geology (Kunsky, 1957) or climate (Parker, 1963) to statements that piping is an 'unusual' form of soil erosion commonly confined to badlands of little significance to agriculture (Hudson, 1971). In the FAO world survey of soil erosion in 1965, piping is ignored in the main text, but is mentioned as a specific problem widespread in Victoria and less so in New South Wales and Southern Australia in an appendix (Beare, 1965).

Yet despite a strong traditional emphasis on surface and rainsplash erosion in soil erosion studies (cp. Hudson, 1971; Morgan, 1979), much recent work is pointing to the role of subsurface soil erosion processes, for example, in Australia (Crouch, 1976) and East Africa (Stocking, 1978).

There is, therefore, growing recognition of the erosional and hydrological significance of piping. The need for more information on these processes is taken up by Newson (1976) and Jones (1978a), but as yet there is no formal quantitative model of hillslope evolution, soil erosion, drainage network development or runoff processes which includes piping parameters.

2. TERMINOLOGY

The literature relating to piping contains a wide variety of terms and a number of different systems of classification which in many ways is as confusing as the variety of landforms and processes they seek to describe. It is therefore unfortunately necessary to review definitions before proceeding any further.

We will begin with some of the more recent definitions of 'piping'.

'The formation of natural pipes in soil or other unconsolidated deposits by eluviation or other processes of differential subsurface erosion' (Chorley, 1978b, 370).

'The natural development of subsurface drainage in relatively insoluble clastic rocks of the dry lands (arid and semiarid regions)' (Parker, 1963, 111).

'Erosion in which the soil is carried away by water running through holes in the ground' (Heede, 1971, 1).

'Formation of underground conduits by the renewal of clay and silt fractions of soils on a grain by grain basis due to percolating water in clastic material' (Jennings, 1971).

The open-ended nature of many of these definitions is immediately obvious. This is partly due to the lack of clear boundaries in nature. However, Baillie's (1975, 9) definition seems to carry this open-endedness beyond the point of usefulness: 'The process of subsurface erosion known variously as piping, tunnelling, percoline drainage and subsurface gullying'.

The dominant theme in definitions of piping is of mechanical rather than solutional erosion and this is expressed in definitions of 'pseudokarst', which describe landforms and processes related to and including piping. Sweeting (1972, 306) states that pseudokarst 'normally' results from solutional processes and was therefore included in the wider definition of karst by Mandel (1967). However, she believes that this definition is too wide and that pseudokarst should be reserved for generally superficial, smaller landforms in non-calcareous and non-evaporite rocks. Corbel (1957) used the term 'sub-karstic' for pseudokarst in basalt. Jennings (1971, 5) said that pseudokarst refers to landforms in finer pyroclastics and loess caused by piping, especially in lavas and 'glacier thaw karst'. Kosack (1952) mapped the world distribution of pseudokarst in a wide variety of unconsolidated sediments, volcanic deposits and non-calcareous rocks, and also mentions it in glacier ice (cp. Halliday, 1960; Shreve, 1972). Kunsky (1957) described

5 types of pseudokarst in Czechoslovakia including descriptions of what might be termed piping in loess and loose sands.
Halliday (1960) did the same in America and Quinlan (1966)
produced a classification of pseudokarst types occurring in
noncarbonate and non-evaporite rocks:

1. mechanical suffosion (defined below), e.g.
    clastokarst (or piping).

2. differential mechanical erosion.

3. thermokarst (soil) and glacier pseudokarst.

4. lava pseudokarst.

He describes piping as occurring in soil, clay, volcanic ash,
loess and sandy gravel.

Parker (1963, 106) states that he is introducing the
term 'pseudokarst' to refer to karst-like features created
by mechanical rather than solutional erosion. However,
Halliday (1960) traced the term back at least to Savarensky
in 1935 (cf. Savarensky, 1940). It has recently been used
explicitly referring to piping and piping-derived landforms
by Feininger (1969) and Löffler (1974), and Khobzi (1972)
used the term 'pseudo-karstique'.

Yet another term is in widespread use in East European
literature and appears to be relatively well-known there
(Jahn, pers. comm., 1975), namely 'suffosion'. The term can
be traced back to Professor A.P. Pavlov of Moscow University
(Pavlov, 1898) and has also been used in French by Tricart
(1965), although Khobzi (1972) remarks that he has failed
to find it defined in any dictionary or encyclopedia. In
fact, Kral (1975) has pointed out that suffosion is defined
in the Kratkaja Encyclopaedia of Geography (1960; 1964).

To Pavlov (1898) and Savarensky (1940), who evaluated
Pavlov's contribution to engineering geology, suffosion
meant mechanical removal of loose particles. However, subsequent workers recognised chemical suffosion as well
(Sokolov, 1960), and an argument ensued as to whether
chemical suffosion should be admitted. Gvozdeski (1954)
proposed that chemical erosion was completely different and
should be grouped with true karst, but chemical suffosion
continues to be recognised.

Both mechanical and chemical forms of suffosion process
were described by Russian, Polish and French workers, for
example, by Berg (1902) in Kazakhstan, Maruszczak (1965) in
Poland, Czeppe (1960, 1965) in Poland and Spitsbergen and
Khobzi (1972) in Colombia. Lilienberg (1955) used the term
'suffosional karst' for the features created by these processes in the Georgian Republic. The equivalence of
'suffosion' and 'piping' has been explicitly recognised in
recent Czech and Polish work (Kral, 1975; Galarowski, 1976).

8

For the most part, this review will be concerned only with mechanical suffosion, except where the two processes are combined. Typically, although not always, piping erosion results in open linear voids or tunnels, and in fact 'tunnel erosion' is a commonly used alternative term despite many attempts by 'the splitters' to draw a distinction between piping and tunnelling. When the process first received a separate name, other than 'subsurface erosion' or the equivalent, in America it was termed 'tunnel erosion' (cf. Bennett, 1939, 108; Cockfield and Buckham, 1946; Buckham and Cockfield, 1950). The term 'piping' seems to have been introduced into soil erosion studies by Fletcher and Carroll (1948), Carroll (1949), Fletcher and Harris (1952) and Fletcher, Harris, Peterson and Chandler (1954), probably from American engineering terminology (Terzaghi, 1943; Terzaghi and Peck, 1948). In Australia and New Zealand also, apart from the picturesque descriptions of 'subcutaneous erosion' by the farmer-naturalist Guthrie-Smith (1921), 'tunnelling' (Downes, 1946) and 'tunnel-gully erosion' (Gibbs, 1945) was the normal term used until recent papers influenced by American soil mechanics literature.

Attempts have been made to distinguish between piping and tunnelling in both America and Australasia. Rosewell (1970, 188) has tried to restrict 'tunnelling' to 'post-construction deflocculation' in earth dams with subsequent removal of the deflocculated material, whereas 'piping' is 'due to the formation of continuous channels through an embankment by the erosive action of seepage' and therefore occurs in 'flocculated clays, sandy or silty materials'. In fact, the distinction that Rosewell was trying to make has become more evident during the last 10 years as piping, and in particular piping failure of earth dams, has become closely associated in many minds with dispersive clays, and as alternative evidence, particularly from studies of piping on natural slopes, has on the other hand shown that it can occur in many other materials. Significantly, Rosewell's terminology has not survived.

In fact, if anything Rosewell's terms have been reversed in meaning by American engineers. Sherard, Decker and Ryker (1972) recognised 'piping' due to dispersion of clays as the main problem in earth dams, but they also observed 'rainfall erosion tunnels' on the crest of dams which occasionally caused failure of the dam if they connect with other weaknesses. Tunnelling or 'internal erosion damage' starts with water entering surface cracks with velocities measured in cm s^{-1} and rarely in m s^{-1}, whereas piping occurs deep within the dam without necessarily following cracks and in flows of mm s^{-1}. However, tunnelling is only really effective in dispersive soils (cp. Sherard, Dunnigan and Decker, 1976). Moreover, in his review of engineering research, Perry (1975) another American engineer, makes no such distinction: any 'progressive internal erosion of the soil by the flow of water along preferred seepage paths such as cracks or sandy lenses' is termed 'piping'.

Associated with tunnelling as per Sherard *et al*. (1976) is 'jugging', a dialect word adopted by the American Soil Conservation Service to refer to the vertical inlet sections of tunnels (Sherard *et al*., 1976; Decker and Dunnigan, 1977; Ryker, 1977). Jugs are the equivalent of Ward's (1966) 'shafts' in New Zealand, 'blow holes' (Atkinson, 1978) or simply 'tunnel inlets' (Ritchie, 1965) and 'pipe inlets' (Jones, 1975). Ritchie (1963) referred to 'pop holes' due to heaving at points where the pipes have become blocked: inlets and outlets may look alike and indeed, reverse roles. Earlier, Fuller (1922) had referred to vertical 'pipes' and sloping 'tubes'. Another curious terminological volteface has occurred within soil conservation literature in Australia, where Crouch (1976) has equated piping only with deep tunnels *initiated* by gullies as opposed to tunnels *initiating* gullies ('tunnel-gully erosion') or shallow smaller forms nearer the surface (v.i.).

Ingles (1968, 43) chose to use the term 'conventional piping' refering to Terzaghi-type failure in non-cohesive materials (cf section 5.2(ii): Terzaghi and Peck, 1948), as opposed to 'tunnelling' which occurs in cohesive materials, particularly heavy clay soils and especially in solodised solonetz, solodics and podzolics in Australia. The latter soils tend to become sticky at their optimum moisture content and thus in engineering practice are frequently compacted dry of optimum and consequently have a higher permeability. Tunnelling begins upstream, with saturation settlement in poorly compacted soil creating a combined crack and discontinuity in permeability near the phreatic line as opposed to piping which begins by hydrodynamic pressure lifting particles at the downstream exit of seepage. However, Ingles's explanation does not fully cover the widesread occurrence of tunnelling in undisturbed solodics (cf. section 4.5).

Many other terms have been used for piping erosion in general, particularly in early papers, and often only used by one or two writers. These isolated descriptions tended to emphasise the karst-like features, for example, the 'sinkhole erosion' of Thorp (1936) in China and Cockfield and Buckham (1946) in Canada, the 'pothole gullying' of Cole *et al*. (1943) in California, the 'soil karst' of Downes (1946) in Victoria, the 'soil caves' of Funkhowser (1952) in Ecuador and the 'pothole erosion' of Kingsbury (1952) in Hawaii. The plethora of terms is indicative of the lack of cross-referencing. The agricultural viewpoint has been expressed in the use of the farmers' term 'under-runners' in New Zealand (e.g. Guthrie-Smith, 1921; Soil Conservation and River Control Council, 1944). Bell (1968) quoted 'run downs' as a local term used by road engineers for sinkholes and pipe orifices in North Dakota. Others have referred to the possible role of animals in 'squirrel-hole gully erosionp' (Sharpe, 1938), and 'rodentless rodent erosion' (Bond, 1941). In Africa, a form of piping involving semi-liquid mud spouts has been called 'bottle flows' (Haldeman, 1956) or 'bottle slides' (Temple and Rapp, 1972). This appears

to be similar to the process of 'heaving' described by
Terzaghi and Peck (1948), although it is now recognised that
dispersive clays may play a role here. The process has been
described by Downes (1946), Marker (1956) and Colclough
(1965) in Australia simply as a normal tunnelling process.

In German, piping has been referred to as 'Subterrane
Abtragung' (Rathjens, 1973), 'Schottbildung' (according to
Khobzi, 1972), 'Materialabfuhur im Untergrund' (Bremer,
1972) and recently as 'Suffosion verzweigte Röhren' by Kral
(1975) and 'Röhrenerosion', literally pipe-erosion, by
Franzle (1976). Riley (1972) records the name of 'tomos'
as a local term in New Zealand, although he takes it to mean
tunnels whereas Blong (1968, 84) took it to refer to collapse
shafts or 'sinkholes'. Trudgill (pers. comm., 1980) has
pointed out that 'tomo' is used to refer to vertical shafts
in limestone.

Finally, a number of classification schemes have been
proposed. Cumberland (1944) divided subsurface erosion in
New Zealand into three types:

1) subcutaneous tunnelling - where storm water enters
   through desiccation cracks or minor slumps and
   reaches a compact, impermeable subsoil.

2) subcutaneous sheet erosion - similar causes to
   (1) but no need for summer cracking. It runs
   beneath the humus rich A horizon and above the loess
   in the Banks Peninsula, N.Z., 'carrying with it a
   film of yellow silt particles'.

3) subcutaneous dimpling - limited to land that has
   been ploughed, where it replaces subsurface sheet
   wash, 'may be a form of slow flowage' (*ibid.*, 100)
   or water removal and slow collapse producing forms
   similar to the 'stepped crescents' of Bennett
   (1939, 282); 'small, hollow depressions, smooth and
   gently concave' (*op. cit.*, 101).

However, tunnelling or piping itself was not split explicitly
into sub-categories until Fletcher and Carroll (1948)
recognised three types in Arizona:

1) Piping that begins at a vertical bank and works
   backwards until caving or sloughing closes it.

2) Piping that begins by inflow into cracks just below
   plough depth from rainfall or irrigation water.

3) Cases where rodents or ploughs open up a route
   through surface soils of highly dispersed sandy loam
   or silt into highly permeable, coarse, poorly
   graded subsoils, resulting in wash-down of material
   until the subsoil is clogged.

Dregne (1967, 38; 1978, 216) recognised the same three

11

genetic types in brief summaries which clearly owed a lot
to Fletcher and Carroll's seminal paper. Ritchie (1963)
divided tunnel erosion in Australia into three forms:

1) Field tunnelling: in the subsoil under 'natural
   paddock'. In all recorded instances there is a
   background of overgrazing or cultivation and pasture cover has failed to provide adequate continuous
   ground cover, resulting in uneven infiltration and
   sheet erosion.

2) Earthwork tunnelling: in earthworks built of unstable soils susceptible to field tunnelling.

3) Tunnelling in strongly self-mulching clay soils:
   in black earths and other soils extremely prone
   to cracking.

He regarded only types 1 and 2 as 'true' tunnelling
which results from the high dispersibility of the soils,
i.e. true tunnelling is not due to cracking (cp. Sherard
et al., 1972, who regarded tunnelling as due to a combination of cracking and dispersive soils, and piping due only
to the latter). Parker and Jenne (1967) provided perhaps
the best general genetic classification available to date:

1) Desiccation-stress crack-piping - the dominant form
   in the western U.S.A., begins by water entering
   a crack and eroding a route down the hydraulic
   gradient to an arroyo or side slope. It may also
   occur due to a) localised subsidence, due to saturation of surficial sediments, forming sinks or
   stress cracks, or, rarely, b) an animal burrow or
   rotted root tube. These pipes have visible inlets
   and outlets, except where slumping has closed the
   ends.

2) Variable permeability-consolidation piping - begins,
   in contrast, at the down-gradient end and works
   upwards from the point where a gully wall or free
   face intercepts a more permeable layer than those
   above or below. It may also occur where uneven
   subsidence and variable consolidation have occurred
   following irrigation in the dryland areas. Vertical
   pipes develop during settlement connecting it to
   the surface.

3) Entrainment piping - unlike the other two, this
   rarely involves creation of an open tube. It occurs
   where newly created large hydraulic heads cause
   channelised subsurface flow with entrainment of
   water-saturated earth materials. Surface subsidence
   may follow.

The latter form is probably what has been termed
'bottle-flows' or 'bottle slides' in East Africa (Haldeman,
1956; Temple and Rapp, 1972; Morgan, 1979, 13) and has
been literally described by Downes (1946) and

12

Marker (1958, 130) in Victoria as semi-fluid clay masses exhuding from lines of weakness, often provided by cracks, ant holes or rabbit burrows. In this extreme form, it seems to be rather rare in nature, although in essence it is the boiling or heaving process championed by Terzaghi and Peck (1948) and may well be present in some form at some time in the creation of other forms of pipe, especially the variable permeability form where vertical concentration of seepage will cause lateral boiling forces.

Crouch (1976) also recognised three distinct forms (Fig. 4) and processes, in his words:

1) Shallow tunnel erosion, developed when:-

   a) The surface soil cracks due to desiccation.

   b) Water infiltrates rapidly down these cracks supersaturating the subsoil.

   c) This supersaturated layer disperses.

   d) The dispersed particles move in the soil water due to a hydrostatic gradient that produces lateral flow.

   e) A tunnel develops when the dispersed soil and water breaches the compact surface.

2) Tunnels in the deep B horizon initiating gullies, developed when :-

   a) The soil cracks deeply.

   b) Infiltrating water concentrates down these cracks and moves along subterranean cracks, eroding the dispersible soil at their base.

   c) When this subsurface water breaks through to the surface further downslope, it flows more freely and hence rapidly erodes a tunnel along its flow path.

3) Deep tunnels initiated by gullies, when either:-

   i) a) The flow of groundwater concentrates on a point in the gully wall.

      b) Concentrated flow removes the dispersible subsoil, initiating tunnel development.

      c) The presence of a tunnel increases the concentration, thereby promoting further tunnel development.

   or ii) a) Gully formation.

       b) Cracks develop from the gully wall due to

13

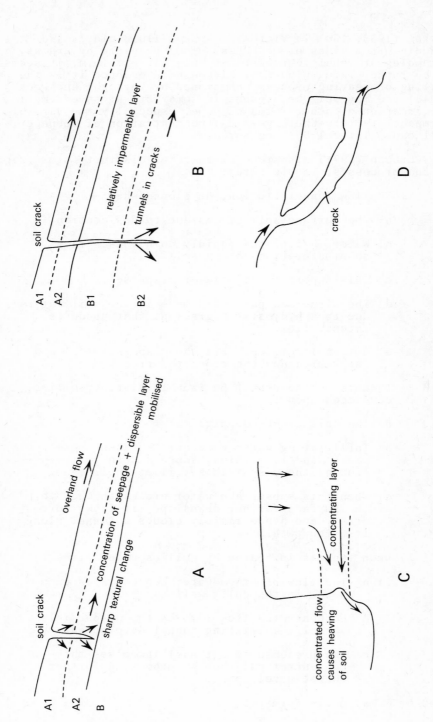

Fig. 4 Varieties of piping process as summarised by Crouch (1976).

14

the soil adjacent to the gully drying out
more rapidly than in the surrounding area.

c) Runoff follows these cracks, eroding the
dispersible soil.

d) The crack widens at its base, causing soil
movement which closes the crack near the
surface (as suggested by Heede, 1971).

In fact, the basis for Crouch's second type was also
present in Newman and Phillips' (1957, fig. 3) diagram.
The last two groups in this scheme are looked at critically
in section 8.1.

The magic trinity of forms has been generally maintained in recent attempts to classify pipes in hydrological
terms. Gilman (1971b), Pond (1971), Bell (1972) Jones
(1975; 1978a) variously recognised ephemeral, intermittent
or seasonal, and perennial or permanent pipes. However,
Jones (1971), Newson (1976), Newson and Harrison (1978) and
Gilman and Newson (1980) only explicitly recognised the
first and last (cf. section 7.1(i)). Jones (1979) has
recently produced a hydrogeomorphic classification combining
hydrological function with approximate location in terms of
landsurface units. This recognises a division between pipes
connected to a stream channel and disjunct piping, with each
division splitting again into core areas on the seepage
slope and the floodplain with varying degree of extension
respectively down or up slope across the valley side-slopes
(cf. section 7.1(i)).

Lastly, Anderson (1979) has used the term pipe (in
quotation marks) to refer to more permeable areas of soil
blocks as revealed by X-ray radiography. Clearly, such
features may or may not result from piping processes in the
normal sense, but they could generate piping processes.

This chapter has been devoted to defining piping and
its related terms. In the process of so doing it has highlighted a number of past disagreements on terminology and,
indeed, a gross redundancy of terms which recent international communication has severely reduced.

Inevitably such a discussion has touched on two major
themes of interest which will be investigated more fully in
the following chapters, namely, the historical development
of research and theories on the causes and nature of piping
processes. The succeeding chapter focuses on placing the
development of ideas and terminology into historical perspective.

3. A BRIEF HISTORY OF RESEARCH

It is extremely difficult to write a coherent and
readable history of research into piping, largely because
so much of it has been conducted in geographical or disci-
plinary isolation. In general the best historical threads
have been within a certain discipline in a certain geographical area, for example, piping is well-known to geomorphologists in Eastern Europe, but almost entirely unknown
to geomorphologists in Western Europe until very recently.
East European geomorphologists hardly referred to any Western
work and *vice versa*. Similarly, geomorphologists, soil
scientists and soil conservationists have in general hardly
been aware of the parallel work of engineers, with the
exception of the now rather dated text by Terzaghi and Peck
(1948) introduced into geomorphological literature by Parker
(1963) (and again *vice versa*), let alone being aware of
parallel work in their own disciplines in many cases. Moreover, hydrological interest in piping is more a matter of
current affairs than history.

As a result, although the first references to piping
can be traced back 80 years or so, with a major increase in
interest in the fields of soil conservation and civil
engineering in the 1940s, it is only in the last 20 years
that international interdisciplinary communication has really
come about. In those 20 years the number of publications
relating to piping have increased many-fold. The Commonwealth Bureau of Soils (1961) bibliography of piping erosion
contained just 14 references, admittedly very incomplete
even for its day. Even in 1970, Lustig (1970, 170) referred
to 'the meagre literature' on piping. Ill-informed though such
comments may have been this could certainly not be said of
the situation today, as will become apparent in this review.

One of the earliest references to piping processes was
made by Von Richthofen (1886) describing landforms developed
in loess. In fact, loess has continued to be a commonly
reported medium for piping, mainly because of its susceptibility to erosion by seepage water and the rapid and large-
scale development of piping within it. Subsequently, Fuller
(1922) and Thorp (1936, 460-463) described dramatic features
in the loess of China. Fuller found 'vertical pipes or
wells' 8-10ft. in diameter (2.4-3.0m) and 15-30ft deep (4.6-
9.1m) connecting with 'downward sloping tubes' 3 feet wide
and 10 feet high (0.9-3.0m) and natural bridges most commonly
on the plateau rims. Open gullies extended back into the
loess either by caving in of the tubes or by the 'eating
back of the loess by the normal process of gullying'. He
considered the forms to be typical of youthful erosion and
limited to areas of rejuvenation, possibly caused by
deforestation, increased rainfall or 'differential elevation
of the loess area', and he suggested that two processes were
responsible: first, solution of calcium bonding and then the
mechanical removal of particles. The features were best

developed in Kansu, probably because this was nearer the Mongolian sources of the loess and therefore it contained more dissolvable calcium. On the other hand, Fuller noted that rainfall was greater in Shansi and Shensi so that features could be quickly destroyed there.

Contemporary with von Richthofen's observations, Cussen (1888, 411) reported piping in alluvial depressions in the middle Waikato basin, New Zealand, consisting of:

> 'numerous funnel-shaped holes...formed by the sub-
> terranean or soil waters in passing along beneath the
> surface of the earth. They create small caverns and,
> finally, underground streams, which draw away the
> loose material from the surface'.

The broad essentials of a theory were already there.

A few years later, Haworth (1897, 17-21) suggested that 'slow creeping of the underground clays and soils', followed by vertical slumping was responsible for the steep arroyos cut 1 to 2 m deep into the grassland plains of western Kansas. And in Russia, Pavlov (1898) described seepage erosion in the steppes, coining the term 'suffosion'. The latter paper appears to have been quite influential in East European literature, although probably not immediately. In contrast, Haworth's ideas were immediately taken up. Follow-ing Haworth, W.D. Johnson (1901) decided that certain hollows in the High Plains of southwest Kansas were caused by settlement of the adobe earth, this time explicitly impli-cating seepage water. First, the water accumulates on the natural, uneven surface. Downward percolation of this water results in settling by mechanical compaction and by chemical solution. He mentions caving in of the roofs of underground channels (*ibid,* 712) and the diversion of surface drainage by subsurface channels (*ibid,* 720). The ideas of Haworth and Johnson were denied by Fenneman (1922), Kirk Bryan (1925), Evans (1927) and W.M. Davis (1927); formidable opponents indeed. In fact, Fenneman was not entirely opposed to Johnson, since both men believed in the extension of gullies or 'draws' by basal sapping, a form of heaving and weakening of the lower banks by seepage water, but Fenneman believed this to be the main process. On the other hand, Bryan, Evans and Davis believed that the large, flat-bottomed gullies and 'charcos' were basically due to erosion by surface runoff. Rubey (1928) returned to support Johnson's thesis of 'sinking of the ground' and provide the first summary of ideas to date. His summary later inspired Buckham and Cockfield (1950) to look again in broader perspective at their earlier study of 'sink-hole erosion' in the Pleistocene White Silts of the Thompson valley near Kamloops, British Columbia (Cockfield and Buckham, 1946). Apart from approximately reusing Rubey's title 'Gullies formed by sinking of the ground', Buckham and Cockfield (1950) observed that their study had been in an arid area (c. 250mm p.a.) whereas Rubey's area was semi-arid, and they speculated that 'the process is also operative in more humid regions than those here discussed, but that its results are completely masked by those of stream erosion' (*ibid.*, 140). Equally significantly, their 1946 paper provided the first direct substantiation of Rubey's thesis. They described 50-100ft diameter (15.2-30.5m) sink-holes draining by diffuse percolation or by discrete passage-ways which outcropped in the middle of gully walls. Many gullies were formed and deepened by the process, with sinks in the gully beds. Like Fuller, they observed that sinks were most frequent near to the edge of the main gully where cracks developed, often extending to 'stepped crescents' (Sharpe, 1938) around the head of the gully.

Contemporary with W.D. Johnson's observations, Gautier (1902) had noted pseudokarst features beneath laterite in Madagascar and began a long chain of essentially descriptive reports in the French language literature of piping features beneath duricrusts in many parts of the world; for example, De Chéletat (1938) in French Guinea, Saurin (1935; 1944) in French Indochina, Segalen (1949) and Chevalier (1949) more generally in tropical Africa, and spilling over into the English language literature, e.g. Hénin (1949), Aubert (1963), Dixey (1920) and Patz (1965) in Africa and McBeath and Barron (1954) in Guyana. Most of these references were relatively isolated. Not until very recently does the term 'piping' seem to have been applied in this context, in Goudie's review (1972; 1973) and in Pitty's (1971, 209) description of features in Hong Kong. Similarly, the term 'suffosion' is markedly lacking in the French works.

In the meantime soil conservationists in America had observed the processes. In a bulletin on soil erosion for the U.S. Department of Agriculture, McGee published a photo-graph of gullying developed by collapse of subterranean tunnels formed during the rapid drainage of floods on the Santa Cruz in Arizona in 1908 and 1909 (McGee, 1911, plate XVI), although he made no other reference to the process in his sixty page text. Sharpe (1938b, 46) referred briefly to 'squirrel hole gully erosion' in his treatise on landslides. In his textbook on soil conservation, Bennett (1939, 108), stated very briefly that: 'the highly friable, very fine sandy soils found in parts of the Great Basin region are peculiarly susceptible to tunnel erosion', and that rodent holes and root holes were enlarged by water to the point of caving in. The suggestion that biotic activity is instrum-ental to pipe development has been made many times since, but one of the most widely quoted rebuttals of this hypo-thesis was published soon afterwards by Bond (1941), who chose the memorable title of 'rodentless rodent erosion' for piping that he observed on Anacapa Island, off the South Californian coast, where there were no endemic rodents.

The first detailed analyses of piping soils came from a small group of soil scientists working for the USDA Soil Conservation Service in Southern Arizona (Fletcher and Carroll, 1948; Carroll, 1949; Harris and Fletcher, 1951; Fletcher and Harris, 1952; Fletcher, Harris, Peterson and Chandler, 1954; Peterson, 1954). Not only did they set the

18

scene for quantitative work with their analyses of infiltration rates, subsoil expansion, and exchangeable sodium percentage (Fletcher and Carroll, 1948, tab. 1), but they also laid down a list of basic factors responsible for piping, including the need for an erodible layer above an impermeable base, and the first classification of piping per se (cf. chapters 2 and 5). Their paper of 1948 was clearly a milestone and had considerable influence on subsequent research. They concluded that the piping in southern Arizona was of relatively recent origin and caused by stream incision following overgrazing, and that the piping soils analysed were all poorly graded and alkaline, high in exchangeable sodium. Peterson (1954) noted that piping was in fact common throughout the whole of the arid west of the United States.

Meanwhile a number of more or less isolated references were being made to similar processes in New Zealand. Most were merely brief mentions in treatises devoted to pioneer descriptions of the regional geology namely Gilbert (1921, 112) in sands in the southern Waikato Heads District, Henderson and Grange (1926, 25) in the middle Waikato basin, and Ferrar (1934, 20) in 'consolidated-sand county' in north Auckland. Ferrar described springs ejecting sand from 'underground channels' at the foot of valley walls and extending upwards to eventually reach the surface and cause sink-holes. Only Taylor (1938, 688-670) was primarily concerned with soil erosion. He says that tunnelling occurs in open-textured and fissured material above impermeable layers, where roots hold the roof together in the Marlborough District (Wairau Valley) and in terraces of water-borne pumice in central North Island. However, he also says that it is 'not common'; a statement that might be questioned in the light of subsequent observations in New Zealand. The best description of all comes from the story of a sheep farm written by Guthrie-Smith. He refers to 'a very remarkable process of subcutaneous erosion, a process akin to the dissolution of a dead beast when first the flesh decays, then the skin shrinks and shrivels, whilst only at the last do the bones protrude' (Guthrie-Smith, 1921, 28). He observed a 'subterranean soakage system' (ibid., 32), with the springs becoming more active in the higher, wetter areas, especially where there was more clay. The 'under-runner' tunnels tended to develop with floors on impervious marl grit and the ceilings just under the matted humus. He observed 'tiny hidden tunnels' branching at 90^0 from the main subcutaneous stream and further lateral branches developing on these, and proposed three stages of development: (i) erosion by percolating rainwater only and deepening of the channel, (ii) erosion by percolation and springs, with vertical deepening, and (iii) erosion by percolation and springs, with deepening and widening of the channel. The sequence is very similar to the evolutionary progression suggested later by Jones (1971; 1975) based on observations in the English Peak District.

Post-war research in New Zealand largely stemmed from

the setting up of the Soil Conservation and Rivers Control
Council which began by commissioning a nationwide survey of
soil erosion by Professor K.B. Cumberland. Cumberland (1944)
divided the country into regions with different erosional
problems, mostly man-induced due to burning, overgrazing or
the introduction of the rabbit. 'Subcutaneous tunnelling'
was particularly in evidence in the foothills and downland
of South Island, where there were:- 'deep, compact, impermeable subsoils devoid of humus and with low infiltration
rates and high dispersion. The areas affected have over
ten percent slopes and are in good pasture with tough,
matted swards. In all cases the pasture has replaced
(usually without ploughing) either indigenous forest,
bracken-fern, with its tough rhizomes, or deep-rooting tussock. In all cases European occupancy seems to have accentuated the already abrupt break between a topsoil with high
percolation and a compact, silt subsoil' (*ibid*. 98-99).
In 1944, the Council appointed D.A. Campbell as Publicity
Officer (McCaskill, 1973) and in August, 1944, the first
popular bulletin was produced to draw public attention to
'The menace of soil erosion in New Zealand' which described
cave-ins of 'under-runners' developing into ravines in the
rolling loess country of the Wither Hills, Marlborough. In
fact, the first of the Council's special research areas,
Wither Hills Reserve, was set up in 1944 to study the severe
tunnel, sheet, slip and gully erosion problems of the area
and to develop counter-measures (McCaskill, 1973, 95). H.
S. Gibbs was particularly instrumental in organising detailed
soil erosion surveys, adopting techniques from the U.S.
Soil Conservation Service (surveys by Gibbs and Raeside,
and Grange and Gibbs described by McCaskill, 1973; Gibbs,
1945). In 1945, Gibbs reported on the detailed situation in
the Wither Hills providing the first detailed description
of soil profiles and chemical analyses. The high level of
soluble salts was singled out as a particular feature, along
with the compact, mottled B_1 horizon with increased base
exchange capacity. Burning, rabbits and overgrazing by
sheep were thought to have initiated the problem. Gibbs
noted evidence of tunnelling in similar soils in eastern
Marlborough, Canterbury, north Otago and Wairarapa districts,
and he was the first writer to quote climatic statistics
for his area in any detail, pointing out the strong seasonal
contrasts in rainfall. A comprehensive reclamation was
begun in 1946 and there is now practically no erosion in the
area, and the frequent floods, which had been a major problem in 1944, have been stopped. The Marlborough Catchment
Board which administers the area has developed a reclamation
technique for the loessial soils based on deep-rooting pasture species (McCaskill, 1973, 106-107). Tunnel-gully
erosion was also a problem in some clays particularly those
overlain with pumice in Hawke's Bay District (*ibid.*, 103),
in coastal areas of Otago (*ibid*. 113) and sporadically
associated with extensive sheet erosion on sandy mudstone
areas in the Rangitikei Catchment Board area (*ibid.*, 117).
Campbell (1950, 198) again described the creation of the
'tunnel gully' in loessial and pumice soils which overlie
impervious strata in his list of 20 different forms of soil

erosion in New Zealand.

Although tunnelling was observed by farmers in Victoria soon after the First World War (Newman and Phillips, 1957), the first published references to tunnelling in Australia appear in the government soil surveys of the 1940s. The Soil Conservation Service of New South Wales reported piping in the Riverina in 1942 (Floyd, 1974). Tunnelling appears in Victoria on a soil erosion map of the state in 1944 and in the reports of Downes (1946; 1949). Downes was aware of Gibbs's work in New Zealand and produced similarly detailed analyses. He particularly noted the flocculated but 'sugary' horizons which tended to 'melt away' on wetting. He considered that the piping was initiated by water accumulating in surface depressions or 'collecting basins', often of bare soil, and slowly infiltrating. By comparative sampling, he showed that these depressions had far higher infiltration rates than their surroundings (Downes, 1946, tab. 2). This was aided by cracks and rabbit burrows. At a more advanced stage the tunnels collapse irregularly in lines, forming gullies 0.6-1.0m deep and temporarily leaving small bridges. Subsidence of areas where clays have been eluviated enlarges the collecting basins. In the final stages the tunnels may reach 2m wide with rectangular cross-section rather than the earlier circular cross-sections, which tallies with Guthrie-Smith's suggestions (cf. section 6.1 (ii)). Soon afterwards, Dickinson (1950) reported tunnelling in Tasmania, though he considered it a rarity. He took his description of the process word for word from Downes (1949) without acknowledgment and added nothing of significance. More recently, Colclough (1967; 1971; 1973; 1978) has described it as a much greater problem in Tasmania than was suggested by Dickinson. However, his work has been mostly concerned with agricultural reclamation measures, which have largely followed the research on the mainland.

The bulk of subsequent work on natural tunnelling was conducted by the Soil Conservation Service of the New South Wales Government. Monteith (1954) created renewed interest in New South Wales with his survey of soils in the Hunter Valley. He associated tunnelling with solodised solonetz of medium alkalinity, dispersive sodium rich clays and columnar cracking of the subsoil. He also made a novel observation of association between tunnelling and gilgai, the tunnels seeming to travel along the edge of gilgai ridges: gilgai also typically occur in alkaline, clay-rich soils and have been explained by differential expansion, contraction and collapse, processes which may also be operative in piping, but they generally lack the element of horizontal removal of material characteristic of piping. There is now an extensive literature on these minor landforms, which shows that, like piping, they occur in both wet and dry regions from the tropics to the arctic (cf. White and Agnew, 1968). However, a quarter of a century later we are still none the wiser as to the relationship between gilgai and pipes in terms of process-response. Could an association of gilgai and piping processes possibly explain some of the more confusing cases such as the 'tabra channels' of Iraq and the Sudan (White and Law, 1968), the Butana grass patterns and 'water lanes' of the Sudan (MacFadyen, 1950; Worrall, 1959, 1960) or the vermicular-arabesque channels of North African desert basins (Smith, 1969) ?

Newman and Phillips (1957) took another step forward when they initiated experiments into reclamation of tunnel eroded land in New South Wales, similar to those already underway in New Zealand, though without specific reference to the latter. They found piping in highly structured Red Brown Earths as well as solodised solonetz in the western Riverina, wherever a compact top soil is found overlying a dispersible layer and a relatively impermeable substrate. The experiments they set up have been reviewed after a 17 year interval by Floyd (1974). Brief mention of the tunnelling problem in New South Wales was also made by Logan (1959), who pointed out that tunnels were sometimes associated with saltings (Condon and Stannard, 1957). However, Foreman (1963) still indicated a general ignorance of the extent of tunnelling in the State. It appeared to be mainly confined to 'drainage lines' but isolated patches were found elsewhere and it was common enough to require caution in planning works involving ponding of water. Charman (1967; 1969; 1970a and b) undertook the necessary field surveys and laboratory analyses to investigate the link with salt affected soils and the distribution of tunnelling in geographical and pedological terms. Crouch (1976) has summarised the main results of the Service's work in his general review of tunnel erosion and continues to study the processes (Crouch, 1977).

Another major problem that the Service addressed itself to in the 1960's was failure of farm dams caused by tunnelling for which they used laboratory models (Ritchie, 1963, 1965; Rosewell, 1970, 1977). According to Charman (pers. comm., 1979) compaction was found to be the most important factor, since typically construction can only take place under dry conditions and in the local climate this usually means that soils are then below optimum moisture content for compaction. Thus there is usually insufficient compaction and small seepages develop: compaction equipment, extra watering and lime or gypsum additives are now common remedies. In fact, early suggestions of a chemical answer came from the Soil Conservation Authority in Victoria (Gyamarthy,1962

Paralleling the work of the Soil Conservation Service, the CSIRO Soil Mechanics Section also became involved in studying piping processes in earth dams during the 1960s, including the first monitoring of the actual development of pipes in the field (Cole and Lewis, 1960; Aitchison, 1960; Aitchison and Ingles, 1962; Wood, Aitchison and Ingles, 1964; Ingles, 1964a and b; Ingles and Wood, 1964; Aitchison and Wood, 1965; Ingles, Lang and Richards, 1968; Aitchison, Ingles, Wood and Lang, 1968; Ingles, Lang, Richards and Wood, 1969), and in 1969 Ingles and Aitchison (1970) applied the experience they had gained to natural soil piping.

Putting the clock back a little and shifting to the African continent, perhaps the first reference there comes from Henkel, Bayer and Coutts (1938) in Natal. In fact, within South Africa piping appears to be most evident in Natal (De Villiers, pers. comm., 1971) and has been the subject of intermittent studies there ever since (Bosazza, 1950; Downing, 1968; Beckedahl, 1977). Henkel *et al. (op. cit.)* observed the development of 'long underground water channels' afresh and described it as 'a new type of subsurface erosion'; the only other reference they could trace was the summary of Thorp's (1936) observations in China provided by Jacks and Whyte (1936, 60). In fact, although their paper has been ignored, except by Downing (1968), their observation of the processes, description of soils and analyses of the infiltration and swelling properties of the soils were remarkably good for their time. They formulated the idea of an erodible layer above an impermeable substrate ten years before Fletcher and Carroll (1948) in America. They described a sandy horizon overlying clay; where the clay swelled to restrict percolation corrasion occurred through the sand, but where wide desiccation cracks developed in the clay quick addition of water before saturation of the surrounding soil led to corrasion of broad channels.

Lester King (1942, 69) observed water enlarging cracks around gullies forming a 'nascent donga' (incipient gully), similar to the process observed in Europe by Walter Penck (1924; 1953, 323), and Bosazza reported it was more widespread, examples being quoted from the eastern Free State, Transvaal and Eastern Cape Province as well. Murdoch (1968) also reported piping in ferralsols in the Swaziland Highveld. Downing (1968) concluded that piping required:

  i) deep mature soils with a) very permeable subsoil, or b) impermeable clay subsoil that cracks prismatically,

 ii) alternate wetting and drying, and

 iii) possibly vegetation aiding (ii).

Only Beckedahl (1977) seems to have been really aware of work elsewhere, such as that of Downes, Terzaghi and Peck, Parker and Stocking. Beckedahl proposed that piping developed when vertical infiltration in high grass hollows saturated the clay underlayer, thus concentrating flow downslope along a zone of higher permeability, using cracks and eroding channels.

Elsewhere in Africa isolated observations were made by Thorbecke (1951) and Clayton (1956) in West Africa, by Berry and his co-worker Ruxton in both the Sudan and Hong-Kong (Ruxton, 1958; Berry and Ruxton, 1959, 1960; Berry, 1964, 1970), Haldeman (1956) in Tanzania and Bishop (1962) in Uganda.

Thorbecke (1951, 38) reported 'interirdische Wasserad-
ern und beite Grundwasserstrome' on pediment slopes in the
savanna of the Central Cameroons. The seepage probably
enters scarp foot depressions around inselbergs and emerges
on the plains as semi-circular springheads 3-7m below the
plain at the head of forested troughs, linear depressions or
'Bergfussniederungen'. Thorbecke found one spring-head
which extended up to 9m headwards in one season. Clayton
(1956) found the same features in Ghana: 'The incised but
wandering paths of the competitive 'spring-heads'are thus a
reflection of irregular courses of underground seepage
along involute prearranged lines of deeper and more saturated
weathering mantle' (ibid., 114). It appeared to be typical
of weathered granite residuum where porosity varies widely
in short distances causing perched water tables, and where
rejuvenation has occurred to initiate sapping. Berry and
Ruxton (1959) observed that seepage above hardpans could
cause landslips, whereas deeper seepage and eluviation particularly where deep root channels (< 30ft or 9m) penetrate
through illuviated or hardpan B horizons into 'zone 1'
debris from granite weathering may cause catastrophic gullying. They observed that lateral eluviation of clays often
emerged in valley-springs (ibid., 61) and generally distinguished between 'weathering profile water' and 'soil water'
or 'rock reservoirs'. The most active removal occurred on
concave slopes. In 1970, Berry actually used the term
'piping' to refer to similar features in the Sudan and was
clearly aware of work in America and Australia. But most,
like Haldeman (1956) and Bishop (1962), appear to have been
quite unaware of other literature on piping. Bishop believed
that the bulk of water flowing into gullies in Queen
Elizabeth National Park, Uganda, was underground; perched
water tables and 100-300ft (30-90m) rejuvenation led to
undercutting of gully heads and temporary tunnels up to 1m
diameter leading from the heads of active gullies, mainly
in unconsolidated Pleistocene lacustrine beds.

More recently, detailed soil analyses have been
employed in the first quantitative studies in Africa by
Stocking (1976a and b, 1978) in Rhodesia, and by the Rhodesian Ministry of Agriculture (Wendelaar, 1976). Following
Australian and American experience, these recent authors
have associated piping with accelerated erosion in dispersive
soils (cf. section 5.1).

In Eastern Europe there seems to have been some follow
up to Pavlov's (1898) observations by Berg (1902) in
Kazakhstan and the observations of dolines, natural bridges
and gullies formed by collapse of underground channels by
Lazinski (1909) in Poland. Most of the subsequent work was
in fact conducted in Poland, where extensive areas of loess
are particularly prone to severe soil erosion, including
'suffosion' (Reniger, 1950). In fact, piping appears to be
present in all the main loess areas of the world, from
Yugoslavia, South Germany, Hungary, Czechoslovakia, Poland
and Austria to Russia, China and New Zealand. Like Fuller
(1922), Kriechbaum (1922) Zaborski (1926) and Malicki (1935;

1938; 1946) maintained that the suffosion of loess was dominantly mechanical. Malicki (1946) noted that although the loess of the Titel plateau on the Danube contained 22% calcium carbonate, suffosion also occurs in loess containing no calcium carbonate. Hence, whilst Sweeting's (1972, 308) recent statement that solution of calcareous cements often plays a part in creating pseudokarst may be true, it is a far from necessary condition. Malicki observed three distinct levels of piping associated with three bands of clay in the loess. Further observations were made in Polish loess and loessial soils by Musierowicz (1951), Maruszczak (1953), Wieckowska (1953), Zolnierczyk (1956), Czeppe (1960), Klimaszewski (1961), Walczowski (1962; 1971), Malinowski (1963), Kühn (1963), Kaszowski et al, (1966), Buraczynski and Wojtanowicz (1971), Kastory (1971), Zaborski (1972), and Galarowski (1976). Walczowski (1971) emphasised the variability in the processes depending on local environment, which he said prevents full generalisation. Marusczczak (1953) maintained that chemical suffosion causes marshes whereas mechanical suffosion is responsible for potholes and open tunnels.

Most of the Polish literature is highly descriptive, no doubt in part reflecting a lack of analytical facilities. Some of the more quantitative work is due to engineering geologists such as Malinowski (1963) who found that the commonly used engineering measurement for assessing the susceptibility of loess to failure was inadequate for cases of tunnel erosion (cp. section 4.6(iii)). As in Western literature, there is the controversy between the roles of saturation or cracking in tunnel formation. Walczowski (1971) believes the tunnels in the Vistula valley grow from exit points in banks or terrace scarps, running through gravelly sand sandwiched between clay below and loess above and eventually collapse to give surface furrows. Near interfluves crack inlets become important, capturing surface water and becoming wider. Czeppe (1960; 1965) and Galarowski (1976) have provided some of the best, detailed studies of the processes, including re-surveying of developing systems (cf. section 8.5), although Walczowski (1971) criticised Czeppe (1960) for including frost cracking in the suffosion process. In fact, Czeppe's observations of the role of frost and of snowmelt flooding in developing piping in Poland and Spitzbergen were significant steps in widening the range of environmental factors. In a highly descriptive and generalised account, Zaborski (1972) also mentions the role of the spring thaw together with intense showers, and he describes how water falling down vertical chimney-like inlet conduits may erode through less permeable layers to create a new level of piping below.

In Russia, suffosion is also a particular feature of loess steppes. Gvozdeski (1954) emphasised that suffosion is purely mechanical, but both he and Lilienberg (1955) in the Georgian Republic recognised cases of combined mechanical and chemical erosion which they termed 'karstic-suffosion'. Sokolov (1960) was also against applying the term 'suffosion' to chemical erosion. Nevertheless, it was again so used by

Alpaidze (1965) for lake depressions around Tbilisi. Returning to the original definition, Bogutski (1968) described
mechanical eluviation forming hollows in loess steppe and
Subakov, Nemchinova and Sumochkina (1967) provide a reasonably detailed account of piping in the 30-60m thick loess
and loess-type loams in the Karshinski Steppe in Uzbekistan.
They consider that both solution and mechanical erosion are
responsible, with local intensification in gypsum layers or
in gravels and clays cemented with calcium carbonate. 10-
30m deep gullies encourage piping which develops along the
curve of the soil water table usually only 3-5m back from
the gully wall, encouraged by cracking near the gully.
Glukhov (1956) and Skvaljeckij, Halmatov and Hasanov (1971)
reported piping in loess banks bordering canals and reservoirs in Central Asia.

In addition, Kunsky (1957) and Kral (1975) have described suffosion in Czechozlovakia, Pinczes (1968) in loess
in Hungary and Sencŭ (1970) in conglomerate in Rumania.
Kunsky put forward a number of quite untenable assertions;
for example, that there is no connection between surface and
subsurface drainage as there is in true karst, that there is
no evolution of underground cavities and still less a pseudokarst cycle, and that pseudokarst forms 'soon' disappear.

In Western Europe, apart from Walter Penck's (1924;
1953, 323) brief observation, remarkably little is said of
piping until the last ten years. Kosack (1952) did a world
survey of pseudokarst forms and Hauser and Zötl (1957)
actually measured sediment discharge from suffosion. Tricart
(1965) briefly mentioned suffosion. But in general, West
European literature showed a marked lack of reference to the
worldwide English language literature until the papers by
Jones (1971) in English and Rathjens (1973) in German.

More has been published on piping by emigré Europeans,
particularly in tropical areas, than within Western Europe,
with the exception of the growth of interest in piping primarily from a hydrological viewpoint in Britain over the last
decade (q.v.). Apart from the African cases already mentioned, the European influence is clearly demonstrated in the
work of Feininger (1969) and Khobzi (1972) in Colombia,
Morgan (1972) in Malaysia, Bremer (1972; 1973) in Brazil,
Baillie (1975) and Sarawak and Franzle (1976) in Africa and
South America.

Last but by no means least, we must return to America
where geomorphological interest was stimulated in the 1960s,
with Brown's (1961) thesis work continuing the approach of
Fletcher and Carroll and at the same time providing the
first review of the state of knowledge based on American and
Australasian work, and with the papers by Brown (1962) and
Parker (1963) bringing it to a wider audience. Parker sees
piping as a fundamental agent of erosion in the 'drylands'
and his subsequent work has sought to elaborate on its
distribution and the processes involved (Parker, Shown and
Ratzlaff, 1964; Parker and Jenne, 1967), in particular making

an important link with the research into the mechanics of piping done by engineers (Terzaghi and Peck, 1948). Smith (1968) obstensibly applied the ideas of Parker freshly to a reinterpretation of periglacial boulder concentrations in terms of piping, without a fact noting that Czeppe (1965) had already done this very thing, using the American term 'piping' although drawing upon East European knowledge on 'suffosion', for active block fields in Spitzbergen. Like Czeppe, Smith published in English in a Polish journal. In the light of Parker's (1963) paper, Smith (1968) clarified his earlier observations of periglacial boulder fields (Smith, 1949; 1953) and expounded a theory for their develop-ment based upon piping processes aided by ice wedge crack-ing and meltwater. He envisaged an accumulation of coarse debris choking the valleys during the Wisconsin glaciation, followed by 'normal' processes after the climatic ameliorat-ion in which subsurface flushing occurred, beginning with localised undermining of the sod and ending with 'stone rivers': the Driftless Area of Wisconsin represents an early stage (Smith, 1949), Hickory Run Boulder Field, Pennsylvania (Smith, 1953) a more advanced stage, and the stone rivers of the Falkland Islands the final stage.

A little ahead of Parker, Mears (1963) had reported on the karst-like features in the badlands of the Chinle Form-ation in the Painted Desert of Arizona, and since then pip-ing has been recognised as a major process in badlands throughout North America (Bell, 1968; N.O. Jones, 1968; Mears, 1968; Barendregt, 1977; Barendregt and Ongley, 1977; Bryan, Yair and Hodges, 1978) and in Israel, particularly by Yair, Lavee Bryan and Adar, (1980), with a brief mention by Gerson (1977).

The history of piping research by civil engineers was very much a separate affair until the 1970's. Only then did the significance of fundamental research into the dispersive properties of clays by soil scientists (cp. Van Olphen, 1953; Richards, 1954) and later by engineers such as Quirk and Schofield (1955) Collis-George and Smiles (1963) and Rowell (1963) become apparent as evidence was brought forward in research by the New South Wales Soil Conservation Service and the CSIRO in Australia (e.g. Ritchie, 1963; Wood, Aitchi-son and Ingles, 1964), at the Israel Institute of Technology (e.g. Kassiff and Henkin, 1967) and by Sherard and associated originally with the USDA Soil Conservation Service(e.g. Sherard, Decker and Ryker, 1972) of piping failures in earth dams which had not been predicted by the application of standard engineering theory.

Engineering concern with piping appears to have begun when Col. Clibborn predicted the collapse of Narora Dam on the Ganges due to piping in 1898 (Fairbridge, 1968, 849). Terzaghi first formulated his theory of piping in Europe in the 1920s (Terzaghi, 1922; Redlich et al., 1929). This was to become the classic approach of engineers until this decade. Terzaghi used his theory to explain a natural devel-opment of piping after severe flooding along the Mississippi

(Terzaghi, 1931), and it became firmly cemented in soil
mechanics literature in his pioneer English texts on the
subject (Terzaghi, 1943; Terzaghi and Peck, 1948; 1966).
Terzaghi and Peck (1948) took a slightly more comprehensive
view than most previous engineers, recognising two basic
forms of piping:-

1. 'Subsurface erosion' in which scouring migrates
   upstream beneath a dam from springs on the tail-
   slope and for which they say 'the mechanics of this
   type of piping defy theoretical approach' (ibid,
   339).

2. 'Heave', involving 'boiling', which they claim
   acts in the opposite direction.

They noted that most piping failures appeared to be due to
'subsurface erosion' (ibid, 506). Nevertheless, it was the
beguilingly simple 'heaving' theory (Fig. 5) which was most
commonly elaborated upon by engineers (Lane, 1935; Casagrande,
1937; Legget, 1939; Bertram, 1940, 1967; Glossop, 1945,
77-80; US Bureau of Reclamation, 1947; Road Research Lab,
1957; Luthin and Reeve, 1957; Capper and Cassie, 1961; Harr,
1962, 124-128; Jumikis, 1962, 332; Sherard et al, 1963;
Linsley and Franzini 1964, 202; Cedergren, 1967).

Soil chemistry had been traditionally ignored by engineers both in the West and in Eastern Europe (Malinowski,
1963; Lehr and Stănescu, 1967). Lehr and Stănescu carried
out simulation experiments and produced graphs of the head
of water required to cause and continue heaving given varying
percentages of silt and silt-clay (op. cit., fit. 4). Small
open pipes were created in soils with less than 15-20% in
the sub-5 μm range, which soon developed turbulent flow, but
more clayey soils merely heaved en masse. Nogushi, Takahashi
and Tokumitsu (1970) did similar, though less informative,
experiments in Japan with a concoction of sand and coal-mine
waste; but they did little more than demonstrate that under
sufficient hydraulic gradient a cavity can grow upwards
through continued settling of the roof without total saturation.

The simulation experiments by Ritchie (1963, 1965) and
Rosewell (1970) for the New South Wales Soil Conservation
Service, and the field observations of actual failure by
Wood, Aitchison and Ingles (1964) for the CSIRO began to
change this and point to chemical bases for dispersion and
piping. At the same time engineers were beginning to realise
the widespread nature of earth dam failures as a result
of reports from around the world; Rao (1960) in India, Cole
and Lewis (1960) in Australia, Jennings and Knight (1956,
1957), Knight (1959), and Pazzi (1963) in
South Africa, Volk (1937) and Jessup (1964) in the United
States, Peterson and Iverson (1953) in Canada, and Aisenstein, Diamant and Saidoff (1961) in Israel.

In November 1964 the CSIRO Soil Mechanics Section and
the Water Research Foundation of Australia convened a

colloquium on the failure of small earth dams. However, the proceedings though available as individual research papers from the CSIRO, were never properly published; a fact which led Perry (1975) to lament that this isolated most of the world from a major part of research into dispersive clays. As details of the Australian research became available (e.g. Ingles and Wood, 1964; Aitchison and Wood, 1965), confidence weakened amongst engineers that they had correctly identified the causes of piping and its remedies. According to Reséndiz (1977), confidence 'nearly vanished' following the additional reports by Sherard (1971) and Sherard, Decker and Ryker (1972a and b), which were the first studies of piping failure in dispersive soils in America. These publications suggested that the standard soil mechanics tests were not adequate to identify piping soils and that, since the mechanics of piping had largely been misunderstood, nor were the standard preventive measures. This view was largely upheld by the symposium on dispersive clays and piping organised by Sherard and Decker in Chicago on behalf of the American Society for Testing and Materials in 1976 (Sherard and Decker, 1977). In the words of Sherard, Decker and Ryker (1972a, 590), the Australian conclusions were 'novel and somewhat surprising to the soil mechanics engineer', although the fundamentals of colloidal flocculation were understood before 1900 and agricultural soil scientists had been aware of piping erosion and of the danger of erosion in clays with exchangeable sodium percentages (ESP) of 15 and over since the 1930s. This realisation called into question the validity of the standard use of gravel filters to protect tailslopes against piping, and renewed experiments were conducted by Sherard and associates at the SCS and by the US Army Waterways Experiment Station, Vicksburg, into the feasibility of filter protection for dispersive clays (Perry, 1975).

At this time, the gap between the natural field scientists and the engineers began to be bridged, with Ingles and Aitchison (1970) referring to the observations of natural piping made by Fuller (1922), Downes (1946) and Parker (1963), and Sherard et al. (1972a) referring to Fletcher and Carroll (1948), Carroll (1949), Parker and Jenne (1967), Bell (1968) and Heede (1971). Moreover, the effects of desiccation or subsidence cracking, commonly blamed for pipe development in the natural landscape, were also actually observed in dam failures. A number of engineers recognised piping on dams caused by rainwater making use of surface cracks rather than seepage of reservoir water (e.g. Sherard et al., 1972a). And in the Baldwin Hills Dam, near Los Angeles, piping developed through the protective clay lining along bedrock faulting caused by the artificial injection of brine in order to extract oil (Jessup, 1964; Hamilton and Meehan, 1971).

Hence Decker and Dunnigan (1977) stated two necessary pre-conditions for piping failure as i) cracks or other large voids and ii) enough clay material to allow enlargement of voids prior to erosion. Indeed, Sherard, Dunnigan and Decker

(1977, 7) go so far as to say: 'There is much experimental and theoretical evidence supporting the conclusion that failures are not initiated by the general seepage through the pores of a well-compacted clay mass', but that cracks are necessary to concentrate flow.

It is apparent, therefore, that the barriers between disciplines, as well as the barriers between workers in different areas of the world, have been breaking down during the 1970s. One of the major barriers has been that between field scientists and construction engineers, for which the breakdown was begun with the papers by Parker(1963) and Ingles and Aitchison (1970). It is clear that both groups have already benefitted from this cross-fertilisation, and it seems inevitable that much more will be gained through closer co-ordination between the disciplines in the future.

This brief historical review has deliberately excluded any mention of recent work on the hydrology of soil pipes. As already pointed out, this is really a major aspect of 'current affairs' and it is dealt with fully in Chapter 7. Suffice it to say at this juncture that whilst interest in soil pipe hydrology really originated amongst field scientists, the first signs of interest appear to have now come from another branch of civil engineering, the engineering hydrologists (Dovey, 1976; Rees, 1979).

4. FACTORS IN THE INITIAL DEVELOPMENT OF PIPING

As stated at the outset, theories as to the nature and causes of soil piping are legion. In fact it is clear that no single factor or group of factors is universally responsible for the development of piping and that many writers have been somewhat incautious in saying categorically that they are.

The most extensive collation of the literature to date is that by Jones (1975) covering over 150 references directly concerned with piping. Crouch (1976) has also produced a very useful and succint review for the Soil Conservation Service of New South Wales based on 23 piping papers, of which nearly half arose from the Service's own detailed research programme.

Most quantitative research has suggested physical causes related to the hydraulic environment and to soil structure and composition. High susceptibility to cracking commonly associated with relatively high silt-clay content and a high percentage of swelling clays such as montmorillonite is most commonly quoted (e.g. Parker, 1963). This is followed by high intensity rainstorms and natural or artificial devegetation aiding erosion, e.g. the frequent torrential rains following summer droughts in British Columbia (Cockfield and Buckham 1946), Colorado (Brown, 1961; 1962), North Dakota (Bell, 1968), and Arizona (N.O. Jones, 1968), or the devegetation caused by pioneer agriculture in Australia (Downes, 1946), Arizona (Fletcher and Carroll, 1948), and Hawaii (Kingsbury, 1952).

This group of factors is likely to be most influential in the development of the stress-desiccation type of piping described by Parker and Jenne (1967) or, by analogy with bedrock caverns, of 'vadose-zone' piping (section 6.3). Perhaps significantly, most of the papers suggesting these causes result from observations in relatively dry grassland areas especially in the western U.S.A. and southern Australia. Other researchers have emphasised the need for steep hydraulic gradients, often associated with the lowering of local base level, and erodible soil layers lying above zones of restricted permeability (e.g. Fletcher et al., 1954). This group of factors may be more universal since they are likely to be present in both 'vadose-zone' piping and water table piping such as would result from boiling or heaving (cf. Terzaghi and Peck, 1948). This group of factors alone would favour Parker and Jenne's (1967) variable permeability - consolidation piping, but it is often linked with high susceptibility to cracking in the erodible subsoils (Fletcher and Carroll, 1948; N.O. Jones, 1968) suggesting that some forms of piping may span the division between Parker and Jenne's stress-desiccation and variable permeability forms of piping.

Associated with the significance of soil erodibility have been suggestions regarding soil chemistry. It has been widely held by many workers that piping was particularly associated with alkaline soils with high soluble salts content, high base exchange capacity and particularly with a high exchangeable sodium percentage.

Yet another group of suggested causes has been biotic, ranging from biotic break-up of the topsoil increasing the surface infiltration capacity to direct development of pipes from animal burrows or plant root holes. These biotic causes have been remarkably popular, and in fact ranked third in frequency of citation in Jones's (1971, 606; 1975, 39) survey of about 30 major papers.

Fletcher *et al*. (1954) provided a very simple and beguiling list of the five necessary factors for piping which has been quoted many times since (e.g. Parker, 1963 : Ward, 1966) :

i)   a source of water;

ii) a surface infiltration capacity greater than the subsoil permeability unless rodents or ploughing break the less permeable surface;

iii) an erodible layer just above the retarding layer;

iv) sufficient hydraulic gradient;

v)   an outlet.

Crouch (1976) has listed the predisposing factors as:

i)   a seasonal or highly variable rainfall combined with high summer temperatures;

ii) a soil subject to cracking;

iii) a reduction or detrimental change in vegetative cover;

iv) a relatively impermeable layer in the soil profile;

v)   the existence of a hydraulic gradient within a dispersible soil layer.

Parker and Jenne (1967, 25) also provided a summary of causes based upon their experience:

A) Silty and clayey materials are susceptible if they:

i)   contain 20% montmorillonite;

ii) desiccate thoroughly, or are susceptible to localised subsidence, or have a stratum of high

permeability relative to lower strata, or
have a temporary perched water table;

iii) contain a high percentage of exchangeable
sodium. However, due to subsidence, loess and
possibly other low-density previously unwetted
earth materials may pipe even in the absence
of montmorillonite and large amounts of
exchangeable sodium;

B) Regardless of location or materials, at least 4
basic conditions are essential:

i) sufficient water either to cause drainage
through cracks or to saturate a layer of higher
permeability than the layers below it;

ii) hydraulic head sufficient to move water
through a subsurface route;

iii) presence of a permeable or deeply cracked soil
or bed rock above gully floor level;

iv) an outlet.

The latter group of essential factors is clearly similar to
the list of Fletcher et al. (1954). Above all, they list
4 minimum requirements: a) enough water must be available
to fill drainable cracks, b) the strata must be montmorillonitic, c) the strata must desiccate thoroughly, if only
seasonally, and d) there must be an outlet for drainage
(op. cit., 4). Furthermore, desiccation-stress crack piping
is favoured by higher sodium to calcium and magnesium ratio,
lower aggregate stability, less vegetation cover and less
slope.

Barendregt and Ongley (1977) also provide two lists of
causes, one general and one local:

Essential conditions, in decreasing order of importance:

i) a high percentage of swelling clay (montmorillonite) and a pronounced dry period to desiccate
these clays;

ii) high intensity rainfall to provide enough water
to saturate a part of the bedrock or surficial
deposits;

iii) a relatively permeable zone above the local
erosional base level-

iv) a steep hydraulic gradient;

v) an outlet above base level (usually appearing
first as seepage zones);

33

Factors favouring piping in their area (Milk River Canyon, Alberta):

i) the presence of swelling clays, especially bentonite;

ii) the abundance of fine-grained sediments;

iii) the presence of soft unconsolidated bedrock as well as alluvium;

iv) large zones of porous material alternating with impermeable members;

v) long, hot dry spells with episodic intense precipitation;

vi) active badland erosion;

vii) incised gullies and a degrading river;

viii) sparse vegetation;

ix) high local relief producing a steep hydraulic gradient;

x) the presence of seasonally large hydraulic heads provided by sloughs on the prairie surface above the canyon walls and by the thick porous sandstones resting on impermeable clayey carbonaceous shale and lignite (Barendregt, 1977, 146).

The following sections will explore the bases of these theories by looking systematically at the results of research as they relate to the broad categories of environmental and soil property factors. These will then be drawn together in the context of a 'general theory', which will include the results of detailed work on mechanical and soil reactive processes.

4.1. The climatic factor

The most frequently cited element of climate has been rainfall. Many have considered absolute amounts of rainfall to be important; hence the common association of piping with 'the drylands'. Speaking of soil erosion in general, Hudson (1971, 27) has suggested that the most important factor influencing water erosion is mean annual rainfall. He argues that in regions of very low rainfall there can naturally be little erosion caused by rain, whereas too much rainfall results in dense vegetation cover, so that most erosion occurs in the range in between, with the exception that tropical rains are heavier and therefore more destructive. Such generalisations are particularly belied by piping erosion, which can now be shown to be active over a very much wider climatic range than hitherto imagined and in which climatic variability appears to have far greater influence than climatic averages. Indeed, most authors have identified some aspect of rainfall distribution and variability as of paramount importance. In some cases snowmelt may be significant and in others temperatures considerably modify the effectiveness of the rainfall and possibly even affect soil response. In the longer term, however, climate is also a factor in pedogenesis and therefore to some degree subsumed in pedological factors.

Kosack (1952) produced maps of the world distribution of pseudokarst which show the main areas falling within a broadly dry climatic spectrum, the BW and BS zones of the Koeppen classification, although his definition is somewhat broader than adopted in this monograph and included pseudo-karst in 'coral chalk' in the Pacific Islands and in polar ice caps and glaciers. Parker (1963, 111) made the first attempt to analyse the climatic distribution of piping by collating about two dozen published reports, most of which fell within what Parker termed 'the drylands', with the exception of sites in New Zealand and Australia (v.i.), Hawaii (Kingsbury, 1952) and Bolivia (Dobrovolny, 1962). Parker and Jenne (1967) therefore concluded that the dominant type of piping is 'stress desiccation cracking'. Dregne (1967) and Lustig (1970) also give the impression that piping is predominantly a process of semi-arid and arid areas, and Cooke and Warren (1973, 141) believe that piping is a 'fairly common and perhaps characteristic feature of certain drylands'.

However, the impression that these are the only or even the main areas of piping is now plainly open to question. Kosack included only the Australian desert in the region of pseudokarst in Australasia and Parker only referred to the work of Downes (1946) in Victoria and Gibbs (1945) and Cumberland (1944) in New Zealand. The latter case he said, 'appears to be exceptional....requiring special localized conditions' (Parker, 1963, 111). But subsequent work, particularly in Australia, New Zealand and Britain has shown that this is not so. Jones (1975, 46-48) produced a collation of 150 references which indicated a distribution extending from periglacial climates to the rainy tropics. He proffered that piping had been reported in most major climatic zones and that further exploration might well reveal piping in most of the zones thus far unrepresented. His list can now be extended somewhat as indicated in Table 1 and Figure 5. Oddly enough, although there are numerous reports of natural piping in the United States, these are still confined to the drier states west of a line from North Dakota to New Mexico, confirming the affirmation of Parker and Jenne (1967) that piping occurs across the Great Plains from Canada to Mexico and from the Cascades to the Sierra Nevadas. But surely Horton and Kirk Bryan observed piping in the more humid east, even if they did not call it such?

Clearly, there are two problems involved in using an inventory of the literature to determine distribution: first, the lack of trained observers in certain parts of the world

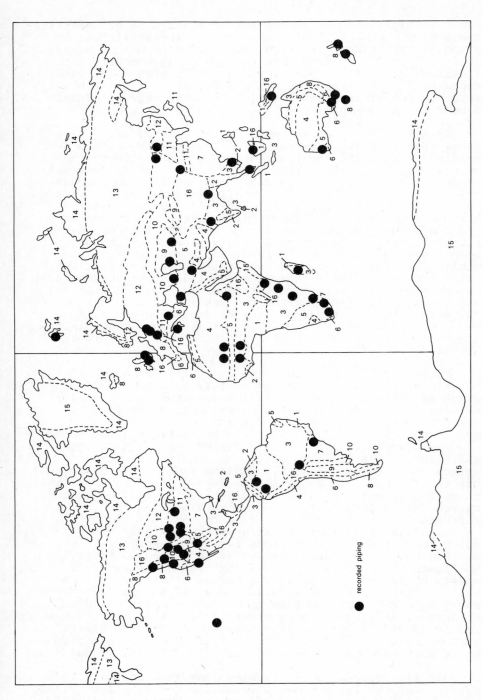

Fig. 5   World Climatic distribution of published observations of soil piping.

and second, 'observer myopia' caused by the natural tendency to study only the best developed cases, the cases causing greatest problems to agriculture and engineering, and the cases of most relevance to the adopted discipline of the observer.   An example of the first problem is probably the lack of reports from tropical arid regions: H.T.U. Smith claims to have recognised features probably due to piping on air photographs of internal drainage basins in North Africa (Smith, 1969) and in the Lake Eyre basin in Australia (Smith, pers. comm., 1971), but there is little other evidence.   The distribution of observations in America may well be a good example of the second problem.

No single climatic feature is common to all the reported instances of piping. However, two features tend to be fairly common, namely, periods of desiccation and periods of intense rainfall. Drew (1972, 207) maintains that most people agree on the need for a semi-arid climate with inter- mittent, heavy rainfall, whilst Rathjens (1973) believes it may be so typical of a savanna climate for it to be incor- porated in a system of climatic geomorphology. Drought followed by heavy convection rainfall has been cited by Cockfield and Buckham (1946) and Buckham and Cockfield (1950) near Kamloops, B.C., with an annual rainfall of 254 mm, in Colorado by Brown (1961; 1962) under 432-508 mm p.a., in North Dakota under 406 mm p.a. by Bell (1968), in Arizona under 279-355 mm p.a. by N.O. Jones (1968), in the Western U.S. generally by Parker (1963) with annual receipts usually below 380 mm and in central Sudan by Berry (1970) with 130- 600 mm p.a.   Jones (1968) noted that 60% of the annual receipts fell in these torrential rains in Arizona and Bell (1968) suggests 80% for North Dakota.   The role of droughts and torrential storms has also been noted in more humid climates; in the wet-and-dry tropical climate of Uganda by Bishop (1962) with 634 mm p.a. and 35-60% of this in heavy storms on about 12 days a year, or in the marine climates of Victoria (Downes, 1946), with 530-580 mm p.a., and New Zealand, as reported in South Island by Gibbs (1945) under 710 mm p.a. and Hosking (1967), 508-762 p.a., and in North Island by Blong (1965), with a mean of 1295 mm p.a. rising on occasions to 1930 mm, or Ward (1966) with 762-1780 mm p.a.. Gibbs reported intensities up to 15 mm hr^{-1} in the summer storms and Ward noted 6 hr intensities of 75 mm with a two-year return period. Recently, Floyd (1974) has related the extension of piping in New South Wales (mean 440 mm p.a.) to exceptionally wet years. Heavy summer rains were particularly effective in initiating piping, because of poorer vegetation cover, cracked A horizons and possibly increased dispersibility of the soil in higher temperatures in the summer, whereas heavy rain in winter did not appear to have an effect.   In South Africa, Henkel, Bayer and Coutts (1938, 241) reported a mean of 865 mm p.a., 81.4% of which occurred during the summer accompanied by marked desiccation. Coutts's experiments had shown that light rain- fall (<12 mm) infiltrated no more than c. 125 mm leaving the subsoils dry and cracked until heavy falls occurred.   A severe case of heavy rainfall was described by Starkel (1972b) in the tropical monsoon climate of Sikkim, where 1091 mm of

Table 1 Climatic distribution of published reports of natural piping.

Climatic type	Geographical area	Principal authors and publications*
Tropical rain forest		Löffler (1974), Baillie (1975), Franzle (1976).
Monsoon tropics		Fontaine (1965), Eyles (1968) Neboit (1971), Morgan (1972), Starkel (1972b), Löffler (1974).
Wet-and-dry and semi-arid tropics	East Africa	Haldeman (1956), Thomas (1960), Bishop (1962), Berry (1970), Stocking (1976; 1978), Wendelaar (1976).
	West Africa	Thorbecke (1951), Clayton (1956), Dregne (1967).
	Sudan	Jefferson (1953), Worrall (1959; 1960), Berry (1964; 1970).
	New Guinea	Löffler (1974).
	Ecuador	Funkhowser (1951).
	India (Rajastan)	Starkel (1972a).
Humid and dry summer subtropics	Brazil	Maack (1956), Bremer (1971; (1973).
	S.E.Australia	Downes (1946, 1949, 1956, 1959), Monteith (1954), Newman and Phillips (1957), Beare (1965), Charman (1969, 1970a and b), Ingles and Aitchison (1970), Floyd (1974), Crouch (1976), Imeson (1978).
	S.W.Australia	Conacher (1975), Conacher and Dalrymple (1977).
	South Africa	Henkel et al.(1938), King (1942), Bosazza (1950), Downing (1968), Beckedahl (1977).
	Turkey	Fletcher et al.(1954).
	Spain	Kunsky (1957).
	Italy	Neboit (1971).
	Yugoslavia	Kunsky (1957).
Mid-latitude marine climates	Britain	Knapp (1970a), Weyman (1971, 1975), Jones (1971, 1975, 1978a, 1979), Stagg (1974, 1978), Lewin et al.(1974), Oxley (1974), Dovey (1976), Newson (1976), Cryer (1978, 1980), Conacher and Dalrymple

Climatic type	Geographical area	Principal authors and publications*
Mid-latitude marine climates cont...	Britain	(1977), Newson and Harrison (1978), Rees (1979), Gilman and Newson (1980).
	New Zealand	Cumberland (1944), Gibbs (1945), Blong (1965b), Ward (1966a and b), Jackson (1966), Hosking (1967), Hughes (1972), Conacher and Dalrymple (1977).
	Tasmania	Dickinson (1950), Colclough (1973).
Mid-latitude	Austria	Zeitlinger (1959), Morawetz (1969).
continental	Czechoslovakia	Kunsky (1957), Kral (1975).
interiors	Poland	Malicki (1946), Czeppe (1960), Kuhn (1963), Malinowski (1963), Pinczes (1968), Walczowski (1971), Galarowski (1976).
	Rumania	Maruszczak (1965), Sencu (1970).
	Russia	Glukhov (1956), Subakov *et al.* (1967), Skvaljeckij *et al.* (1971), Klimontov (1972).
	China	Fuller (1922), Thorp (1936)
	North America	Bennett (1939), Bond (1941), Fletcher and Carroll (1948), Fletcher and Harris (1952), Fletcher *et al.* (1954), Brown (1961; 1962), Parker (1963), Mears (1963), Parker and Jenne (1967), N. O. Jones (1968), Barendregt (1977), Barendregt and Ongley (1977), Bryan *et al.* (1978).
Polar and sub-polar		Czeppe (1965).

* including research theses.

an annual 4050 mm fell, reaching a maximum intensity of 760 mm h^{-1}, and initiating mud-flows, slumps, slides and piping.

Downes (1946) stressed the importance of a combination of years of serious drought, especially affecting sunny slopes, heavy storms and the difficulties of land management with growing seasons varying from 3 to 10 months in Victoria. Newman and Phillips (1957) suggested a similar combination in New South Wales (c. 410 mm p.a.). Indeed, devegetation where due to climatic fluctuations, to human error or to a combination of the two has been frequently cited as a cause of accelerated piping erosion (cf. section 4.3). In fact, Colclough (1978) suggests that timber cutting and clearing for cultivation were the major factors in Tasmania where rainfall is moderate (500-650 mm) and evenly distributed, apart from a drier late summer.

Nevertheless, desiccation cracking may also be a cause in some more humid areas such as Britain, especially in peaty soils (Newson, 1976b; Gilman and Newson, 1980). The latter workers quote 2192 mm p.a. with measurable rain on 231 days a year in their mid-Wales research area, although 44% falls between October and January. Heavy rainfall (>50 mm) may occur 1-7 times a year, and some summer storms of 100 mm in a day have been reported (with estimated 10-year return period) which they found to be important to pipe formation in combination with cracking caused by summer droughts, especially in the light of evidence gathered during the great drought of 1975-76. Jones (1971; 1975) studied piping in a basin in the English Peak District with a mean annual precipitation ranging from 942 to 1018 mm across the area, an estimated mean annual evapotranspiration of 554 mm and a period of general soil moisture deficit between April and October reaching maximum deficits of 40 mm. Some evidence was found of pipes developing from summer desiccation cracks which penetrate the B horizon of the lessivé brown earths exposed in the streambanks, but this appeared to account for no more than c.10% of the pipes and he concluded that hydraulic 'boiling' under saturated conditions was probably a more common cause (Jones, 1978a, 13).

Ward's (1966, 71) conclusions in North Auckland are interesting here. Excepting the area described by Blong (1965) and possibly some of Cumberland's (1944),piping in New Zealand has been considered to be best developed where precipitation is under 380 mm p.a. and particularly in areas with a prevalence of high intensity aperiodic rainstorms. His study area fulfilled the latter but not the first criterion. Nevertheless, although pipes are undoubtedly extended during high intensity storms, he concluded that 'the localisation of a continuous or near-continuous supply of subsurface seepage water is more critical in the development and perhaps the initiation of piping, than are large water flows during storms'.

Löffler's (1974) observations in New Guinea also tend to argue against the need for a combination of desiccation and heavy storms, since he reported piping in both the tropical savanna with 2000-2500 mm rainfall strongly seasonally distributed and in the wet tropical forests with c.5000 mm evenly distributed throughout the year and a uniform temperature of about 27°C.

Both Bell (1968, 250) and Smith (1968, 200) have suggested that a periglacial climate may substitute for highly expansive clays and desiccation cracking in creating fissures in the soil. Smith suggests that original zones of higher permeability may be caused by melting of segregations of ground ice, leaving voids as well as providing meltwater. Moreover, cryostatic and hydrostatic pressures developed during seasonal refreezing could assist by forcing water along lines of weakness. However, he notes that observational proof is lacking. Bell provided some observational proof of piping promoted by crack systems left by ice wedges during contemporary freeze-thaw cycles in North Dakota and of piping developed in blocks of stagnant ice continuing into the underlying silt (we might call this 'superimposed glacial pseudokarst').

In some areas snowmelt replaces high intensity rainfall as the eroding agent, for example, in Spitzbergen (Czeppe, 1965) in Montana and the Falkland Islands (Smith, 1968), in North Dakota badlands (Bell, 1968), the Colorado Rockies (Heede, 1971), Poland (Galarowski, 1976), and in Albertan badlands (Barendregt and Ongley, 1977). Heede (1971) actually observed that cloudburst storms were not capable of generating any pipeflow.

Although many have suggested that climatic variability may be a factor, only N.O. Jones (1968, 149) in Arizona seems to have suggested that climatic change may have initiated arroyo and pipe development: 'A secular trend may be more important than unusual occurrences in producing major changes in vegetation, because only a series of unusual occurrences can eliminate new seedlings. Such ecological changes might be more important than any feasible degree of overgrazing.'

4.2. The role of biota.

Biotic activity has been a popular explanation for pipe development, varying in degree of explanation from complete responsibility to minor modification, but the published literature has generally settled on the latter.

Sharpe (1938) and Bennett (1939) both suggested that rodents were important in America. Carroll (1949, 4) maintained that 'the type of piping with which we are most familiar begins with the burrowing action of gophers and other animals'. He quotes a 'unique' case on the Old Santa Cruz West Branch near Tucson, where water drains through animal holes and shrinkage cracks in clay loams overlying gravelly-sand. The sandy layer is thin and soon clogs and stops the piping. More destructive piping has been caused by gopher holes along the banks of the San Pedro River, near Benson, Arizona, laying waste large areas of farmland. The gopher holes penetrate a highly dispersible subsoil, the improved drainage causes shrinkage cracks and a positive feedback occurs.

Cockfield and Buckham (1946, 8) also maintained that animal burrows were generally the initial points for washouts and presumed that the pipes probably began from such points of discharge and grew headwards. More recently,

41

Hosking (1967) and Baillie (1975) have also held that biotic
holes are important. Czeppe (1960) suggested that moles
and rodents were important near the surface, but that deeper
piping either followed root holes or developed on interfaces
between loam and clay debris or bedrock.

However, the early suggestions in America led Bond
(1941) to try to dispel doubts about the essentially hydrological nature of soil pipes by studying piping on Anacapa
Island, California, where no burrowing animals were known
to exist. Fletcher and Carroll (1948) also concluded that
burrowing animals are only a minor factor, in so far as they
can create inlets for water, but that they are neither a
necessary nor a common cause.

Newman and Phillips (1957, 165) took a more serious view
of the effect of rabbits in Australia: 'Rabbits are associated in all cases with tunnel-eroded areas and there is
little doubt that in some instances they have actually
initiated tunnel formation.... However, the major role they
play in tunnel development is their destruction and deplet-
ion of existing vegetation. Rabbits will not cause tunnel
erosion where other pre-disposing causes are not present'.
Occupation of existing tunnels undoubtedly encourages secondary tunnels. In presenting the results of the tunnel-
control experiments set up by Newman and Phillips, Floyd
(1974) has repeated that rabbits use existing pipes and he
believes that increased rabbit infestation between 1957 and
1970 aggravated the piping problem, but he emphasises that
tunnel formation came first. Jones (1971; 1975, 151) mentioned burrowing crabs possibly creating short pipes (<1 m
long) on English salt marshes by digging through from salt
pans into creek banks. Berry (1970) also mentions some
animals associated with pipes in the Sudan. Newson (1976b),
Jones (1978a) and Gilman and Newson (1980) considered and
largely discounted the possible role of moles in upland Wales
on grounds of the known distribution of *Talpa* and the geometry of the pipe networks (cf. section 6.3). An important
distinction is that mole runs are designed not to fill with
water (Newson, 1976b) or to exit into streams (Jones, 1978a).
However, moles may use pipes and even modify them. Weyman
(1975, 12) suggests that some pipe systems, such as those
studied by Kirkby and Weyman (1972) in the East Twins basin
on the Mendip Hills, Somerset, U.K.,' are little more than
randomly oriented mammal burrows in which the downslope
routes are used by water', a view followed by Waylen (1976,
106), whereas others, such as those studied by the Institute
of Hydrology on Plynlimon, mid-Wales, U.K., 'show all the
normal properties of channel networks', but this seems to be
a gross overgeneralisation for both implied end-members of
the sequence (cf. section 6.3).

Quite independently of any other literature on the
topic, Banerjee (1972) has considered a biotic origin for
pipes in ferrallitic soils in West Bengal. He suggests that
holes left by rats and particularly by termites and other
insects are smoothed and enlarged by water at depths of

50-150 cm. He also considered formation by iron replacement of tree roots, but found that the local fossil wood was siliceous and retained a clear wood structure. His argument for the prime role of insects is based on finding pipes of only a few millimetres diameter together with clear insect holes. Nevertheless, he points out that the vertical extensions and honeycombing typical of termites are absent and that termites tend to be only in the 'upper horizons'. In the light of the presence of these pipes above an impermeable layer of plinthite, i.e. in a good position for perched water table development, the present author would tend to question this supposed origin.

A curious variant on the theme of animals using pipes is the report by Kunsky (1957) of loess caves being used in Bohemia by the early Christian Church and later by other persecuted sects in Moravia and Austria between the 16th and 19th centuries. Most recently they have been used for growing mushrooms! Berry (1970) also mentions the use of water wells dug into pipes for the rainy season in the Sudan.

Whilst animals may contribute to piping, vegetation generally affords protection against severe piping erosion, which only tends to develop when that vegetation is removed, often due directly or indirectly to human activities (cf. next section). Many authors have linked piping with sparse vegetation cover, and there can be a positive feedback in so far as once piping has developed the improved drainage can cause vegetation to become sparser. Peterson (1954) attributes bare areas 'several hundreds of feet' wide along gullies the western U.S.A. to excessive drainage by pipes tributary to the gullies. Heede (1971) and Beckedahl (1977) also associated piping with strips of poorer vegetation. However, Fletcher et al. (1954) indicated that piping was marked by more succulent vegetation in Arizona. Jones (1975) noted the common association between pipes and flush vegetation along percolines in England and Wales, although lines of poorer vegetation marked the course of pipes in rather sandy acid brown earths (> 30% over 250 mm) overlying stony sand colluvium in the Brecon Beacons. However, soil biota, animals or decayed roots, may encourage concentration of throughflow in the soil through linear 'biopores' or through more porous 'pedotubules' (pseudomorphs of roots as described by Brewer and Sleeman, 1963), which could develop into piping (Jones, 1975, 59-62), although we have no direct proof of this.

Lastly, an odd case of 'negative activity' by ferric iron bacteria was noted by Jones (1971, 604) in which the bacteria had completely blocked a 10 cm diameter pipe in a gleyed lessivé soil. Similar blockages have been reported in agricultural drains in the Netherlands and elsewhere.

However, attention was focused on human activity rather than on natural biotic activity by Bond's (1941) observations of 'rodentless erosion'. In this remarkably influential and frequently quoted note, Bond reported that piping is

especially common in heavy clay soils where rodents are
relatively rare and that on the rodentless Anacapa Island
surface cracks were small in well vegetated areas but where
sheep had bared the ground, cracks were larger and tunnelling had developed. Human activity has been increasingly
implicated since the 1940s.

4.3. The role of human activity.

      Human activity has been blamed for the development or
accentuation of piping erosion in many parts of the world.
The most commonly cited elements of human interference have
been clearing of the land for agriculture and overgrazing,
but piping has also been caused by irrigation and by construction work. In general terms, we can divide the problem-
atical human activities into 1) those which affect soil
stability and 2) those which affect the local water balance,
although in many cases strong feedback links occur between
soil stability and water supply. We must also remember
that as the processes of piping erosion have become better
understood and as this knowledge has been communicated to
the agricultural and engineering community at large, so
Man's role has been slowly changing from that of a primary
factor in the generation of piping erosion more towards a
prime factor in preventing, halting and destroying piping.
This latterday role will be discussed more under the heading
of 'control and prevention' at the end of chapter 8, but
there is necessarily some overlap between this and our main
concern here, because most of the research which has implicated Man has been directed towards a conservationist goal.

      In early work in South Africa, Henkel, Bayer and Coutts
(1938) reported that their study area in Natal had never
been cultivated, although veld burning could have been a
factor. However, Man's role first become definitely apparent
in the soil conservation surveys of the 1940s, especially in
New Zealand, America and Australia. Guthrie-Smith (1922,
33-40) had described tunnelling in Hawke's Bay Province,
N.Z., as a natural process, but when Cumberland (1944, 98)
made his classic survey of soil erosion in New Zealand, he
decided that it was culturally induced. Lowdermilk (1936)
had already referred to cultural interference causing the
clogging of channels into and through the soil by increasing
the mobility of the fine fraction leading to increased
flooding in America, and Cumberland found that piping was
always associated with the replacement of natural vegetation
by pasture during European colonisation (cp. chapter 3).
Where ploughing had taken place, 'subcutaneous dimpling'
occurred rather than tunnelling or the more diffuse 'subcutaneous sheet erosion'. Fletcher and Carroll (1948) also
reported that ploughing, particularly combined with deep
rooted crops resulted in downwashing of the surface material
through coarse subsoils creating surface depressions in
Arizona. Carroll (1949, 7) restated the farmers' belief
that growing alfalfa or other deep-rooted legumes for more
than 3 years on a field encouraged more rapid percolation
and subsequent initiation of pipes and sinkholes. Fletcher

and Carroll also noted that in every one of their four study
areas in Arizona severe piping seemed to follow overgrazing
in the latter part of last century. Brown (1961; 1962)
reached a similar conclusion in Colorado. Lack of vegetation
cover allowed rainfall to compact the surface resulting in
quick runoff and stream incision, lower soil moisture levels
and cracking. However, although heavy grazing up to c. 1920
was responsible for initiating piping by devegetation, it
did not explain the chemical and physical structure of the
soil, which was already predisposed to erosion. Severe
droughts in the 1930s and 1955 aggravated the problem in
some areas of Colorado, as described by Downes (1946) in
Victoria.

From his survey of the literature, Brown (1962, 220)
concluded that 'in each instance....piping is associated
with prior land denudation,....(e.g. by) grazing'. Parker
(1963) also states that overgrazing is the commonest cause
of piping and Ritchie(1963, 111) has reiterated this view
from New South Wales: 'In all recorded instances of (field)
tunnelling, there is a background of overgrazing or cultivation and pasture cover has failed to provide adequate continuous ground cover, resulting in uneven infiltration and
sheet erosion'. Neboit (1971) has similarly blamed overgrazing in Southern Italy.

Downes (1946) listed the factors favouring pipe development in Victoria under the headings of soil types favouring, topography favouring and 'acquired conditions favouring' piping. In his view it was necessary to add irregular
infiltration and high runoff to favourable soil and topography and these were acquired characteristics caused by
vegetation loss and development of an impermeable surface.
Deforestation after 1870 and the very irregular growing
season which made grazing management difficult, possibly
aggravated by the 1902 drought, were the main triggers to
erosion. Marker (1958; 1976) took up Downes's theme. Early
records show that the number of springs increased soon after
settlement, probably largely because removal of the natural
vegetation made more water available. Recently, these
springs have become seasonal perhaps because the increased
number of springs lowered the water table more (Marker,
1958, 134), but Marker also mentions recent rejuvenation
caused by increased runoff. The change in land use accentuated the periodicity in water supply régime which also
increased erosion hazard. Furthermore, spring development
accentuated leaching which resulted in salt deposition downslope, the original vegetation, particularly tussock grass,
being replaced by a few salt-tolerant species unable to protect the soil, and the creation of 'saltings'. Charman
(1967) also noted that piping is sometimes associated with
saltings. Marker (1958, 130) also mentions 'semi-liquid
clay' erupting under pressure along lines of weakness,
particularly through rabbit burrows, and this reminds us
that one factor, though perhaps a minor one in direct terms
was the human introduction of the rabbit to Australia (section 8.7). Newman and Phillips (1957, 159) list overcropp-
ing, overstocking and rabbit infestation as contributary

45

here, and Stapledon and Casinader (1977) cite deforestation.

Conacher and Dalrymple (1977) have referred to a number
of examples of human interference in England, Australia and
New Zealand. Like Marker, they trace the development of
'salt scalds' in the semi-arid wheatbelt of S.W. Australia
to forest clearing, and they observed small pipes conducting
throughflow into the scald areas (*ibid*, 51; Conacher, 1975,
48). Plough pans are formed in the area by repeated ploughing to depths of 8-13 cm, which promoted downward translocat-
ion of fines and organic materials, and by compaction from
stock and farm machinery, which persists at depths just
below the limit of ploughing. Compaction and pan formation
encourage lateral movement of soil water (Conacher and
Dalrymple, 1977, 52-54). In contrast, in southern England
ploughing of seepage slopes (unit 2) has caused widespread
destruction of near-surface soil pipes as well as other
diagnostic pedological characteristics at the same time as
creating an Ap (plough pan) horizon (*ibid.*, 58), whilst
ploughing on mid-slopes (unit 5) may actually enhance
impermeability below the Ap horizon and result in the development of unit 2 characteristics (*ibid*, 60). An unfortunate
example of pipe destruction by ploughing in Wales is also
described Gilman and Newson (1980) (cf. section 8.6).
Meanwhile, deforestation in North Island New Zealand has
resulted in a change of process-response but not of landsurface unit type. Here clearing has accelerated mass movement and increased the rate of colluvial deposition on the
lower slopes with lateral subsurface seepage and piping play-
ing a significant role (Conacher and Dalrymple, 45). Crozier
(1969) has also attributed slope instability, severe polygonal cracking and piping in Eastern Otago to deforestation
in the last 120 years.

Czeppe (1960) studied three areas in the Upper San
basin, Poland, with contrasting settlement and cultivation
histories. One was intensively cultivated, one was partly
depopulated and largely covered by forest and uncultivated
fields, and the other was a completely depopulated area no
longer cultivated. Deforestation and ploughing appear to
have caused excessive soil creep and creation of a dusty,
structureless loam reaching 2m thick at the base of side-
slopes, which overlies a 0.5m thick clay loam resting on the
argillaceous shale bedrock. Proof of downslope migration is
given by clay intercalations, oriented boulders and increasing thickness downslope. Piping developed after the land
had not been cultivated for 10 years, causing landslides on
the valley slopes, and was aided by field rodents on the
agricultural terraces. Preferred locations for piping were
i) within the colluvial loam using animal and root holes,
ii) at the base of the loam above the relatively impermeable
clay underlayer or at the interface between this and bedrock,
the upper interface being largely a product of cultivation.

Floyd (1974) reported on the most extensive experiment
to date on methods of pipe control on agricultural land and
the relationship between pipe development and cultivation

techniques. Fuller discussion of his results will be deferred until a later section (section 8.7).

Aghassy (1973) has described two interesting cases of pipe initiation by human activity in Israel, one due to Bedouin cultivation techniques in the southern coastal plain during the first half of this century, the other due to the construction of a narrow-gauge railway in the same period. Apparently once the British brewing industry discovered the Bedouin grew barley in the area, the economic incentive led the arabs to plough nearer and nearer to the margins of the Nahal (Wadi) Besor. They used camel-pulled ploughs which worked straight up and down the wadi sideslopes. This led to rilling and piping and the development of badlands (cf. section 8.1). The second case occurred due to oversteepening of hillside cuttings during the building of the Ottoman Railway before the First World War in Gaza. Rill and pipe networks 2-5 times the size of the plough-induced networks developed until conservation work after 1948.

Parker and Jenne (1967) reported pipe initiation caused by interference with natural drainage due to highway construction in America. Piping systems and associated landslides have been described in glacial sediments disturbed by road construction near Binghamton, New York (Coates, 1977; 1979).

Parizek (1970; 1971; 1973) has noted a similar instance in America due to highway cuttings. Unstable slopes may result due to piping failure, because the cuttings act as groundwater drains. This changes seepage forces by either increasing the driving moments or reducing the resisting moments. The probability of failure is increased when the cuttings intersect the water table especially in residual or transported soils, such as alluvium, colluvium and glacial drift, and in highly weathered bedrock. In Poland, Kühn (1963) describes catastrophic building collapses in the town of Klodsko where a combination of natural rainfall and damaged sewage pipes has created a network of tunnels in the loess soils underlying the old town centre.

Interference with surface water flow may initiate piping as well as interference with subsurface flow. Crouch (1977) has analysed a less dramatic case in New South Wales in which construction of an open drain through a residential area led to tunnel development which threatened buildings 12m away in little over a year. He showed that the soil was highly dispersible yellow solonetzic and should have been identified and precautions taken. Colclough (1965) has also mentioned discharge from road culverts initiating piping in Tasmania. Harris and Fletcher (1951) described over-zealous irrigation as the cause of piping on a farm in Arizona in an internal report for the Soil Conservation Service. More recently, Parker and Jenne (1967) have considered irrigation to be a probable cause in part of Wyoming and Nebraska, and Subakov et al. (1968) have blamed irrigation for initiating piping in loess and loessial loams in Uzbekhistan. Glukhov

47

(1956) and Denisov *et al,* (1960) reported canal bursts, the
former particularly concentrated around sand or sand-clay
interlayers in loess, causing pipes 1 m in diameter and failure of major canals in Tashkent and Tadzhik S.S.R. Commissioning of the 20 km long Khodza-Kola Canal was delayed for
5 years by these failures. Skvaljeckij *et al.,* (1971) have
reported slumping of 'dry' loess over a wetted substrate
along the edges of the Selburski reservoir in Tadzhikistan
and the development of piping through cracks 5.7 m deep between the slumped terraces, parallel to the banks, despite
a period of carefully graduated filling extending over a
period of six years. A seasonal fall of 11 m in water level
during the summer probably contributed to the stress. Fuller
(1922) blamed contour ploughing for increasing the amount
of water infiltrating into the loess in China and initiating
pipes. Stocking (1976) has similarly described misguided
post-war construction of contour banks in Rhodesia in order
to reduce erosion by overland flow, which by ponding up the
water created a hydraulic head which initiated piping failure of the banks.   Indeed, piping failure of farm dams built
of earth is a common experience, described by Rao (1960) in
India, Pazzi (1963) in South Africa, Burton (1964), Rallings
(1964;1966) and Carmichael (1970) in Australia, the Ministry
of Water Development (1967) in Rhodesia and Cole *et al,*
(1977) in Thailand. Despite distinguishing between 'field
tunnelling' and 'earthwork tunnelling', Ritchie (1963) notes
that both forms result from highly dispersible soils and he
makes no distinction in terms of processes. The problem of
piping in major dams is discussed elsewhere (section 5.2).

4.4.   The pedogeomorphic context.

The aim of this section is to analyse the location of
piping in the landscape on the basis of reported observations and in the light of the theoretical catenary framework
put forward by Dalrymple, Blong and Conacher (1968) which
has recently been very fully elaborated in Conacher and
Dalrymple (1977).

Dalrymple *et al.* (1968) suggested a catenary sequence
of landsurface units, the nine unit landsurface model
('NULM'), which in its latest form is illustrated in Fig. 6.
Conacher and Dalrymple (1977, 3) regard the component units
as process-response subsystems, defined in terms of characteristic responses to pedogeomorphic process combinations
and intensities.   They note that although the model was
originally put forward primarily in geomorphic terms, subsequent concentration on establishing identifying criteria
has focussed on hillslope processes which are at least partly
pedological.   They also point out that 'position on the landsurface catena is often, but not always, an important deter-
minant of process', and that 'most processes operate on most
landsurface units, and the units are essentially identified
by relative differences amongst their responses on different
parts of the landsurface catena' (*ibid.* 101).   Soil piping
is seen as a characteristic response to eluviation by lateral
subsurface soil-water movement in four of the units (*ibid.,*

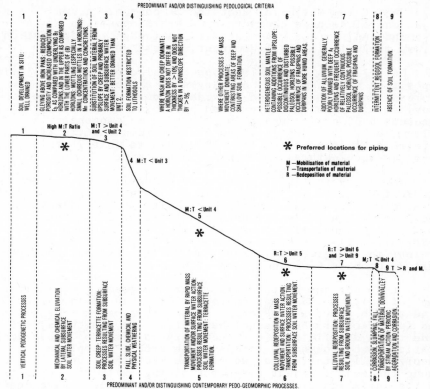

Fig. 6 Preferred locations for soil piping according to the nine unit landsurface model (NULM) of Dalrymple, Blong and Conacher (1968) and Conacher and Dalrymple (1977).

101-109):

Unit 2 The Seepage Slope : a zone of predominantly lateral subsurface seepage immediately below the interfluve zone, which is dominated by vertical infiltration of rain-water. Collection and transmission is not spatially uniform, resulting in percolines and surface depressions caused by soil subsidence and piping. It is 'frequently the zone of stream/valley initiation by subsurface processes'.

Unit 5 The Transportational mid-slope : piping is not characteristic of the intervening units 3 and 4 where soil creep, fall and rockslide predominate, but on the mid-slope subsurface processes may once again become significant in transporting material downslope in combination with soil creep, flow, slump, slide, raindrop impact and surface wash. It is a water-transmitting and shedding zone. Cultivation can easily accelerate transportation of material and create Ap horizon plough pans (with a noted relationship to piping, cf. section 4.6(i)). B horizons are frequently more highly structured than in adjacent units with marked and persistent interpedal cracks.

Unit 6 The Colluvial Footslope : commonly a smooth concave zone of seepage characterised by redeposition of mass
movement and surface wash material, although transportation
across the unit is still dominant, partly subsurface.

Unit 7 The Alluvial Toeslope : characterised by alluvial
deposition from upstream, but may also exhibit surface
depressions resulting from subsidence due to subsurface
eluviation. Alluvial layers and palaeosol horizons are
frequent and fairly continuous creating inhomogeneity.
Dalrymple *et al*. (1968) stated that if this unit is left as
a terrace by downcutting it will return to unit 2 status,
with seepage features becoming more dominant.

Jones (1978c, 1979) identified essentially the same
landsurface units in his fourfold division of piping in
terms of hydrological significance, with piping on Units 2
and 5 being generally disconnected from the stream and
therefore potentially slower in supplying discharge (cf.
section 7.1(i)). However, of course, not all units need be
present in any given hillslope, and direct connections from
unit 2 to units 5, 6 and 7 can occur.

It is perhaps worth adding that the geomorphic environment of units 2 and 7 may be very similar in so far as both
tend to lie above a steepening slope, the convex creep slope
and fall face or the channel bank free face, which may provide the hydraulic gradient necessary to develop the hydrostatic and hydrodynamic forces which create the piping.
Theoretically, therefore, we might expect these units to be
among the most frequent sites for piping, with pipe outlets
on the units immediately downslope. Pipe outlets have been
reported on the upper sections of valley side-slopes, but
exact classification in terms of NULM is not clear. Conacher
and Dalrymple (1977, 25) note that Yair's (1973) pipes and
rills on convex hillslope units in Israel are in fact on a
low-angled unit 5, not unit 3. However, Barendregt and
Ongley (1977) explicitly recorded piping on units 3, 4, 5,
6 and 7, emerging on 8 (Ongley, pers. comm., 1980), in
Canadian badlands.

Perhaps more significantly, Dalrymple *et al*. (1968) and
Conacher and Dalrymple (1977) do not recognise piping as a
diagnostic feature of unit 8, the channel wall, or unit 9,
the channel bed. This may seem particularly odd since piping has so often been recognised by outlets in gully or
channel walls, but it must be remembered that the units are
defined in terms of process-response, particularly of dominant process-response. Lateral corrasion, saturation slumping
and fall are recognised as the main processes forming this
unit. Nevertheless, in a minor way piping may modify the
streambank by inducing slumping. The writer has observed
minor piping developed in the tension zones around bank
slumps on stream channels and along salt marsh creeks with
surface inlets on the partly slumped floodplain or marsh
surface. It is presumed that this piping is post-slumping,
but it could play a significant part in the final demise of

the partially slumped bank, particularly by draining flood
waters through the disturbed zones as described by McGee
(1911), Terzaghi (1931), and Leopold and Miller (1956).
Heede (1971; 1974; 1976) reports shearing off of piped gully
walls on Alkali Creek, Colorado, as a major geomorphic
process causing migration of the gully bed: some of these
pipes had inlets on the 'sliding surface of the gully
slopes' (Heede, 1971, fig. 3). Again, channel walls may be
locally modified by pipe-induced gullying and the growth of
a tributary. Lastly, the classic mechanism of 'boiling'
described by Terzaghi and Peck (1948, 510) and translated
into the field context in channel walls by Parker (1963)
is seen as beginning by eluviation on the channel walls.
Undoubtedly many pipes are formed by this process.

So far as unit 9, the channel bed, is concerned, whilst
Conacher and Dalrymple's view that downvalley transportation
of material by stream action with subsidiary periodic aggradation and corrasion is undoubtedly true of the 'model'
case, some piping processes may yet be found. 'Boiling'
has been recorded in the beds of open channels (Terzaghi
and Peck, 1948, 52; Blench, 1962; Martin, 1970). The seepage responsible may be effluent seepage or induced by the
passage of waves over the bed. Perhaps more obvious in a
geomorphic sense is the sequence of lowering of channel beds
by collapse of the roofs of underlying pipes. This has been
described by Cockfield and Buckham (1946) in Canada, by Bell
(1968) and Hamilton (1970) in the United States, by Banerjee
(1972) in Bengal and by Aghassy (1973) in Israel, all quite
independently. This process will be taken up more fully in
section 8.1.

A review of the frequency of observations of piping in
the literature might be used both to confirm and extend the
observations of Conacher and Dalrymple. Unfortunately, very
few other authors have explicitly noted the landsurface unit
observed in terms of NULM nomenclature, but of 36 relatively
clear field references, 5 mention the seepage slope or upper
slope of the hills, 10 the valley side-slopes, most likely
the transportational midslope, 4 appear to refer to the
colluvial footslope, 28 the alluvial toeslope and channel
walls and 5 the channel bed. On this tentative basis, the
predominant location of piping landforms appears to be on
NULM units 7 and 8. However, observations in Britain suggest
that here at least the seepage slope is more common than is
suggested by the literature at large, especially where the
seepage slope is covered by upland peat. Hence, Jones (1979)
includes the seepage slope as a core area of piping which
is not directly connected to stream channels in his theoretical hydrogeomorphic schema. Possibly this applies to more
humid locations in general (cf. Downes, 1949, in Victoria;
Conacher and Dalrymple, 1977, in New Zealand) and the imbalance in the literature may be partly due to the high number of reports from arid and semi-arid environments. A
further complication, as pointed out by Conacher and Dalrymple (1977, 35), is that the predominant process on landsurface units can change with a change in geomorphic environment.

Hence, in particular unit 7 can be converted to unit 2 by stream rejuvenation and terrace formation. This situation appears to apply in Morgan's (1972) study area in Malaysia where pipe outlets on the terrace scarps issue anastomosing water courses which drain across the floodplain to the channel. Jones (1975) describes a similar situation in mid-Wales on terraces of solifluction material.

Downing (1968) also appears to describe a gradual process of change of landsurface unit in Natal, which is unique in so far as piping is the agent of change. On vleis (bottomlands) in Natal pipe resurgences cause the build-up of cone-shaped mounds of sediment which gradually coalesce and are stabilised by vegetation. This 'aggrades' the floodplain, forcing the river away, and presumably replaces an alluvial toeslope (unit 7) with a colluvial footslope (unit 6) (see also Beckedahl (1977)).

A relatively large body of quantitative information exists on slope angles associated with piping. However, although individual observers report specific slope angles associated with piping, including limiting angles in some instances, a wide variety of slopes is actually recorded from less than 1^0 in the plains of central Sudan (Berry, 1970) to 35^0 on shale badlands in the Negev (Yair, Lavee Bryan, and Adar, 1980), although the latter were only temporary pipes that tended to self-destruct during the period of their artificial rainfall experiment. Nevertheless, clearly the physiographic, hydraulic and pedological context is more important than any specific surface slope angle.

For piping to occur on low slopes there must be either a plentiful supply of water, possibly seasonally, steep hydraulic gradients, perhaps caused by an adjacent free-face or gully wall, or very favourable soil profiles (sections 4.5 and 4.6) either singly or in combination, e.g. occurrences on slopes of 1 : 500 ($0^07'$) in the deep alluvial soils of the Butana grassland plain in the Sudan (Worrall, 1959; 1960; Berry, 1970) or the common floodplain piping. Newman and Phillips (1957) observed that piping did not occur on slopes of less than 2% (1^0 20') in New South Wales and Downes (1946) reported a lower limit of 4^0 in Victoria, both in solonetzic soils developed on Lower Palaeozoic metamorphic rocks. Colclough (1965) reported piping 'on a wide range of slopes' in podzolic soils in Tasmania, but supposed an undisclosed minimum 'to allow water to flow and exert pressure'. N.O. Jones (1968) indicates a minimum slope of c. $1^044'$ in the sediments of the San Pedro Valley, Arizona. Other typical slopes quoted are 10% (5.7^0) (Cumberland, 1944) and 17-29^0 (Gibbs, 1945), both in loessial soils, and 15-18^0 (Blong, 1965) in loose pumice and fluvial beds overlying less permeable ash beds, all in New Zealand, compared with 10-20^0 slopes on flysch in Upper Austria (Zeitlinger, 1959), or generally over 13^0 (Knapp, 1970a and b) and 9-18^0 (Jones, 1975, 291) both in peaty gley podzols in Wales. Ward (1966) gives an *upper*-limit of 12^0 for the best pipe development in loessial soils with weakly nutty/prismatic clay loam subsoils in New Zealand, and Stagg (1974) also noted that pipes occurred mainly on the flatter slopes in the East Twins basin, Mendip. Baillie (1975) reported pipes mainly on slopes of 15-20^0 in Sarawak, with only a few on slopes over 20^0 and none near 30^0, because more surface runoff occurs on steeper slopes. Although the local relief in Feininger's (1969) quartz diorite batholith area in Colombia is generally about 150m, with slopes of 20-35^0, he noted that pseudokarst only developed in the clayey bouldery saprolite where local relief was less than 50 m. One possible reason for an upper limit is that processes like soil creep and mass movement may more effectively destroy pipes developing on steeper slopes as respectively suggested by Feininger (1969) and Conacher and Dalrymple (1977). It is also possibly true that most collecting areas from which pipes may run are concave (cf. section 7.1(i)). But it is the hydraulic grad- ient, not the landslope, that is of paramount important (Jones, 1971) and this is clearly controlled largely on the relative slopes of adjacent slope segments.

Slope aspect has not generally been considered signifi- cant for pipe development, although a number of New Zealand workers have noted a relationship (Cumberland, 1944; Gibbs, 1945; Hosking, 1962, 1967; Hughes, 1970, 1972). Downes (1946) also noted that slopes facing the northern sun in Victoria suffered worse from desiccation and piping, follow- ing devegetation by overgrazing. Hughes (1970; 1972) proved a relationship by quantitative analysis of tunnelling inten- sity in the loessial soils of the Banks Peninsula, in which tunnelling is most common on west and north-west facing slopes and least likely on south-east or south-south-west facing slopes. This he attributed to greater desiccation on north-to-west facing slopes resulting in poorer surface protection by vegetation, rain pan formation and greater overland flow which then enters desiccation cracks and init- iates tunnelling. In contrast, Lear (1976) has found no relationship between intensity of desiccation cracking in the sides of salt pans on the Dovey Marshes, mid-Wales, and pipe drainage, the principal hydraulic gradient to the main creek being at right angles to the most desiccated sides.

A geomorphic factor of greater importance than slope or aspect to the development of piping is the stability of the landsurface system. Although there appears to be little current development in many British pipes (cf. Section 8.5), and both Jones (1971) and Gilman and Newson (1980) have suggested that they may be in a quasi-stable equilibrium at present, the initial development of piping tends to be associated with systems which are, perhaps temporarily, in disequilibrium. This in particularly true of areas where pipes have been observed developing and where some authors have inferred cycles of pipe development. This point is taken up at greater length in section 8.1.in terms of the processes operating, but at this juncture it is relevant to observe that 1) piping is commonly associated with a wide range of other erosion processes and rarely if ever in iso- lation (cp. soil erosion maps of Downes, 1949) and 2) piping

has been commonly associated with general degradation of
the landsurface, whether due to human interference with the
surface soils and vegetation cover, biotic or climatic destruction of the same or by the natural rejuvenation of drainage. Stream incision was noted, for example, by Fletcher
and Carroll (1948) following overgrazing. Carroll (1949)
and N.O. Jones (1968) reports active downcutting on the San
Pedro River, Arizona, and the development of high, arched
outlets or else of new lower level pipe networks as the pipe
systems try to keep pace. Jones (1968, 13) reported that
incision of 15 ft. (4.5m) occurred between 1890 and 1935,
associated with major floods in 1926-7, and a further 5 ft.
(1.5m) up to 1965. Piping of river terrace scarps (e.g.
Morgan, 1972; Jones, 1975) may indicate a similar cause.

Bell (1968, 253) suggested that a cycle of piping is
started in the badlands of North Dakota by regional uplift,
which encourages water to use desiccation cracks, ice wedge
cracks, joints, or subsidence cracks over beds of burned or
burning lignite and seep through the sediments.

However, relative rates of rejuvenation and pipe activity are quintessential. Excessive lowering of the water
table might kill piping, as suggested in the case of apparently fossil pipes and caves in the mottled zone beneath
fersiallitic duricrusts in Sierra Leone by Bowden (1979; 1980;
pers. comm.).

4.5. The role of soil chemistry.

The role of soil chemistry has been the subject of a
certain amount of controversy. N.O. Jones (1968) concluded
that it was largely irrelevant to pipe development in Arizona,
whereas Heede (1971) proved a difference in pH of 7.6 against
8.9 between non-piped and piped areas as statistically
significant. In particular, during the past decade increased
attention to the behaviour of dispersive clay soils has
focussed a considerable amount of literature on the role of
the sodium ion.

4.5.(i)   *Piping in soil systems influenced by sodium salts.*

Many research workers have associated piping particularly with these soils; hence the relatively high ranking of
high exchangeable sodium percentage, high base exchange
capacity or high soluble salts content among the reported
casual factors. Although it is now apparent that piping is
much more widespread than alkaline or saline soil systems,
it is also clear that in some areas alkalinity does have a
major effect upon soil erodibility and permeability and hence
upon susceptibility to piping erosion.

Fletcher and Carroll (1948) reported piping in poorly
graded, alkaline soils with jointly high exchangeable sodium
and calcium in South Arizona, although they later concluded
that the association of sodium and lime was coincidental and
that exchangeable sodium merely contributed to the severity
of piping but was not a necessary factor (Fletcher, Harris,

Peterson and Chandler, 1954). Piping was found in Arizonan soils with exchangeable sodium percentages (ESP) ranging from 0 to 90%. However, Brown (1961: 1962) found extremely salty soils in piped areas in Colorado. In two of his three areas sodium dominated, but in the other, where piping was significantly less developed, the soils were gypsiferous with associated differences in structure (cf. section 4.6 (iii)). ESP values calculated on the basis of Brown's table (1962, tab. 1) range from 22% to 62% in the two saline areas (Table 2). He suggested that accelerated erosion occurred after periods of drought, resulting from soil dispersion caused by ionic imbalance between the soil and the fresh thunderstorm rainwater. This suggestion was also made by Ingles and Aitchison (1970) and co-workers from studies on Australian earth dams (cf. section 5.1).

Similarly, Parker (1963) observed piping in generally sodic systems in Arizona and New Mexico: only one of his sites was not sodic, a sandy, badland site that was never-theless alkaline. He took samples from the surface above pipes, and from the roof, wall and floor of pipes at 8 sites in soil, alluvium and badlands. The samples showed 1) sod-ium, lime and gypsum contents indicated alkaline material at all sites, 2) higher total soluble cation concentration in the walls and roof than in the floors of pipes, possibly explained by cation exchange during capillary water movement, 3) higher ESP in roofs and floors than in walls, 4) salts generally leached from the upper profile and deposited on the roof and walls, 5) surface materials highly dispersible, but 6) the dispersive effects of sodium were partially re-pressed by higher salinity in the roof and walls of pipes. In addition, Parker also found 1) that the cracking potent-ial was high in all areas and greatest at the depths at which pipes developed, 2) that the dominant clay species was mont-morillonite, and 3) that the pipe floors had low permeabi-lity. Later Parker, Shown and Ratzlaff (1964) found total sodium domination (ESP ~ 100%) in montmorillonitic clays at the Officer's Cave pseudokarst in Oregon, and Parker and Jenne (1967, 6) noted that higher ESP leads to greater volumetric changes in montmorillonite during wetting and hence to greater fracturing of cements and clay coatings. And the greater the ESP at any given soluble salts content the more readily will montmorillonite disperse and flush out.

Bell (1968) noted that piping in the Badlands of North Dakota occurred in an environment rich in sodium sulphate and montmorillonite/beidellite/illite clays. Oxidation of pyrite and marcasite in lignite beds adds sulphate to the capillary water in the silts, thus:

Na_2 -resin + $FeSO_4$ → Fe-resin + Na_2SO_4.

The silty sediments 'melt away' and piping develops where dispersed clay is available (*ibid*, 246).

Heede (1971) conducted a paired survey between piped

Table 2.

A. Physical and chemical characteristics of piped soils in Colorado based on Brown (1962, tab. 1).

Site and depth (mm)	ESP*	CEC+	Exchangeable ions (me/100g)					Soluble Salts++	Dispersion Ratio (%)	Bulk density (mg m⁻³)	Shrink-age	Text-ure
				Ca	Mg	Na	K					
Silt 0-300	22.1	22.6	52.4	29.0	17.8	5.0	0.6	64	76.6	1.6	34.2	Silty clay
- 600-900	33.3	23.2	67.7	38.2	21.4	7.5	0.6	25	77.7	1.7	37.9	Clay
Bayfield 0-300	49.4	23.1	52.8	27.0	13.5	11.4	0.9	90	62.0	1.6	32.5	Silty Clay
- 600-900	62.0	23.4	49.8	22.2	12.5	14.5	0.6	56	94.6	1.6	31.8	Loam
Nunn 0-300	6.7	25.3	83.8	67.7	13.0	1.7	1.4	4.8	38.7	1.2	20.2	Silt
600-900	7.0	27.3	86.6	69.2	14.1	1.9	1.4	4.8	41.6	1.2	20.4	Loam
> 1.8m	4.0	35.3	83.1	64.9	15.2	1.4	1.6	-	-	1.3	25.0	

* Exchangeable sodium percentage, added by reviewer.

+ Cation exchange capacity (me/100g): Ca, Mg, Na, K all quoted in me/100g.

++ CaSO₄ me l⁻¹

B. Comparison of soils with and without pipes (Heede, 1971, tab. 1)

Soils	ESP	SAR	pH (1:5)	Gypsum me / 100 g.	Conduc- tivity at 25 C. mmhos/ cm.	Moisture at satur- ation (%)	Sand %	Silt %	Clay %	Textural Class	Number of Samples
Without pipes											
Mean	<1.0	0.4	7.6	0.7	0.5	45.1	25.4	26.8	47.8	Clay	19
Standard error of the mean	~0.0	.07	.14	.06	.05	1.35	2.53	1.33	2.13		
With pipes											
Mean	12.0**	10.7**	8.9**	1.6**	1.7*	38.7	30.7	23.5	45.8	Clay	13
Standard error of the mean	2.27	2.0	.17	.19	.37	3.16	4.05	2.39	2.87		

Differences between soils with and without pipes

** highly significant

* significant

57

and non-piped clay soils in Colorado and concluded that exchangeable sodium percentage (ESP) and sodium adsorption ratio (SAR) were statistically higher in piped than in non-piped soils (Table 2). Mean sodium ion concentration was 12.2 me 1^{-1} (SE = 2.6). Highly significant differences were also found in pH (but this was clearly a reflection of the ESP) and gypsum content (but the amount was too low to effectively reduce the dispersion caused by the strong sodium influence). Conductivity was significantly higher in piped soils, but not sufficient to call them saline. From analyses of the shale parent material it appeared that this was the source of the sodium (Table 21). In Heede's view: 'Most of the piping soils of the study area can be classified as soils with a tendency towards the formation of solonetz', i.e. the ESP values suggest that displacement of calcium and magnesium and adsorption of sodium is in progress although as yet the soils exhibit no or very weak horizons and some alluvial strata (*ibid.*, 12). Nevertheless, columnar structure is normal below the surface soil layer and some of the piped side slopes show more conventional solonetz features. In Rhodesia, Wendelaar (1976) noted that piping is a characteristic feature of dispersive soils with a high sodium exchange percentage. Stocking (1976a and b; 1979) describes a novel cycle of pipe development in soils developed on the Karroo Sands in Rhodesia, which is dependent upon the accumulation of sodium ions in areas of bad drainage with seasonally fluctuating water tables. The poor drainage helps to counteract the more normal preferential dispersion of sodic clays and the fluctuating water table encourages the weathering of sodium-rich feldspars. Deflocculation of clays causes clogging of pore spaces and the formation of what Purves and Blyth (1969) call a 'subsurface dam wall'. Piping develops beneath this 'dam wall' whenever gullying steepens the hydraulic gradient sufficiently, tapping the large reservoir of subsurface water impounded by the 'dam'. Stocking concluded that the sodium was derived from weathering of granite surfaces overlain by thick wind-blown sediments in which the sodium accumulated under poor drainage conditions during the Pluvial Period. His results suggested that piping occurs where the ratio

$$\frac{\text{Exchangeable Na}}{\text{Total Cation exchange}} > 50\%.$$

Where the percentage was lower, gullies were common but not pipes. Stocking (1979) noted that soils displaying severe gully and pipe erosion often have a sharply defined sodic horizon in or very near the present zone of water table fluctuation which impedes drainage and causes sodium accumulation in a positive feedback situation.

Accumulation of cyclical salts has been associated with piping in S.E. Australia. Here, Downes (1954; 1956) attributes accumulation mainly to salty rainfall during the recent Arid Period, when rainfall was about 50% of the present. The most intense effects are found in the present-day zone of 500-750 mm annual rainfall and has led to widespread solods

and solodic soils (some formerly termed red and yellow podzolics). Present-day rainfall is responsible for varying
amounts of desalinisation by leaching. Desalinisation results in the eluviation of sodium clays down-profile and
possibly to the breakdown of the clay complex, creating a
less permeable B horizon, with the characteristic columnar
structure of the solonetz stage breaking down to the small
columnar and prismatic structures in the solodised solonetz
and solod stages.

In the semi-arid West Australian wheat belt, Conacher
(1975) considered throughflow to be responsible for excessive salinisation in areas known as 'salt scalds'. Clearing
of the area for agriculture resulted in increased throughflow and percoline generation through reduced transpiration
losses; up to 56% of the rainfall drains as throughflow in
one area. Pits dug in scald areas revealed small pipes,
up to 1.2cm diameter, at a number of different levels in
the profile which carried a major proportion of the throughflow. Evaporation of seepage water emerging from above the
plough pan on the lower slopes results in the build-up of
soluble salts, often highly toxic to plants, creating the
bare, waterlogged scalds (Conacher and Dalrymple, 1977, 51-
55). In this case, the roles are in part reversed and piping
seems to be creating the saline soils.

A considerable amount of research has been conducted
into erosion problems in saline soils by the Soil Conservation Service of New South Wales, both in terms of soil conservation and of piping failure in earth dams (cf. sections
5.1 and 8.7). Monteith (1954) reported piping in solodised
solonetz in the Hunter Valley. These soils have a columnar
structure in the B horizon and medium alkalinity, which encourages dispersion, whereas high alkalinity could lead to
clay flocculation. The trials set up to study tunnel erosion by Newman and Phillips (1957) in the Riverina were mainly
concerned with piping in solonetzic and solodic soils. They
reported a compact surface layer above a subsoil showing
extensive cracking during dry periods, a highly dispersive
layer which becomes super-saturated and mobile when wet and
a relatively impermeable B horizon in soils susceptible to
piping. Pipes were usually developed in the dispersive,
silty A_2 horizon, but no pipes developed where the A horizon
was over 28 cm thick probably because this lessens cracking
and delays saturation of the dispersible A_2 horizon. Pipe
formation appeared to be independent of the dispersible
horizon where the B horizon was highly structured and deep
tunnelling occurred into the gullies. Floyd (1974) reported
on the results of the experiments in reclamation set up by
Newman and Phillips (cf. section 8.7). He reports dispersal
indices of 5.7-7.3 in the loamy A horizon of solodised soils,
falling to 1.5 in the mottled clay B horizon.

High on the tablelands close to the Snowy Mountains,
Charman (1967) found that tunnel erosion occurred only in
small patches and was not a serious threat in brown and grey-
brown podzolic and alpine humus soils. But the severity of

piping erosion in the sodium-rich soils of the coastal
areas of Grafton, Taree and Kempsey led him to undertake
detailed field mapping and laboratory analyses in these
areas (Charman, 1969; 1970a and b). Charman states that
piping is a significant erosion problem in New South Wales
and that it is more widespread than previously supposed. The
proportion of sites examined that were subject to piping in
the three areas varied between 26% and 37%, but the percentage of soils considered susceptible to piping on the basis
of Ritchie's (1963) dispersal index and a volumetric expansion test (cp. Appendix I) varied between 37% and 50%.
Charman (1970a, 77) argued that the discrepancy is due to
infiltration capacities being generally good in the area.
This, he says, leads to fewer subsurface weaknesses through
which dispersed soil can move so that a more diffuse subsurface erosion, indicated by slumping of subsoils into
gullies, is more common than tunnelling. Perhaps we could
add the view that whilst Charman's laboratory tests indicated
potential for piping, whether that potential is activated
in the field or not really depends upon other factors such
as hydraulic gradients.

True halomorphic soils are not common in the coastal
areas due to the high rainfall, but Charman believes that
the continued leaching through of cyclic salt leads to subsoils high in exchangeable sodium and consequently with high
dispersibility. The influence of salt seems to be limited
to cation exchange effects producing dispersibility and is
not sufficient to dominate the soil morphology or cause
serious salting problems (Charman, 1969, 345). Hence none
of the susceptible soils in the Kempsey area were solodised.
Although some salts may be derived from weathering of sedimentary parent material in the area, most of the present
sodium content of the soils comes from current rainfall. In
fact, Monteith (1954) calculated that current rainfall could
produce an accumulation of 10 tons of salt per acre in
10,000 years. This contrasts with Downes's area in Victoria,
west of the Great Dividing Range, where rainfall is much
lower. Charman (1970a, 75) suggests that the higher rainfall
in coastal areas has taken the soils through the solonisation-
solodisation process at a much faster rate than inland.
Charman (1969, 332) stresses that 'the fact that clay colloid
particles with a high percentage of large monovalent ions
(mainly sodium) on the exchange complex are highly dispersed
in water, is of paramount importance in the tunnelling problem'. All that is then required for tunnelling to begin
is the infiltration of water and some subsoil weakness.

However, Charman had hoped to find a correlation between
great soil groups of the Australian classification (Stephens,
1962) and the incidence of tunnelling, but this did not
materialise. Examples were found of piping in all soil
classes except krasnozems and coastal sands. Nevertheless,
the main areas of piping occur where, according to Charman
infiltration and leaching are sufficient to create sodium
dominance of the clay material. Charman (1970a, 75-76)
pointed to problems of classification of some of these soils,

particularly as to whether some of the podzolic soils were once subject to a solodisation process. He concludes, however, 'that processes that are not clearly reflected in soil morphology are affecting the suitability of these soils for earthworks'.

Fig. 7 is an analysis based on the laboratory data published by Charman (1969, tabs. 1 and 2; 1970a, tabs. 1 and 2; 1970b, app. I and II), in which each soil profile has been classified according to the highest level of susceptibility to piping recorded within the profile. It reveals that i) the highest number of all susceptible soils fall in the yellow podzolic group, followed by grey-brown podzolics and yellow solodics (including solodised solonetz), ii) the first two groups also contain the highest number of 'susceptible' and of 'highly susceptible' respectively, iii) however, these results are partly due to the high number of samples taken from the soils of the first two groups and, in fact, only 55% of yellow podzolic soils were at all susceptible, iv) only yellow solodic and podzols were 100% susceptible. Charman (1970b, 267) decided that the acidic podzols in the eastern or seaward area 'can be regarded as the end product, beyond the solodization process where sufficient exchangeable sodium is maintained only by the percolation of rainfall charged with cyclic salt' and that 'in other places Yellow Podzolic soils are not dispersible because rainfall, possibly in combination with other more localized factors, is sufficient to reduce the sodium to a safe level'. One is tempted to suggest that perhaps the similarity in terms of the strong textural and structural contrast between A and B horizons in both podzolics and solodics (cf. Stephens, 1962), as well as in dispersion properties, is important to the piping process. One might also question whether the high dispersibility of some podzolic soils is necessarily due to the influence of sodium (cf. section 5.2(i)). Unfortunately, in general, Charman maps and discusses only 'susceptibility to tunnelling' without recording the actual incidence of tunnels, which he himself admits is not the same thing (v.s.), so useful though it is there is a slight problem of disentangling fact from theory in the analysis.

More recently, Craze (1974) has used the same methods as Charman in a survey of an area in the Southern Tablelands of New South Wales. His results confirm those of Charman in so far as soils susceptible to tunnelling were generally confined to duplex or texture-contrast soils, i.e. yellow podzolics, yellow solodics, yellow solonetzics and solonchaks. 8 of the 14 profiles (57%) had a susceptible horizon. Only one solonchak was sampled. Floyd (1974) also states that although tunnelling occurs on solonetzic, solodic, podzolic soils and Red-Brown Earth, it is most severe in the first two, especially in type Dy 3.5 of the Northcote (1971) classification, and is not found in Red Earths.

Colclough (1965; 1967; 1971; 1973; 1978) reports on tunnelling in nearby Tasmania, although strictly from the

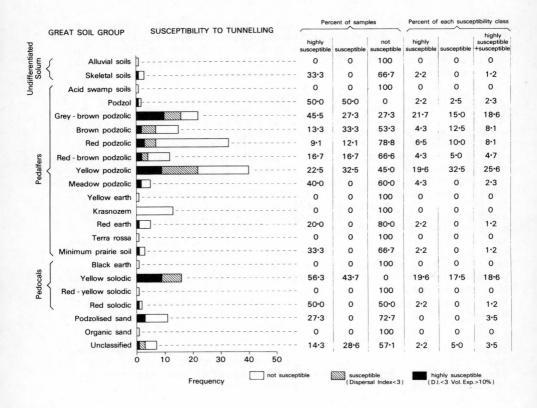

GREAT SOIL GROUP	SUSCEPTIBILITY TO TUNNELLING	Percent of samples			Percent of each susceptibility class		
		highly susceptible	susceptible	not susceptible	highly susceptible	susceptible	highly susceptible +susceptible
Undifferentiated Solum							
Alluvial soils		0	0	100	0	0	0
Skeletal soils		33·3	0	66·7	2·2	0	1·2
Acid swamp soils		0	0	100	0	0	0
Pedalfers							
Podzol		50·0	50·0	0	2·2	2·5	2·3
Grey - brown podzolic		45·5	27·3	27·3	21·7	15·0	18·6
Brown podzolic		13·3	33·3	53·3	4·3	12·5	8·1
Red podzolic		9·1	12·1	78·8	6·5	10·0	8·1
Red - brown podzolic		16·7	16·7	66·6	4·3	5·0	4·7
Yellow podzolic		22·5	32·5	45·0	19·6	32·5	25·6
Meadow podzolic		40·0	0	60·0	4·3	0	2·3
Yellow earth		0	0	100	0	0	0
Krasnozem		0	0	100	0	0	0
Red earth		20·0	0	80·0	2·2	0	1·2
Terra rossa		0	0	100	0	0	0
Minimum prairie soil		33·3	0	66·7	2·2	0	1·2
Pedocals							
Black earth		0	0	100	0	0	0
Yellow solodic		56·3	43·7	0	19·6	17·5	18·6
Red - yellow solodic		0	0	100	0	0	0
Red solodic		50·0	0	50·0	2·2	0	1·2
Podzolised sand		27·3	0	72·7	0	0	3·5
Organic sand		0	0	100	0	0	0
Unclassified		14·3	28·6	57·1	2·2	5·0	3·5

Frequency: 0 10 20 30 40 50

□ not susceptible   ▨ susceptible (Dispersal Index<3)   ■ highly susceptible (D.I.<3 Vol. Exp.>10%)

Fig. 7 Distribution of susceptibility to piping in relation
to Great Soil Group in soils analysed by Charman
(1969; 1970a and b) in New South Wales. Compiled
by present author from raw data published by Charman.

point of view of a Soil Conservation Officer writing for a
farming audience, giving examples of practical reclamation
projects rather than the detailed soil analysis and mapping
undertaken by Charman on the Australian mainland. Colclough
(1973) states categorically that 'tunnel erosion occurs in
soils that have a high sodium content in the subsoil',
mostly in podzolic soils developed both on sandstone or mudstone.

In a recent study of erosion in the New England Tablelands of New South Wales Imeson (1978) has referred to soil
pipes often on the lower slopes in dispersible soils with a
relatively impermeable B horizon of blocky structure and
50-60% clay. He states that soils with high exchangeable
sodium percentage seem to concentrate in a narrow band along
drainage courses. He noted that the high turbidity of side
streams in dry weather suggests that water of low electrolyte
content is draining down from podzolic soils upslope and
coming into contact with solodic soils, where sodium had
accumulated from downslope wash on the lower slopes, causing
the dispersal of clays. Most horizons were susceptible to

dispersion except the A_1. Aggregates from the A_1 horizon were the only ones that did not slake on wetting; organic content and therefore cation exchange capacity were higher and sodium content lower in this horizon.

Sherard and Decker (1977) point out that only since about 1972 has much information been available on dispersive clays in the United States, and only by 1976 were tests generally included for this in surveys since standard engineering tests are not appropriate. According to Decker and Dunnigan (1977): 'they are high in exchangeable sodium and many have high pH values. They slake down rapidly and completely, with the clay fraction going into suspension even in quiet water. They puddle or run together and have very low permeability rates when wet but crack severely when dry'. Decker and Dunnigan say the Russians call the soils solonetz and the Americans alkali, black alkali or slick spot soils, 'the latter due to the spotty or localized occurrence of small areas in large bodies of normal soils'. The spotty distribution is thought partly responsible for some dam failures, since not all parts of the borrow area are tested before used in construction.

Sherard, Decker and Ryker (1972) reported piping and 'jugging' (vertical tunnelling) caused by rainfall infiltration on the surface of earth dams in Oklahoma, Mississippi and Arkansas. They concluded that i) all affected dams contained dispersive clays, ii) dispersiveness is related to the total dissolved salts and sodium content of the pore water, and that iii) small cracks caused by hydraulic fracturing, drying etc., probably start the process.

N.O. Jones (1968, 100-104) analysed 27 samples from the San Pedro Valley, Arizona, for total salinity (conductivity) and calcium and magnesium and calculated ESP assuming sodium was the only other significant cation. He found high salinity and high ESP in the main pipe system (37% of his samples), but only moderate or low-moderate salinity and low ESP in the other systems (Fig. 8). The spread of results led him to conclude that there is no simple correlation between soil chemistry and piping.

Although geomorphologists have reported that soil piping may sometimes be found in true karst areas (Jennings, 1971, 30 and 32), calcium cations are inhibitive to soil dispersion. Coumoulos (1977, 56) has suggested that the reason why no earth dams have yet been reported as failuring through piping in Greece is the high calcium content of local river water. Where soil piping does occur above karstic bedrock it may be aided by voids created by solution below or solution of residual rock fragments within, according to Jennings (*ibid*), though he emphasises the need for swelling clays and high sodium ratios in the exchange complex. Presumably, rainwater feeding soil piping *above* karst has generally not had the opportunity to take up calcium from bedrock.

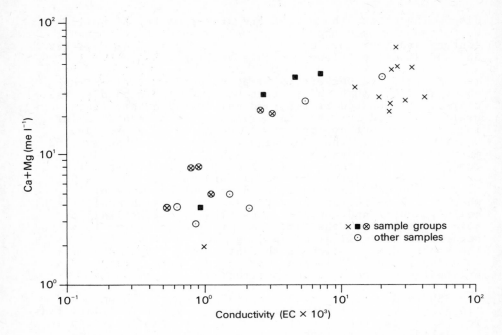

Fig. 8   Salinity of piped soils in Arizona analysed by N.
          O. Jones (1968, fig. 27).   Assuming that Ca, Mg
          and Na are the only cations significantly affecting
          conductivity of the saturation extract, the graph
          shows groups of samples from different pipe systems
          ranging from high salinity with high Na/(Ca + Mg)
          ratio (X), through moderate salinity with low Na/(Ca
          + Mg) ratio (■), where there was a high gypsum
          content, to low-moderate salinity with the low
          Na/(Ca + Mg) ratio (⊗).   The spread and separation
          of the groups suggest that no single chemical pro-
          cess can be responsible for pipe development.

        To conclude this section we must mention the interest-
ing case of piping of coastal salt marshes.   It might be
supposed that sodium ion concentrations in salt marshes
would be so high as to maintain a 'stable deflocculated
state', but as indicated in Fig. 16 seepage of salty
water through the marshes is likely to cause flocculation;
seawater is likely to have a total ionic concentration of
1.1×10^6 me 1^{-1}. Piping has been reported on salt marshes around
the British coast recently by Pethick (1971), Jones (1971;
1975, 147-154), Lear (1976) and Kesel and Smith(1978), but
no chemical analyses have been reported.   Jones (unpub.) had
analyses of concentrations of soluble Ca^{++}, Mg^{++} and Na^+
cations performed on 5 samples from Brancaster Marsh, Norfolk,
and the Deveran marsh, Cornwall. The Brancaster site was an
old marsh only flooded in extreme tides, whereas the marsh
on the Fal estuary, Cornwall, was very young and regularly
flooded by tides.   The results given in Table 3 show higher
ESP in the young marsh and a tendency in both areas for less
dispersible material forming the pipe roofs; these were true

Table 3. Chemical and physical analyses from piped salt marshes
(Jones, unpub.)

	Total soluble cations		Sodium ion concentration	Soluble sodium percentage	Mean weight diameter of aggre-gates +	% aggregate stabil-ity +	% composition*				
	ppm	me l^{-1}					organic	gravel	sand	silt	clay
Brancaster pipe roof	11000	220	5,200	47.3	5.23	94	10	2	51	27	22
pipe wall	11750	235		44.3	3.96	73	6	1	12	46	42
pipe bed	15350	307	5,800	37.8	4.95	86	5	0	24	48	28
Fal pipe roof	16100	340	12,350	76.7	5.19	93	56	0	7	73	20
pipe bed	11400	228	8,300	72.8	2.42	94	41	37	28	58	14
									(18)*	(36)*	(9)*

+ Analysed as per Kemper and Chepil (Black et al, 1965, 499–510) and Kemper (Black et al, 1965, 511–519).

* Sand, silt and clay percentages of less than 2mm fraction, except those in parentheses where gravel is a significant fraction. Those in parentheses are percentages of the less than 2 cm fraction.

pipes not bridged creeks (cp. Yapp *et al,* 1917; Steers, 1960).

The detailed theory of pipe initiation in saline end alkaline soils is taken up in section 5.1.

4.5(ii). Piping in soil systems exhibiting base-deficiency.

Despite the strong emphasis on the role of salinity in pipe development shown by the bulk of the literature, it is now clear that piping can occur in, and is indeed wide-spread in, soils in which the sodium ion can have no signi-ficant effect upon soil structure and erodibility. Ingles and Aitchison (1970) therefore appear to be wrong in sug-gesting that piping is not common in strongly acid soils (pH <5) because of the strong flocculation caused by hydrogen ions.

Curiously, in fact the pH values reported by Ingles and Aitchison in their work on earth dams subject to piping ranged down into this acid zone: 4.8-8.2 (Ingles, 1964a), 4.4-7.9 (Ingles, 1964b), 7.1-8.2 (Ingles and Wood, 1964), 4.1-9.1 (Wood, Aitchison and Ingles, 1964), 5-7 (Ingles and Aitchison, 1970).   But by and large their own work and the literature supports their view.   Published pH values for piping layers range from 9.4 in lignitic materials in North Dakota (Bell, 1968) to 3.9 in lessivé brown earths in the English Peak District (Jones, 1971), but the majority lie between pH 6 and pH 9 : pH 7.7-9.1 (N.O. Jones, 1968, 98) and 7.9-8.6 (Brown, 1961) in Arizona, pH 8 in eastern Oregon (Parker, Shown and Ratzlaff, 1964), 6.2-6.8 in Hawaii (Kings-bury, 1952) and pH 5.4-6.3 increasing towards the B horizon in the Wither Hills, New Zealand (Gibbs, 1945).   Heede (1971) went so far as to find a statistical difference between a mean of 8.9 in piped soils and 7.6 in adjacent un-piped soils.

Nevertheless, a glance at Charman's (1969) pH measure-ments shows 4 out of 19 susceptible soils with pH values of 5.0 and another with pH 4.6 (grey-brown podzolic).   In fact, over half of the soils with pH 5.0 were susceptible.   These were red, red-brown, and brown podzolics and a podzol. Charman noted that in this district (Kempsey) none of the soils were solodised and suggested that salt influence 'is not sufficient to dominate the soil morphology and charact-eristics, or causes serious salting problems in lower drain-age lines' (*ibid.,*345).

In fact, there seems room for doubt that salt influence may have any significance in these cases. Disregarding any such influence, the type profiles of all these podzolics (Stephens, 1962) seem to possess some of the fundamental properties associated with piping processes for 30 years or more, namely, increased clay content in the B horizon which has a nutty to blocky structure (cp. section 4.6(iii)). The only possible exception would be in the case of the true podzol (Stephens, 1962, 18) but Charman's analysis shows that in this particular case the B_2 horizon, which was the susceptible horizon, contained c.60% clay. Table 4 combines Charman's results with Stephens's (1962) description of the soil types. It shows a clear disregard for chemical status and a concentration in profiles with marked textural contrasts, particularly podzolic and solodic.

In Britain, Dalrymple (pers. comm., 1968) observed piping podzols with pH 3 developed on Bagshot Sands in Berkshire. Jones (1975, 124) recorded values for peaty podzols and brown earths in mid-Wales of 4.3 and 4.2 in the piped layers, and Newson (1976) reported pH 4. in similar peaty podzols.

In addition to these British examples, Baillie (1975) concluded that piping is one of the major erosion processes in the forested uplands of Sarawak, where soils are highly leached and acid as a result of continuously high temperature and rainfall. Many of the soils are cambisols, on slopes often over $20°$, but some of the deeper, better drained soils are acrisols, both of which contain the necessary textural contrasts. Cumberland (1944, 185-186) also mentioned piping in the yellow-grey podzolic loams of Hawkes Bay, New Zealand.

Reports of piping in base-deficient soils therefore clearly give the lie to the oft voiced opinion that sodium-generated dispersion is the fundamental cause of piping. Alternative mechanisms, and indeed an alternative to any form of destruction of soil aggregation, are discussed in section 5.2.

4.6 <u>Soil morphology and composition</u>.

It is now evident that soil piping occurs in a far wider range of soils than has been supposed by many authors. Different soil properties and different external factors combine in different cases, and in the demise of one factor another may become important. Nevertheless, a few of the soil factors appear to be almost universal and these are related to soil structure, texture and clay mineral composition, particularly contrasts within the soil profile which affect hydraulic conductivity, erodibility and susceptibility to desiccation. In fact, the spatial variation of soil properties, both vertical and horizontal inhomogeneities, may be as important as representative point values, if not more so, and much of the bewildering variation in the soil properties of piped soils from different areas may be explicable in terms of relative contrasts at each location; a factor often referred to (e.g. Fletcher and Carroll, 1948) but rarely quantified, especially in the horizontal plane.

4.6(i). Texture.

Considerable emphasis has been placed upon textural analysis in the quantitative literature, perhaps partly due to the influence of engineering theory, despite the fact that properties such as structure, porosity, erodibility and

Table 4. Summary of Charman's analyses of tunnelling susceptibility reorganised in terms of Great Soil Groups as described by Stephens.

(Data from Charman, 1969, 1970a and b; classification and description based on Stephens, 1962).

	Sample size	% susceptible	Profile
I Solum Undifferentiated			
Alluvial soils	1	0	alluvial layers only
Skeletal soils	3	33	stony-gravelly, organic accumulation at surface, shallow.
II Solum differentiated. A. Pedalfers			
i) Dominated by acid peat or peaty eluvial horizon.			
Acid swamp soil	1	0	organic surface, gleyed mineral layer
ii) with organic, sesquioxide, and sometimes clay illuvial horizons.			
Podzol	2	50	A_{00} thin, A_0 rare / A_1 accumulated) sandy / organic)/sandy / A_2 grey) loam / B_1 organic + iron oxides (pan) / B_2 variable sand-clay
iii) with clay and sesquioxide illuvial horizons.			
Grey-brown podzolic	22	73	A coarse-medium texture / B usually clay, nutty-blocky structure. / C varied.
Brown podzolic	15	47	A_{00} very poorly developed, no A_0 / A coarse-medium texture / B clay-clay loam, nutty-blocky structure. / C variable.
Red podzolic	33	21	A_{00} thin-moderate, little A_0 / $A_1 + A_2$ coarse texture / B friable red clay: granular-nutty structure / C variable
Red podzolic	12	33	between last two
Yellow podzolic	40	55	A_{00} thin, no A_0 / A_1) / A_2) coarse / B yellow clay, granular-angular nutty structure.

68

Meadow podzolic	5	40	Little A_{00}, A_0 A sand-sady, loam-loam B coarse structured with concretions. C.

Acid to neutral, lacking pronounced
eluvation of clay.

Yellow earth	1	0	A friable sandy loam-clay loam. B clay loam-clay (gradual change) vesicular massive - weak blocky. C.
Krasnozem	13	0	Little horizonation A. Clay, organic accumulation, well-flocculated giving loamy tilth. B, C. Clay, granular-nutty, very porous, friable.
Red Earth	5	20	A sandy-loamy B slightly finer texture, vesicular structure.
Terra Rossa	1	0	A) granular-nutty, little B) horizon differentiation in clay-clay loam numbers (as this case)
Minimum Prairie Soil	3	33	A organic, loamy-clay crumb-nutty B clays of nutty-coarse blocky structure (moderate texture contrast).

B. Pedocals

Black Earth	1	0	A clay-clay loam, granular-cloddy B granular-coarse structured clay.
Yellow solodic	16	100	A light texture B clay-strong textural contrast C
Red-yellow solodic	1	0	ditto
Red solodic	2	50	ditto

C. Others (not described by Stephens)

Podzolised sand	11	27	
Organic sand	1	0	
Unclassified	7	43	

drainage status are of more direct relevance to the detailed development of piping. In general, there appears to be a preference for soils with a moderate to high silt-clay content, notwithstanding Glossop's (1925) outdated theory which actually suggested the reverse (section 5.2(ii)). Most reports are of piping in soils developed from fine sedimentary rocks or within alluvium or loess.

Cumberland (1944), Newman and Phillips (1957) and Floyd (1974) mention piping in deep clay-loam subsoils developed on shattered schists. Fuller (1922), Cumberland (1944), Malicki (1946), Malinowski (1963), Hosking (1967), Hughes (1972) and many others mention piping in loess and loessial soils. Gibbs (1945) noted that piping was abundant in clay loam soils in the Wither Hills, New Zealand, but absent from the adjacent stony soil, where gullying developed instead. Blong (1965a) observed that piping was more common in fluvial and pumice deposits than in adjacent ignimbrites, perhaps because of their greater porosity. Feininger (1969) reported that piping was common (literally thousands of pipes) in clayey bouldery saprolite developed on quartz diorite but was not found on the surrounding metamorphic rocks, probably because they were more homogeneous and weathered to give fewer residual boulders. Jones (1971) remarked that piping was abundant in soils developed from shale bands of the Millstone Grit series, but virtually absent upstream on the Rivelin Grit beds. However, Colclough (1973) states that piping in Tasmania occurs mainly in podzolic soils developed on either sandstone or mudstone, without distinction, in the presence of high sodium concentrations in the subsoil (section 4.5(i)). Collation of reported pipe size and soil texture and physiographic and hydrological environment suggests that the best developed piping does occur in soils with high silt-clay content (Table 5), which may aid piping in a number of ways, viz. providing cracking potential, easily eluviated particles and stronger roofing to prevent destruction. Carroll (1949, 4) pointed out in the early days of quantitative analysis that cohesion is necessary to maintain an open pipe and he deduced that the danger of piping increased with decreasing grain size, aided by the presence of humus and mycelia.

The one glaring exception to the 'rule' is provided by observations of piping processes in periglacial rubble deposits. Smith (1968) says that these are coarser and more heterogeneous than previously-described examples; obviously barring the earlier description of the same features by Czeppe (1965). Smith states (*ibid.*, 199) that bearing this in mind, one might expect a somewhat different end-product, i.e. stone rivers with fines removed and water flowing beneath the boulders.

Some authors are also of the opinion that, going the other way down the grain size scale, too much clay can inhibit pipe development, contradicting the logical conclusion of Carroll's remarks. Bell (1968, 247) states: 'It is important to note that clay soil is not amenable to piping',

although he says it aids pipe development by collapsing over
underlying sediments to provide surface orifices to pipe
systems. Sherard (1952) also concluded that finer soils
with higher clay contents tended to have the greatest resistance to piping. Coarser soils were more susceptible, but
very fine, very uniform sands were the most susceptible.
However, recent work on dispersive clays clearly makes this
an outmoded view, and ironically, Sherard himself would be
amongst the first to agree. Among others, Zaslavsky and
Kassiff (1964; 1965) found that safety from piping varied
inversely with the size of aggregates or particles washed
out, and Kassiff, Ingles, Sherard and others have demonstrated how dispersive clays can suddenly increase erodibility
(cf. section 5.i(i)).

There has also been a marked lack of reported quantitative analyses of variations within soil profiles, despite
the long-standing suggestions that pipes need an impermeably
substrate (Fletcher and Carroll, 1948) and fine material to
bind the roof (Carroll, 1949) and prefer layers with higher
cracking potential (Fletcher and Carroll, 1948; Parker, 1963).
Only Parker (1963), Brown (1961; 1962), N.O. Jones (1968)
Heede (1971) and J.A.A. Jones (1971; 1975) have reported
quantitative analyses of textural variations in the profiles
although qualitative field descriptions have been more numerous. Gibbs (1945) appears to have given the first description of a piped profile, which consisted of a loamy A_1,
a silty A_2 of loose, granular structure with mottling, a
clay loam B_1 with hard nutty structure and strong mottling,
a compact silt loam B_2 and a structureless silt in the C
horizon on loess parent material. Newman and Phillips (1957)
also describe soils based on field examination of 50 profiles, which had a strong textural contrast between A horizons
varying from loamy sand to clay loam (with the A_2 sometimes
slightly heavier or more silty) and a clay B horizon (Floyd,
1974, described the same soils). The textural variations
found by both qualitative and quantitative analyses are
generally consistent with the hypotheses of Fletcher and
others (v.s.) in so far as erodibility, strength and permeability can be inferred from them, although the impermeable
substrate is not universal as is clear from the data collected by Parker (1963).

Both Heede and Jones carried out spatial surveys of
piped and non-piped soils in their same respective areas and
found no significant difference in textures between piped
and non-piped profiles. Heede (1971, 6-7) compared 13 piped
and 19 non-piped samples. Jones (1975, 135-138) concluded
from analyses of 60 samples grouped in 17 profiles from the
Burbage Brook that both classes of profile showed a gradual
transition, most marked between A and B horizons, from loams
down to sandy loam sub-soils, with a very slight tendency
for higher (<5%) silt-clay contents around the actual pipe
level generally in the B horizon, which may be due to lessivage. In a further sample of the top, side and bed of 24
pipes, Jones found significantly finer roofs compared with
beds (e.g. beds had more in the 2-1 mm and 1-0.6 mm ($\alpha = 0.001$)

Table 5.   Summary of reported pipe size in relation to soils, geomorphic location and climate.

Size	Material	Geomorphic location	Climate	Source
max. 0.6-4.6m	mainly alluvial soils	river terraces	semi-arid	Jones (1968)
'few metres'	shales and erosion glacis	canyon walls	semi-arid	Barendregt (1977) Barendregt and Ongley (1977)
.05-6.1m 0.6-1.8m 0.15-0.38m	alluvial soils derived from shale with blocky B horizon	alluvial valleys	semi-arid	Brown (1961; 1962)
max. 3m	clay loam/silt clay loam	badlands on alluvial terraces	semi-arid	Bell (1968)
0.91-1.22m	Pleistocene White Silts	river terraces	arid highland	Buckham and Cockfield (1950)
0.91m	loess	flat upland, usually <45m from plateau rim	semi-arid	Fuller (1922)
0.15-0.96m	silt loam	hillslopes and gully walls	marine	Gibbs (1945)
25mm-0.91m	loess soils, often hardpan	hillslopes and stream-banks	marine	Hosking (1967)
<0.76m	cracking A horizon, prismatic B, impermeable clay C on sedimentary rocks	valley heads, seepage lines, landslides	marine	Ward (1966)
0.15-0.35m	brown earths with clay loam and C horizon	$10-20^{0}$ slopes	humid continental	Czeppe (1960)
<0.30m	alluvial and volcano-alluvial sediments	terraces and plateau edges	savanna to wet tropical	Löffler (1974)
<0.30m	clay soils on shale and in colluvium	-	tropical rainforest	Baillie (1975)

80-120mm	solonetz and solodic soils	cultivated slopes, banks, dams	marine	Floyd (1974)
max. 400m) mode 50-100mm)	peat gleys and podzols	highland mid-slopes	marine	Bell (1972)
max. c.0.30m) mode 12.50mm)	pumice soils	hillslopes in 'maturely dissected' relief	marine	Blong (1965)
<100mm	shale badland surface	mid-slopes	arid	Yair et al.(1980)
mean 90mm (ephemeral) 220-900mm (perennial)	lessivé brown earth on alluvium/collovium	streambank	marine	Jones (1971; 1975)
14.41-mm) (ephemeral)) 130mm) (perennial))		mid-slopes	marine	Humphreys (1978)
60mm	estuarine alluvium	salt marsh	marine	Lear (1976)
c. 50mm	estuarine and coastal alluvium	salt marsh	marine	Jones (1975)
mean 46mm	peaty podzol	highland mid-slope	marine	Morgan (1977), Newson and Harrison (1978) Gilman and Newson (1980)

and 600-250 μm ($\propto$ = 0.01) ranges and less in the
20-5 μm range ($\propto$ = 0.01)). Comparable though less significant differences were found in samples from the Afon Cerist
in Wales (*ibid.*, 140). Similar analyses by Parker (1963,
tab. 1) based on 8 semi-arid pipes also suggest higher silt
and/or clay contents in roofs than floors in five of the
cases: Parker himself noted that 'at six sites the texture
of the samples is controlled by silt and clay whereas at
the other two sand is chiefly responsible (*ibid.*, 109-110).

4.6 (ii). Erodibility

One of the major requirements for the development of
piping listed by Fletcher *et al.* (1954) and by numerous
others has been the presence of an erodible layer above an
impeding layer. Erodibility is a particularly difficult
property to measure, partly because despite considerable
advances in our understanding of the phenomenan over the past
decade there are many aspects still incompletely understood
(cp. Paaswell, 1973). But it is nevertheless remarkable
that more attempts have not been made to take some form of
measurement of erosive potential in piping studies until
latterly. Most measurements to date have been made by
engineers concerned with the chemistry of dispersive clays
in earthworks.

Downes (1946) analysed water-stable aggregate distribution in Victoria. All the B horizons were highly dispersible, heavy clays with no water-stable aggregates over 0.25
mm diameter and a mean dispersion percentage for 5 samples
of 80.4%. Farmers described them as melting away when wetted
and Downes' consequently called them 'sugary'subsoils. The
A horizons also had a poor structure with only 42.3%
aggregation. Brown (1961, tab. 2 and 4; 1962, tab. 1) presented data from Colorado and from the U.S. Soil Conservation
Service, Tucson, Arizona. All the Colorado soils were
highly erodible by Middleton's (1930) index, although he
did not sample for variations within the profiles. The Arizonan data suggest generally greater dispersibility in mid-
profile, with, for example, percentage of aggregates larger
than 0.1 mm falling from 26% to 6% and returning to 17%
down profile in one of the worst affected areas in the San
Pedro Valley.

Beckedahl (1977) applied dispersion tests to piped clay
subsoils in South Africa and concluded that there was greater
dispersion above and at pipe level than below.

Dispersal indices have been a regular feature of soil
analyses by the New South Wales Soil Conservation Service
since Ritchie (1963) developed the index to enable tunnelling
soils to be identified (Charman, 1969, 1970a and b; Crouch,
1974). An index of 3.0 or less is taken as indicating
susceptibility to tunnelling and a combination of this with
volumetric expansion of 10% and over indicates a high susceptibility to tunnelling (Ritchie, 1963; Charman, 1969, 1970a
and b). with a reinforcement of erodibility by an inferred

high cracking potential. In the soils of New South Wales, dispersivity is closely related to exchangeable sodium percentages and the results of these analyses have already been discussed in section 4.5(i). From Charman's tables (*loc. cit.*) it is clear that most dispersible samples are from the A_2 and B horizons of solodic soils. Craze (1974) noted dispersibility indices of 3 or less in 2 topsoils and 8 subsoils. Low indices (of 3-4) were quite common in the Reedy Creek Catchment, but the majority of expansions were less than 10%. These low volume expansions together with low or non-plasticity, characterise the clay type as kaolinite with low cohesiveness. Tunnelling is not a problem in undisturbed soils with this level of dispersibility but deep gullying is found where vegetative cover becomes sparse.

Jones (1971; 1975, 126-134; 1978a) performed water-stable aggregate tests under simulated subsurface wetting conditions (Kemper and Chepil, 1965, 499-510; Kemper, 1965, 511-519) in lessivé soils. Analysis of percentage aggregate stability and van Bavel's mean weight diameter (MWD) in profiles ranging up to 1m upstream and downstream of bank outlets on the Burbage Brook, England, showed that pipes did not occur in the zones with greatest aggregate stability (>90% stability). In one case, piping was found in a lense of lower stability (Jones, 1978a, fig. 2); in another piping occurred at the upper edge of a more dispersible and relatively impermeable layer (Fig. 9). Comparison of 14 piped and 7 'blank' profiles showed no overall difference in MWD over 72 samples with means of 3.8 and 3.9 mm respectively. However, in terms of aggregate stability, the piped profiles appeared overall to be slightly less stable ($\bar{x}$ = 68% for piped, 86% for unpiped, t = 1.95, $\propto$ just short of 5%). Analyses-of-variance indicated comparable variability between and within profiles in each group. However, comparing the two groups, piped profiles had much greater overall variability in aggregate stability ($\propto$ = 0.005). The high between-profile variability in piped profiles was due to the presence of two types of piped profile, namely, those with relatively high stability throughout and those with very diminished stability usually at just one point in the profile. Frequently the latter case occurred where outlets were low in the bank, running on or in colluvial/alluvial deposits. Piping occurred in the top 20cm of the profile where aggregate stability of the roof was greatest. On average greatest stability within the profile occurred at the roof of pipes (86% against 76% above and 75-78% below): this is an interesting variant of the 'erodible layer' hypothesis. A wider sample of the roof, wall and beds of 20 randomly selected outlets taken as part of the study of pipe geometry (cp. section 6.2) showed average aggregate stabilities of 73% (roof), 60% (wall) and 55% (bed), which proved to be significantly different in all three comparisons between roofs, walls and beds ($\propto$ = 0.01 by Wilcox on test).

Early work on erodibility by engineers concentrated on Atterberg Limits and grain-size. Sherard (1952) noted that all dams sampled with plasticity indices of > 5 failed within

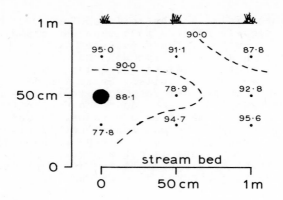

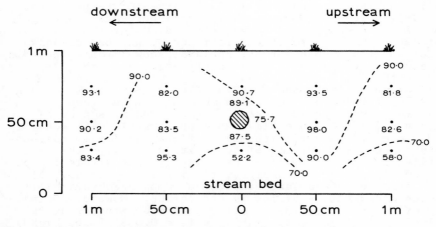

Fig. 9 Percentage aggregate stability around pipes in
the English Peak District.

a few years of filling and that the higher the liquid limit
for a given plasticity index the greater the resistance to
piping. Gibbs (1962) later established for the American
Bureau of Reclamation that soils with plasticity indices of
< 10 and liquid limits < 30 were generally erodible. Cole
and Lewis (1960) took a similar view in Australia, but it
was pointed out in discussion by Aitchison that dispersivity
may also be a factor in Victoria. That discussion marked a
significant turning point in engineering outlook. Only
Reséndiz (1977) appears to have challenged current views
that dispersive clays are important in dam failures. He
maintains that Skempton's activity coefficient is 'an appropriate index to identify soils susceptible to piping' and
that coefficients of 0.3-1.0 are necessary (though not sufficient). However, Sherard and Decker (1977, 470) appear to
have had the last say by pointing out that most clayey soils
used fall in this range, and in particular that their own
tests with high-sodium clays have involved conversion from

dispersive to nondispersive states with no change in Atterberg limits and activity (cf section 4.6 (iv)).

Sherard, Decker and Ryker (1972, 679) pointed out that although classical engineering theory had considered fine cohesionless silts and silty sands to be most susceptible, clay is actually more so. Neither cohesive nor cohesionless soils commonly provide the necessary conditions for piping as they see it, namely i) walls to cracks which can be eroded at the water velocities available, and ii) sufficiently cohesive walls and roofs to support an open pipe. They point out that the idea that cohesionless soils were particularly susceptible was due to Lane's 'weighted creep ratio' theory (Lane, 1935; Terzaghi and Peck, 1948) developed in relation to seepage beneath concrete dams. However, in this case the concrete provides a roof, so the mechanisms are completely different.

4.6(iii). *Soil structure and hydraulic conductivity*

In view of its obvious importance, remarkably little detailed attention has been paid to soil structure and permeability. There is a particular lack of measurements relating to permeability and most of the information available on soil structure comes from Australasia.

The common view in Australia is that piping/tunnelling is associated with duplex soils, also called 'texture-contrast' soils, with less permeable B horizons and A horizons with granular or crumb structure. The depth and severity of piping in these soils depends largely upon the properties of the B horizon, particularly the degree of cracking or development of interpedal voids. Crouch (1976) recognises three distinct types of tunnelling.

1) Shallow tunnels occurring in the A_2 horizon or the top metre of the B horizon, where water is arrested by the sharp textural, structural and permeability contrast at the A_2/B interface. This was described by Downes (1946), Fletcher and Carroll (1948) and illustrated by Newman and Phillips (1957, fig. 3).

2) Deep tunnels in the B_2 horizon, where cracks have penetrated the relatively impermeable B_1 layer. According to Crouch, these are usually confined to steep slopes as described by Gibbs (1945) and Hosking (1967). This form initiates gully development and has been called 'tunnel-gully erosion'. Newman and Phillips (1957) reported this deep piping in highly structured or cracked B horizons.

3) Deep tunnels initiated by gullies and usually assisting the expansion of gullies, as described by Parker (1963), initiated either by cracks in gully walls or by concentration of water on a point in the gully wall.

Contrasts in structure and permeability are clearly paramount in the first two of these. The columnar structure typical of the B horizons of soils in the solonetz and solodised solonetz range has been particularly associated with piping in New South Wales (Monteith, 1954; Newman and Phillips, 1957; Floyd, 1974). Downing (1968) describes piping occurring in soils in Natal with impermeable clay subsoil that cracks to form prismatic peds. Ward (1966) describes piping in Northland with generally structureless A horizons, B horizons with well-developed prismatic structures becoming angular blocky at depth, and impermeable, massive-structured clay C horizons. Piping is developed above the impermeable, unstructured Bg or C horizons.

In Colorado, Brown (1961; 1962) described profiles at 3 pipe sites as having a loose, granular A horizon. At two of the sites the B horizon was blocky or slightly blocky, followed respectively by massive and strongly blocky C horizons. At the other both B and C horizons were massive. Piping was severe at sites with massive B or C horizons, but less intense where the C horizon was blocky. His descriptions indicate that piping generally occurred just above or just within the massive layers and that the area with less intense piping had much smaller and shorter tunnels developed in the weakly blocky B horizon; presumably permeability was not as great here as in strongly blocky subsoils and the blocky C horizon allowed more vertical drainage and reduced lateral seepage in the B horizon. These pipes were only 15-38 cm across, less than 90 cm deep and extended less than 5 m back from the gullies compared with pipes with diameters up to 1.8 m, extending up to c. 20 m back from gully walls in the severely affected areas.

Jones (1975) studied piping in lessivé brown earths in England, which had somewhat similar but less marked structural differences with slight clay enrichment of the B horizon (B_t). But he noted two significant features: macropore distribution and subsoil cracking. Cracking occurred during the summer, although only c. 10% of pipes appeared to have originated from cracks. Macropore distribution, however, showed a more consistent relationship to piping, being above and around but not below the pipes, and reflected in permeability contrasts (*ibid.*, 106; Jones , 1978a, fig. 3) (Fig.12).

Beckedahl (1977) has described piping in the Drakensberg which occurs in profiles dominated by clay horizons. The pipes are actually developed in a mottled clay containing sandstone pebbles which he says forms a zone of higher permeability above a heavy compact clay; this also happens to be the zone of greatest erodibility (section 4.6 (ii)).

Few measurements are available of hydraulic conductivity. Fletcher and Carroll (1948, 547) confirmed the existence of an impeding subsoil, with surface infiltration rates of 0.016 mm s^{-1} above 0.0002 mm s^{-1}, and Harris and Fletcher (1951) found impeding layers with conductivities of 0.0016 and 0.0006 mm s^{-1} below piped layers with conductivities of

0.0092 and 0.0141 respectively. Fletcher *et al.* (1954, figs. 8 and 9) illustrate two typical profiles of soil permeability, one with greater permeability at the surface and gradual though irregular lessening of permeability down-profile, the other with restricted permeability at the surface and maximum permeability in mid-profile. In the first example, they believe that piping will occur if all 5 fundamental factors are favourable (cf. chapter intro.),regardless of cultivation techniques or rodent activity. In the second, they believe that piping will develop only if rodents or other activity allows water in at an accelerated rate.

Heede (1971, 10) found that samples taken from soils with ESP > 1.0 (piped) in gully banks in Colorado were between 1.6% and 11.8% as permeable as samples from unpiped soils at the same depth (ESP < 1.0). This suggested that wetting and subsequent swelling of the sodium-rich layers is slower and that cracks remain open longer, whilst decreased infiltration increases flow through the cracks. From Heede's diagram (*ibid.*, fig. 10) typical layer permeabilities appear to be of the order of 3.0×10^{-6} mm s^{-1} in piped soils and 5.3×10^{-6} - 4.9×10^{-5} mm s^{-1} in non-piped soils, which are all remarkably low. Löffler's (1974) rough estimates of matrix permeability based on the texture and structure of piped soils in New Guinea range from slow $(0.00069 - 0.0028$ mm $s^{-1})$ to moderate $(< 0.0167$ mm $s^{-1})$.

Crouch (1977, 65) reports a very low permeability $(0.000064$ mm $s^{-1})$ in the tunnelled B horizon of a yellow solonetzic soil in New South Wales.

In Britain, Jones's (1975, 96-104) study of variations in the profile revealed generally greater permeabilities and that piped profiles had significantly lower matrix permeability (α = 0.001, n = 53) than unpiped profiles, 1.66 mm s^{-1} against 5.85, due to a marked reduction in permeability in the lower part of the profile in piped cases. These pipes fell into two basic categories: namely, i) cases where permeability remained restricted below pipe level and ii) cases where there was a slight increase in permeability below the restricting layer in the pipe floor, but never regaining the rates measured above the pipes (Fig. 10). Measurements of the spatial variation within the profile over a distance of 1 m from 3 pipe outlets also suggested that the outlets tended to locate near the edge of high-to-medium permeability layers (*ibid.*, figs. 4.1.2-4; Jones, 1978a, fig. 1), which can be compared with the results of point counts of macropore distribution (v.s.).

Laboratory measurements of horizontal hydraulic conductivity above and below 12 pipe outlets revealed an impermeable substrate in every case (Fig.10a), and this was confirmed by parallel piezometer measurements of layer infiltration capacities in the field. Interestingly, there was little contrast in porosity within the piped profiles and the differences in hydraulic conductivity therefore indicated greater organisation of pore space rather than lower bulk density.

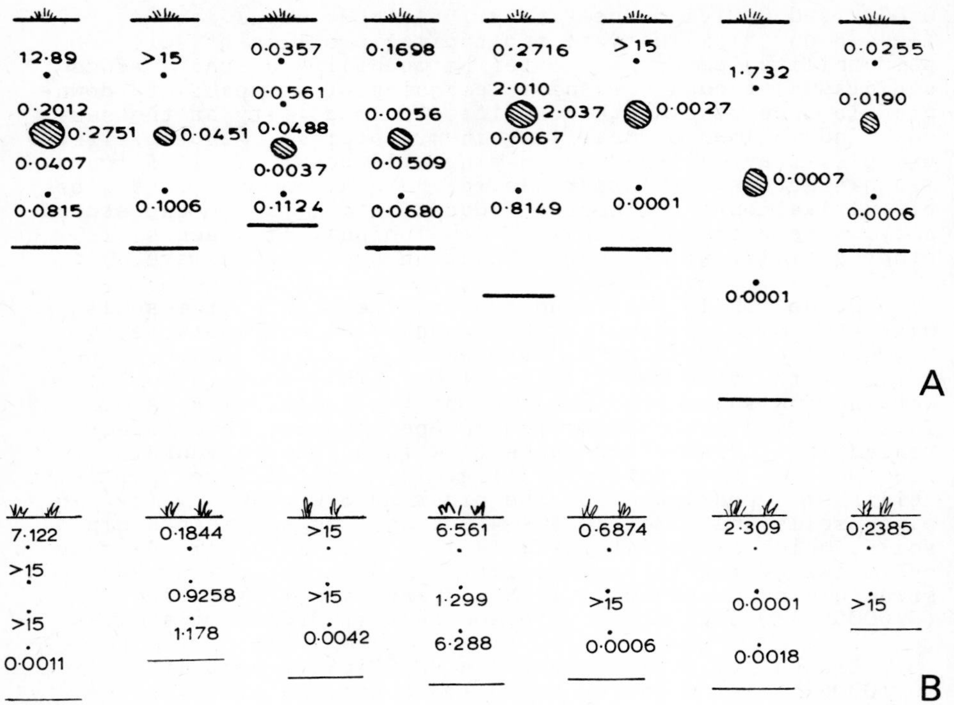

A

12·89   >15   0·0357   0·1698   0·2716   >15   0·0255

               0·0561   2·010   1·732   0·0190

0·2012   0·0451 0·0488   0·0056   2·037   0·0027
  0·2751   0·0067

0·0407   0·0037   0·0509

0·0815   0·1006   0·1124   0·0680   0·8149   0·0001   0·0007   0·0006

                                                      0·0001

B

7·122   0·1844   >15   6·561   0·6874   2·309   0·2385

>15

        0·9258   >15   1·299   >15   0·0001   >15

>15

        1·178   0·0042   6·288   0·0006   0·0018

0·0011

Fig. 10   Horizontal saturated hydraulic conductivity in
           piped and non-piped bank profiles on Burbage
           Brook, Derbyshire.

Average bulk density was slightly lower in the piped profiles
(1.00 Mg m^{-3} against 1.12), but neither this nor the associated difference in total porosity was significant (Jones,
1975, 108). Two profiles in Yuma County, Arizona, analysed
by Harris and Fletcher (1951, tab. 1 and 2) had porosities
of 58-42%, compared with values of c. 50-70% measured by
Jones (1975), and bulk densities of 1.4-1.2 Mg m^{-3} again
without clear zonation. Brown (1961; 1962) measured higher
bulk density in the severely piped areas than in the other
areas, 1.6-1.7 against 1.2-1.3 Mg m^{-3}. N.O. Jones (1968)
reported densities of 1.15-1.52 Mg m^{-3}. He comments that
'low density is a consistent feature of the areas subject to
piping' and that the horizon with pipe development seems to
be associated with the low density soils in many cases (*ibid.*
97). This is not corroborated by Brown's or J.A. Jones's
measurements. However, Brown does comment that because of
the high sodium content his heavy soils were more erodible;
the lower bulk density in the area of less severe piping
was associated with gypsiferous soils with higher flocculation, and (by calculation from his tables) low ESP.

     In fact, bulk density and total porosity measurements
are not particularly useful, partly because they take no
account of the organisation of pore space and partly because

the permeability of a horizon is often controlled by
megascopic voids (interpedal voids, planes, metachannels
and orthochannels etc., cf. Brewer 1964, 366) which tend
to escape, or even be avoided, in standard core-sample techniques. The latter problem also applies to hydraulic conductivity determined from laboratory samples.

In Eastern Europe, attention has been directed towards
measuring 'macro-porosity' in loess on the theory that the
collapsing or sagging of loess on wetting is the major
cause of subsidence (cp. Jennings and Knight, 1957, in S.
Africa). However, Malinowski (1963) pointed out that the
current Macroporosity Index[†] was inadequate since the problem was more complex than thought hitherto because of the
importance of suffosion.

Measurements in peaty podzols represent a special case.
Stagg (1974) observed shallow, discontinuous, ephemeral
piping near the base of the peaty horizon in the Mendips.
He describes the profile as a black, stoneless peat A horizon
with some mineral matter, a 100 mm thick Ea horizon, very
stony grey loam with weak blocky structure, above a clay
loam B/C horizon with weak angular-blocky structure. Infiltration capacities obtained for four cases of the peat surface varied considerably from 0.05 mm s^{-1} to 0.005 mm s^{-1}.
Vertical infiltration rates on the subsurface horizons were
.001 mm s^{-1} at base of peat (peat/mineral horizon) and 0.003-
0.008 mm s^{-1} in the underlying mineral horizon. The surface
of the peat was therefore reasonably permeable at some points
and there was a constricting layer at the peat/Ea interface.
However, most pipe discharge probably entered through cracks
and inlets (cp. Weyman, 1971) so it is the permeability of
the Ea surface that is important.

Jones (1975, 248-250; 1978c, fig. 9) mapped horizontal
saturated hydraulic conductivity around a seasonally active
pipe system developed in peaty gley podzols at Maesnant,
Plynlimon, based on 26 samples. He found a mean conductivity
of 0.0094 mm s^{-1} in the surface peaty layer compared with
0.6087 mm s^{-1} in the Eg : the best developed pipes were found
in the upper part of the Eg. However, he also observed that
the mineral horizon was less conductive within 1-2 m of the
trunk pipe than further away (.0125 against 1.3001), and
overall the peat and mineral conductivities were only distin-
guishable at the 10% level. The bouldery, stony clay solifluction deposits beneath could not be adequately sampled but
appeared to provide the traditional impermeable underlayer.
In the Hiraethog peaty gley podzols of the Upper Wye, Plynlimon, Morgan (1977) reported a saturated hydraulic

[†]Macroporosity Index = $\dfrac{\text{subsidence increment on wetting}}{\text{height before wetting}}$

in samples under a load of 3 kg m^{-2}. Unstable if index >0.02.

conductivity of 0.015 mm s^{-1} in the horizontal and one-tenth
of this in the vertical in the stony silty clay B horizon
underlying the pipe zone, although she comments that water
is transmitted through the interpedal voids in the prismatic
structure. In the piped Oh horizon, cracks 0.5 mm wide and
20 cm apart were observed during the 1976 drought which
aid pipe development (cf. section 4.6.(iv)).

Bell (1972) has described the general situation on the
Upper Wye. The podzol profiles consist of 10-40 cm of a
peat and peat/mineral A horizon, an E layer (leached ash
grey clay pan 5.20 cm thick) above an iron pan, and a B/C
horizon of brown silty clay with slatey mudstone flakes
above (supposed) boulder clay or what he termed 'preluvium'
(loose slatey material probably of periglacial origin).
Pipes are found at the peat/E horizon interface and on the
colluvial mantle/boulder clay interface, i.e. at the major
sites of contrast in conductivity in the regolith. He
noted that pipes on the boulder clay interface are often
beneath prominent shrinkage cracks in the clay pan and he
suggested that they probably evolved from the most suitable
downslope drainage path afforded by the crack system. The
cracks are probably formed in successive dry years and then
persist in both the peat and the clay pan. This idea has
been reiterated, particularly in the case of the peat, by
Morgan (1977), Institute of Hydrology (1978) and Gilman and
Newson (1980).

Another special case occurs on salt marshes. Kesel
and Smith (1978) have described piping on the Nigg Bay
Marshes, Scotland, from immediately below the root zone (20
cm) to over 75 cm depth in bedded silts and sands, but not
in the underlying gravels. Jones (1975, 153) reported a
somewhat similar situation on a young marsh in the Fal estuary, Cornwall, with piping at the interface between a 12
cm thick overlayer consisting of 73% silt and an underlayer
with 37% gravel (Table 3) with presumably greater permeability. Conversely, Jones's analysis of a case on Brancaster Marsh, Norfolk, showed piping at the interface of a
granular surface horizon and a massive silt-clay layer (c.
80% silt-clay). Unfortunately, we have no hydraulic conductivity measurements for any of these cases, and it is
not clear whether the Nigg pipes follow the sedimentary
bedding. It is possible that in some cases the apparent
lack of an impermeable substrate can be explained by a confined aquifer situation, as suggested by Jones for the Fal
case, in which pipes are created at the top of the aquifer
in more erodible material.

To date, Jones (1975, 115-124; 1978a) is the only one
to have reported measurements of reduction - oxidation status
in piped soils. He found that redox potentials were not
simply related to hydraulic conductivity. An interesting
comparison can be made between plots of hydraulic conductivity, bulk density and total porosity, macropore distribution and redox potential in a lessivé brown earth illustrated
in Fig. 11. They suggest that the pipe runs in a lenticular

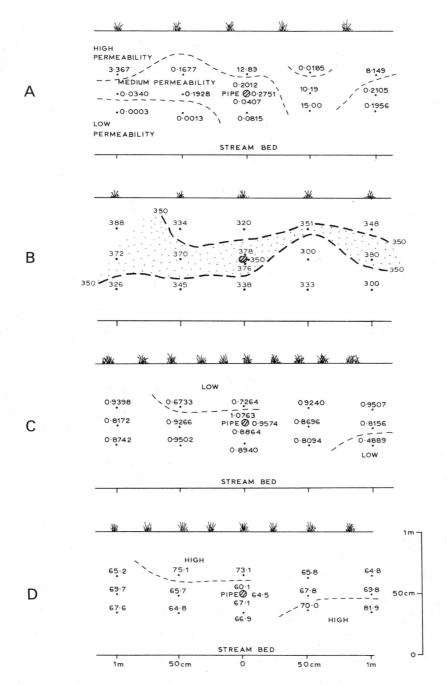

Fig. 11   Variation in horizontal saturated hydraulic conductivity, total porosity, bulk density and redox potential around a pipe on Burbage Brook, Derbyshire.

zone with slightly better aeration, approximating to a similar lenticular zone of medium permeability and moderate bulk density and porosity. The lower redox potential of many of the samples from the A horizon, which has a generally lower bulk density, and higher porosity and *matrix* hydraulic conductivity might be explained by greater aeration in the piped zone due to the presence of the pipe itself draining the upper B horizon and possibly by a concentration of macro-pores (which may not be well represented in core samples used to determine hydraulic conductivity) around pipe level. A macropore plot of a similar pipe outlet which was higher in the profile, so that the 30 x 30 cm counting template included part of the A horizon, shows some reduction in macropores in the top soil (Fig. 12), which may be related to the horizon contrasts developed by lessivage.

However, Jones also found cases in which pipes occupied zones with greater reduction and he suggests that the para-dox can be explained assuming that the redox potential reflects the average water supply situation, so that irres-pective of hydraulic conductivity if pipes flow only infre-quently then a high redox potential may exist even in poorly conducting soils (Fig. 13). Using size of master pipe in each of six pipe groups as a surrogate for frequency of pipe-flow, Jones was able to demonstrate a significant correlation coefficient of -0.65 (Kendall's τ, with conductivity part-ialled out) between pipe size and redox potential (Jones, 1978a, 7). This agrees with the result that pipe diameters were larger in percolines outfalls (cf. section 6.1(i)), which was obtained from a much larger sample (138).

Comparing redox potentials in piped and unpiped profiles Jones (1978, 124) found that the only real difference was a significantly greater oxidation (+ 100 mV) in the lower part of profiles with piping ($\propto$ = 0.05), which could be explained by the fact that the pipes will also to some extent drain and aerate the underlayer, as was illustrated in Kirkham's (1951, figs. 3-4) analyses of the effect of land drains.

Returning from the microscopic to the synoptic view. The main unifying factor between all soils subject to piping, according to Baillie (1975), is a profile with a permeable layer overlying an impermeable layer. In fact, Baillie lists the main conditions for pipe development as, i) this permeability contrast, ii) sufficient gradient on the imper-meable layer to drain and iii) material clastic enough for particle or aggregate detachment but cohesive enough to prevent premature collapse. He noted that there are large areas of duplex soils in arid, semiarid and subhumid regions, characterised by abrupt changes in the profile from sandy to heavy texture, and that piping develops in these espec-ially where a high ESP leads to dispersion. In the FAO/ UNESCO (1968) classification, these are solonetz or planosols. There are also large areas of vertisols (self-mulching soils) many of which are characterised by severe cracking at the surface in the dry season : Worrall (1959) observed features

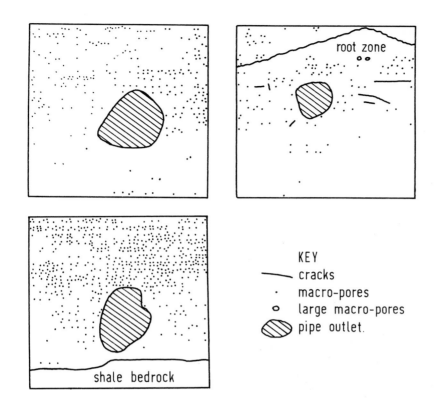

Fig.12   Distribution of macropores around pipes on Burbage
          Brook (after Jones, 1975, fig. 4.2.1. and Jones,
          1978, fig. 3).

similar to piping in vertisols in the Butana grassland of
Sudan and, in fact, published a photograph of a plaster
caste of a vertical pipe (*ibid*.). Baillie also suggested
that increased clay content and decreased permeability with
depth is often found even in transported materials where
piping occurs, quoting the observations of Terzaghi (1931)
in river alluvium, Burton (1964) in marine alluvium and
Fuller (1922) and others in loess.   In the humid and subhumid tropics, large pipes are often found where ferricrete
is underlain by a pallid zone of less permeable saprolitic
material, the strong ferricrete providing an ideal roofing
material (Baillie, 1975; cp. section 8.6).   In his own study
area in the tropical rainforest of Sarawak, piping was common
in acrisols and cambisols which like ferrasols all develop
widespread throughflow because of increasing clay content
and reduced permeability down-profile.

      These observations are independently supported by those
of Franzle (1976), who contrasted the acrisols, luvisols,
porous ferralsols and podzols of tropical and sub-tropical
rain forests with the vertisols and luvisols of savanna

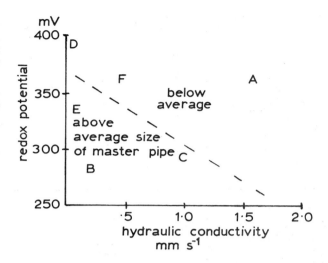

Fig. 13   Mean redox potential of piping layer against mean
          horizontal hydraulic conductivity for pipe groups
          on Burbage Brook (after Jones, 1975, fig.4.3.5. and
          Jones, 1978, fig 4).

regions. Piping is an important process in the former group
because of the high surface infiltration capacities but low
field capacities, whereas surface planation processes are
supposedly more important in savanna areas where low interception storage is combined with low surface infiltration
capacities. High field capacities are also supposed to
limit subsurface processes in the latter group. Nevertheless, Franzle's statement does seem to be rather sweeping
and whilst his observations in rain forest ferralsols are
supported by Bremer (1973), piping has also been widely
observed in savanna soils (e.g. Clayton, 1956; Rathjens,
1973, Löffler, 1974; Stocking, 1976).

     Therefore, with some exceptions, subsequent experience
does tend to support the contention of Fletcher and Carroll
(1948) that an impeding layer is necessary, at least for
strong development of piping. Similar support is also given
by studies of 'bedrock piping' in badlands (section 8.1).

*4.6(iv). Clay minerals : cracking potential and
           dispersivity.*

     Particular attention has been paid to clay mineral
species both because of their relevance to cracking potential
and because of their significance for soil dispersion and
erodibility in alkaline soils. Fletcher and Carroll (1948,
545) made the suggestion that piping is associated with
highly expansive clays and this view has been emphatically
upheld by Barendregt and Ongley (1977, 238): 'We assume that
the swelling and cracking of montmorillonitic clays is essential to the initiation of piping'.

86

The main species reported in areas of natural piping are listed in Table 6. Generally, montmorillonite and illite dominate in reports from the western USA, but less expansive clays are reported from Britain, New Zealand and New Guinea. Yair *et al.* (1980) reported proportions of 50% montmorillonite, 40% kaolinite and 1% illite on shale badlands which developed piping. Nevertheless, expansion tests showed a maximum of only 4% swelling, which was 'not well understood'. They suggested that the kaolinite inhibited swelling, particularly in the presence of large quantities of calcium (gypsum); this would imply that the material was calcium-montmorillonite rather than the sodium-montmorillonite found earlier by Bryan *et al.* (1978) in Canada. The kaolinite and chlorite dominance reported at an English site by Jones (1971) may be related to the apparently insignificant role of cracking in pipe initiation there. Jones (1975, 141) reported similar results from a streambank site in mid-Wales, but with illite slightly more dominant than chlorite. Gilman and Newson (1980) quote the work of Stewart, Adams and Abdulla (1970) suggesting similar composition in slope soils in mid-Wales, implying that this obtains in the piped slopes they studies on Plynlimon. However, much of the hillslope piping in Britain occurs in a peaty surface horizon, where as Gilman and Newson point out desiccation cracking of the humic material is important to piping (cf. section 6.3). Desiccation cracks in peat do not completely close due to structural changes and oxidation, and in blanket peats there may also be a substantial surface subsidence on drying. In peaty podzols desiccation cracking is more prevalent in the organic horizon and clay minerals appear to be subsidiary. Gilman and Newson (1980, tab. 3) quote data from Adams and Raza (1978) indicating shrinking rates on air drying of : 57.0% Oh, 19.4% Eg, 8.5% Bg and 4.0% Br. Pipes in their area are basically developed within the Oh with only minor erosional excursions into the Eg. Nevertheless, Morgan (1977) has reported prismatic structures in the mineral layer there and that water is conducted along the interpedal voids.

Volumetric expansion tests were made by the U.S. Soil Conservation Service for Brown (1961, tab. 2) in Southern Arizona (San Pedro) and indicated values between 7.9% and 20.2% in piped subsoils. Brown (1962) also found high shrinkage and heavy soils characteristic of all three piping sites studied in Colorado. Volumetric expansion tests have become a standard form of analysis in the New South Wales Soil Conservation Service, since Ritchie (1963) identified expansion as a basic diagnostic of susceptibility (cp. section 4.5(i): Charman, 1969, 1970a and b; Craze, 1974).

Percentage of clay as well as species may be important. Again, American cases tend to have high clay content: 52-24% sub-5μm in Arizona (Brown, 1961), 46% clay in Colorado (Heede, 1971). However, values of only 10-18% clay (< 2μm) are reported around streambank outlets in Britain (Jones, 1971; 1975, 140). Jones (1975, 107; 1978a) noted that cracking commonly occurred during the summer in the clay-enriched B horizons of lessivé brown earths on Burbage Brook.

Table 6. Major Clay minerals reported in piped soils.

Clay species	Cases associated with piping.
montmorillite	Kingsbury (1952), Brown (1961), Parker (1963), Mears (1963), Parker *et al.* (1964), Ward (1966), Parker and Jenne (1967), Bell (1968), Gile and Grossman (1968), and Heede (1971), from Brown; Bryan *et al.* (1978), Yair *et al.* (1980)
smectite	N.O. Jones (1968)
beidellite	Bell (1968)
bentonite	Bell (1968), Barendregt (1977), Barendregt and Ongley (1977)
illite	Parker (1963), Mears (1963), Parker and Jenne (1967), N.O. Jones (1968)
mixed layer complexes, particularly montmorillonite-illite	Parker (1963), Parker *et al.* (1964) Parker and Jenne (1967), N.O. Jones (1968)
kaolinite	Parker (1963), Bell (1968), J.A.A. Jones (1971; 1975), Yair *et al.* (1980)
halloysite	Löffler (1974) with kaolin
metahalloysite	Ward (1966)
chlorite	Parker (1963), J.A.A. Jones (1971; 1975).

Although only a little over 10% of pipe outlets appeared to have developed from cracks, these pipes tended to be larger than average (13.0 cm cp. 8.9 cm average diameter) and lower in the bank (mean level 15% bankfull cp. 44% average); differences significant at 0.1% level. Pipes developed from cracks therefore probably developed with some advantage, although they were not found on percolines, where 'boiling' would be a more likely process, and where the largest pipes were found. Jones (1975) noted that in contrast piping in the nearby peat bog and in peaty podzols in Wales was dominantly crack-generated, the former largely by desiccation cracks, the latter by mass movement cracks.

Carroll (1949) has also pointed out that cracks may be generated by a sudden lowering of the water table, as this will increase the effective load on the lower layers in proportion to the difference between drained and submerged unit weights (cf. Terzaghi and Peck, 1948). Such cracks are typically much larger than desiccation cracks and were observed to be used by piping along the San Pedro River (Fletcher *et al.*, 1954). Peterson (1962) and Heindl and Feth

(1955) argued against this saying that such fissures
('earthquake cracks') tended to be scarce wherever true
piping was widespread. N.O. Jones (1968, 119-122) found
pipes using large fractures in this area and noted that
there had been a lowering of the water table of 4.6-9.1 m
since 1980, but he was unable to verify Carroll's suggestion
of differential subsidence. Some of these cracks did not
extend to the surface, but were buried by 0.3-0.6 m of sediment perhaps due to subsequent sedimentation or current
infilling. Jones (*ibid*, 119) distinguished 4 types of fissures i) surficial desiccation cracks in soils and alluvium
which reform seasonally, ii) regular joint systems in the
St. David Formation (Plio-Pleistocene silts and silt-clays),
iii) large scale orthogonal polygon fracture patterns in
alluvium, and iv) marginal cracks around gullies and sinks.
The pipe systems were largely controlled by these fractures,
with linearity more marked in the deeper, bedrock systems
than in the shallow systems following desiccation cracks in
the soil.

Engineers concerned with the failure of earth dams have
also recorded clay species both from the point of view of
cracking potential and dispersion. Kässiff and Henkin (1964)
reported dominantly montmorillonite clays in failed loess
dams in the Negev, with a little illite. Ingles (1970)
stated that the majority of serious failures occur in montmorillonites and illites; kaolinites do suffer from it but
not nearly so frequently and marine clays or clay deposited
under saline conditions are notoriously bad. Most of the
Australian failures are in the montmorillonite group, but
wind-blown desert sand and loam are also susceptible.

It is clear, however, that most clay mineral analyses
have not been sufficiently specific in reporting varieties
of montmorillonite present. As pointed out by Chilingar
(1970) in discussing the paper by Ingles and Aitchison (1970),
conversion of the sodium-montmorillonite clays they had
reported to calcium or aluminium-base montmorillonites would
significantly reduce the swelling problem. Ingles (1970)
also pointed out that montmorillonites may have substitution
in their tetrahedral layer instead of the octahedral layer
which will cause them to behave like sand and not to disperse.
In other words, varieties need to be specified, not just
species.

Sherard, Decker and Ryker (1972, 611) saw no reason to
doubt the Australian conclusion on the importance of montmorillonite in dispersive soils. Analyses of 9 clay specimens
from Venezuela, Oklahoma and Mississippi showed a similar
mix of minerals with the smectite group dominant: montmorillonite 30-50%, illite 10-25%, kaolinite 10-25%, quartz 20-40%.
They speculate that the main relationships linking sodium
and dispersibility only apply to soils 'with at least a
moderate content of montmorillonite'. However, they could
find no significant difference in clay mineral composition
between piped (rainfall-tunnelled) and non-piped areas of
dams. Moreover highly dispersible clays (56%-80% by SCS

Dispersion Test) were found in soils dominated by halloysite and plotting in the 'stable flocculated' zone on Fig.5.1.1.

Reséndiz (1977) has argued by logical deduction that clay activity is 'the most significant variable likely to affect the susceptibility to piping' since it relates directly to both permeability and swelling potential. According to him, higher clay activity is likely to give lower suscepti- bility to piping because greater swelling potential should result in small cracks being self-healing. He collated all the published data he could get to prove his theoretical deduction (Fig. 14). Piping was almost universally assoc- iated with activities of 0.3-1.0, which generally indicates illite dominated soils, though kaolinite-montmorillonite mixes may have similar activity. According to Del Castillo (1973) illite predominates in piped dams in Mexico and America. Reséndiz (1977, 350) argues that activity coeffic- ients above 1.0 indicate clays rich in montmorillonite, which 'are generally too expansive to permit open cracks in the presence of water and too impervious to allow seepage velocities large enough to induce piping'. He also notes the absence of piping in clays with activities below 0.3, which are either silts or kaolinitic and halloysitic clays. These low plasticity soils 'hardly possess enough cohesion to keep cracks or pipes open' and their grain sizes are too large for transportation by seepage water (kaolinite or halloysite in the range 0.5-1.0 µm would require permeabil- ities over 10^{-4} cm s^{-1}).

Reséndiz concluded with a broadside at 'Sherard's criterion', actually the SAR/TDS graph used by many engineers in Australia, America and Israel. He added Mexican data from his compatriot, Del Castillo (1973), to emphasise the spread of piping failures which appears to pay no attention to the mapped zones of dispersion and non-dispersion (Fig. 15), in contrast to the clarity of distribution of piping failures in terms of clay activity.

Notwithstanding this evidence there seem to be a number of problems with Reséndiz's view:

1) the overwhelming consensus of opinion and observat- ions of piping in the natural landscape associates piping with highly expansive, especially montmorillonitic clay species. Montmorillonite is also reported as dominant in Australian piped dams (e.g. Aitchison and Wood, 1965), although Reséndiz noted this and seemed to question their data showing activities of c. 0.8 with 80-90% montmorillonite and 20-10% kaolinite 'which is hardly believable' (*op.cit.*, 350). He noted that the only Australian dam with activity greater than 1.0 (A = 1.1) failed after seepage of very pure water with 'the lowest ever reported' TCC of 0.6 meq 1^{-1} for a case of piping.

2) Sherard and Decker (1977, 470) pointed out that most clayey soils used by engineers fall in the activity range 0.3-1.0 anyway and in particular that their own tests with high sodium clays have involved conversion from dispersive

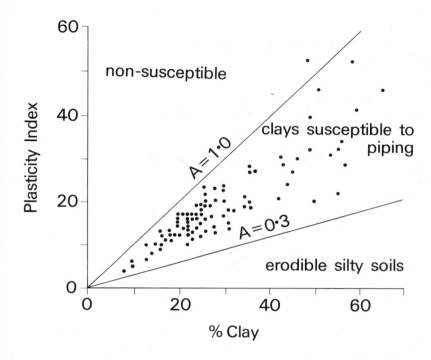

Fig. 14   Clay activity in artificial embankments failed by
          piping according to Reséndiz (1977, fig. 4).
          o - Sherard's data, + - Del Castillo's data.

to non-dispersive states with no change in Atterberg limits
and

3) the question of whether highly expansive clay is
safe because it will close cracks quicker or whether it is
more susceptible because it cracks more when dried must
receive a different answer in different circumstances. The
writer is inclined to the view that Reséndiz's fail-safe
mechanism can only apply to cases of what he himself describes as 'thin cracks' (*op. cit.*, 348).   It is also true
that such a property cannot necessarily be relied upon in
every case of 'thin' cracking.   Sherard and Decker (1977,
469) noted that erosion depends on relative rates of erosive
and swelling actions and that at present no laboratory test
caters for this aspect.

There is, therefore, considerable evidence that highly
expansive clays may frequently be responsible for initiating
piping by cracking, but they are by no means the universal
cause.   Dispersive clays have more recently become the focus
of attention.   This topic was taken up in the discussion of
the role of soil chemistry in section 4.5(i) and will be a
primary point of interest in the following chapter in the
analysis of current understanding of the mechanics of piping.
However, as will be seen, these too lack the causal univer-
sality commonly claimed for them in latter years, despite

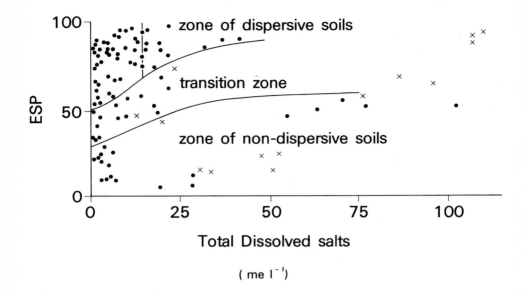

Fig. 15   Comparison of Sherard's dispersion criteria with
          recorded piping failure according to Reséndiz
          (1977, fig. 5).
          o = Sherard's data, + = Del Castillo's data.

compelling engineering evidence (e.g., Ghuman *et al.*, 1977;
Philips, 1977; Forsythe, 1977). It might, for example, appear
enigmatic that the engineer Villegas (1977) should analyse
the residual soils of a quartz-diorite batholith in Colombia
and pronounce them dominantly kaolinitic, non-dispersive,
highly resistant to annual rainfalls of up to 6000 mm, and
safe for dam construction without protective filters, where
in the same climate and soils Feininger (1969) discovered
extensive natural pseudokarst.

Moreover, cracking is as much a histic as an argillic
property, as illustrated by studies of piping in peaty gley
podzols (e.g. Gilman and Newson, 1980), and it seems likely
that well-decomposed peat may behave in a similar fashion
to dispersed soil in blanket bogs, as evidenced by many
reports of deep-seated piping here (e.g., Thomas, 1956;
Johnson, 1957; Bower, 1961; Radley, 1962; Ingram, 1967;
Tomlinson, 1980; Cruickshank, 1980; cp. section 8.2).

5. TOWARDS A GENERAL THEORY OF PIPE INITIATION

It is a difficult if not impossible task to try to subsume all possible modes of pipe initiation within one general theory. Nevertheless,there seems to be sufficient observational evidence and consensus of views to allow some form of theoretical summary under two headings: first the broad conceptual framework and second the detailed mechanisms. Under the broad conceptual framework we seek to summarise the context in which pipe initiation can occur in the natural landscape, without detailed recourse to the predisposing factors of climate, human interference, vegetation history, geomorphic environment etc., that have been analysed in the previous sections. Both Jones (1975, 55-62) and Gilman and Newson (1980) have attempted this with remarkably similar results. The second subsection of necessity draws heavily on research conducted by civil engineers and soil conservationists in relation to artificial landscapes, but an attempt has been made to limit discussion of their work to aspects and processes of direct relevance to piping initiation in the natural environment.

Since the views of Gilman and Newson are readily available in Monograph No. 1 of this series, the next subsection essentially follows the writer's previous summary, noting only any significant divergence of view or extra observation by the former authors.

5.1. General conceptual framework

A basic requirement is the irregular horizontal concentration of rainwater within the soil profile. This may occur either at an irregular basal surface to the soil or at any similar surface of low relative permeability within the soil profile or regolith, or by lenses of greater permeability in the profile, or else by cracks or biotic voids. Once concentration has been achieved both upslope and downslope extension may occur. For piping initiated in percolines or hollows, accelerated weathering and destruction of soil structure by persistant saturation will increase both surface and subsurface height differentials between seepage zones and intercols, enlarge the soil-water catchment and improve the capture of surface runoff. This will result in headward sapping and downslope erosion by inertia as and when more efficient drainage develops, especially piping through 'boiling' or interpedal voids. In other cases, where cracks or biotic voids already provide 'efficient drainage' in the form of concentrated flow or 'combining flow' (Liao and Scheidegger, 1968), whether within percoline hollows or elsewhere, similar extension and enlargement and pipe development may occur. Extension and integration of networks must depend upon the balance between opposing components of the hillslope system, particularly soil creep and other forms of mass movement, as indicated by Dalrymple, Blong and Conacher (1968) and Conacher and Dalrymple (1977)

for soil pipes on valley-side slopes and in the case of
infilling or growth of hollows outlined by Carson and Kirkby
(1972, 394), just as for streamheads (Kirkby and Chorley,
1967).

Piping is only an end-member of a continuum of forms
of interflow or throughflow, in which it is difficult to
assign classificatory boundaries. In their stochastic model
of 'flow through porous media', Liao and Scheidegger (1968)
divided the forms of flow into 'splitting flow', such as
diffuse seepage, and 'combining flow', such as channel flow,
which they wrongly supposed to be absent in the regolith.
But both forms do occur within the soil and it is not so
easy to divide them within the soil. Difficulties of definition arise partly because of semantic problems caused by
the wide range of scale involved. At the smallest end of
the pipe spectrum, on pipe headwaters and tributaries, water
flows along an irregular network of cracks, voids and
channels of various origins. True splitting flow, as described by Liao and Scheidegger (1968), comprises threads of
water which have positive finite probabilities of passing
one or other side of soil aggregates through the pore space.
But the typical pore space does not consist solely of a
statistically normal distribution of structural packing pore
sizes, as theory has often supposed (for example, Childs,
1969, 94-95). Abnormally large pores commonly exist, meso-
pores, 30-100 µm in diameter (Jongerius, 1957) or macropores,
which may originate from the larger packing pores, biopores
or the boundaries of soil peds, along desiccation or subsidence cracks. Since only linear pores greater than 30 µm in
diameter can materially assist drainage (Jongerius, 1957),
it follows that these are the most likely sites for attrition
and modification by water flow, increasing the anomalous
nature of these voids. (Jones (1975, 59) also speculated
that the 'hole', which Culling (1960) deduced from kinetic
solid state theory must migrate upslope associated with localised accelerated rates of soil creep, could be used by water
as random foci for concentration.) A degree of combining
flow is therefore already present where these voids exist.
A measure of combining flow may even be achieved without the
existence of complete voids. Pedotubules, pseudomorphs of
roots filled with loose material (Brewer and Sleeman, 1963),
may act in a similar way (Gaiser, 1952). Similarly, linear
lenses of lower density, disturbed and fractured soils may
be generated by mass movement which concentrate seepage in
a zone of greater permeability (and possibly greater erodibility). At times flow velocities in these zones may become sufficient to initiate true piping.

These rudimentary networks will be mostly discontinuous
and often show very little evidence of water flow. At this
point a second problem in boundary definition becomes apparent : at what stage may this rudimentary network by called
a 'pipe' network? It is convenient to retain the established
definition of soil pipes as 'subterranean channels developed
as a consequence of the movement of water in currents'
(Parker, 1963, 103). In other words, they are sculpted by

water and one would expect their geometry to reflect hydraulic conditions. Jones (1971; 1975, 61) introduced the term 'pseudo-pipe' to refer to features which act like pipes in so far as they carry elements of combining flow, but do not owe their form to flowing water. Conceptually, it is a useful distinction, since the beginning of any regularity in form marks a major step forward in the organisation of a system. However, in practice it is difficult to decide how much modification by erosion is needed before calling a feature a 'pipe'. One suggestion might be that a feature is a 'pipe' once statistically consistent relationships can be established in the hydraulic geometry. On the other hand, Jones (1975, 61) noted that it is quite probable that most of pipes he examined on the Burbage Brook would probably fail this test and yet be recognisably shaped by water, so that, as we are seeking a quantitative definition, perhaps surface roughness, bed roughness or channel curvature statistics may provide sounder bases. Conceivably, one would also expect differences such as increased frequency of capacity flow in pseudo-pipes compared with true pipes and an increased proportion of laminar flow compared with turbulent flow.

As in any naturally growing population, one would expect randomly distributed variations in size and viability amongst the branches of these networks. The more viable will tend to become the larger by capturing drainage from the others and excavating their beds to a level at which they are fed by ground-water as well as stormwater. Some soil pipes might even capture parts of the ground-water spring network in this manner. During periods of network extension the most viable branches of the lower capacity networks are likely to be absorbed by adjacent higher capacity networks, for example, larger pseudopipes transformed to true pipes or the larger pipes into surface stream channels. Alternatively, pipes pseudopipes and percolines may achieve a quasi-equilibrium in the drainage system (cf. section 6.3).

5.2.   The mechanics of pipe initiation.

In focussing on the detailed 'mechanics' of piping, it is increasingly obvious that we are looking at the mechanics of a physico-chemical system. Given a concentration of water within the soil as outlined above and given an outlet and therefore a hydraulic gradient to generate flow, i.e. the fundamental necessities of Fletcher and Carroll (1948), the basic requirements for the onset of piping appear to be destruction of the soil structure (dispersion) and sufficient permeability to generate the required level of shear stresses for attrition and transportation.

Up to the 1960s engineering theory concentrated almost exclusively on problems associated with permeability, and avoided the complication introduced by soil cohesion. Indeed, cohesive soil was considered to be relatively unsusceptible. But cohesion is an almost universal property of natural soils and the developments of the 60's and 70's have forced

95

engineering theory in the direction of soil physics and soil chemistry, making it much more relevant to the natural landscape.

The following discussion is approximately divided into two subsections, first the weakening of soil bonding and then the mechanics of flow and transportation. However, systematic links and interactions mean that a complete separation is neither possible nor desirable.

5.2(i). The process of dispersion

Aitchison (1960) was among the first to suggest that piping processes could involve dispersion of clays at the exit point. Subsequently Aitchison, Ingles and Wood (1963) defined the boundary conditions for dispersion (implicitly in alkaline soils) in terms of the amount of sodium in the exchange complex and the ionic concentration of the seepage water (Fig. 16). The Soil Mechanics section of CSIRO devoted considerable attention to the problem in earth dams in Australia during the 1960's, including an almost unique series of observations of the actual development of piping in the dams (Cole and Lewis, 1960; Aitchison, 1960; Aitchison and Ingles, 1963; Wood, Aitchison and Ingles, 1964; Ingles 1964a and b; Ingles and Wood, 1964; Aitchison and Wood, 1965; Ingles, Lang and Richards, 1968; Aitchison, Ingles, Wood and Lang, 1968; Ingles, Lang, Richards and Wood, 1969). Based upon this work, Ingles and Aitchison (1970) have provided a very useful account of the chemical processes that may be associated with the destruction of soil granulation and the onset of piping in natural alkaline soils.

The basic requirements appear to be 1) substantial porosity or 'voidage' and 2) weakening of interparticle bonds, followed by 3) a displacement force. In soils bonded by clays, concentrated salt solutions or partially soluble salts like gypsum, dispersion may be accomplished by the dilution of the bonding salts by throughflow of low salinity water or by cationic exchange on the surface of the clay micelles or bonded particles. Destructive cationic exchange involves a replacement of bonding divalent cations such as Ca^{++} and Mg^{++}, by monovalent ions of Na^+, K^+ or bicarbonate (HCO_3^-) in percolating waters, thereby increasing the repulsive forces of the Gouy layer of the micelles. The critical stage at which instability appears in clays (Fig. 16) seems to be a complex function of the soluble sodium percentage and the clay mineral species or varieties present according to the work of the dam engineers in Australia (Wood *et al.*, 1964; Aitchison and Wood, 1965; Ingles and Aitchison, 1970). Clay complexes with exchangeable sodium percentages (ESP) greater than 12% can spontaneously deflocculate in low salinity water causing dispersion (Ingles *et al.*, 1969).

Parallel work by Israeli dam engineers confirms the significant relationships between soil chemistry, clay types and seepage in loess dams in the Negev (Kassiff, 1956; Aisenstein, Diamant and Saidoff, 1961; Henkin, 1967; Kassiff,

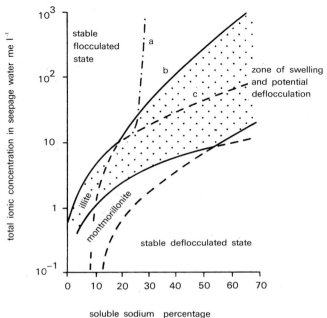

y-axis: total ionic concentration in seepage water me l⁻¹

10^3 — stable flocculated state — a

10^2 — b — c — zone of swelling and potential deflocculation

10

illite / montmorillonite

1

stable deflocculated state

10^{-1}

x-axis: 0 10 20 30 40 50 60 70

soluble sodium percentage

Fig. 16   Stability against piping in alkaline soils related
to total cationic concentration of seepage water,
soluble sodium percentage of the soil and clay
minerals, after Ingles and Aitchison (1970, fig.3),
Wood (1964, fig. 2) and Aitchison and Wood (1965,
fig. 2). Montmorillonite deflocculates to the
right of the line 'b', illite to the right of line
'c'. Kassiff and Henkin (1967, fig. 1) observed
the boundary line 'a' in dominantly montmorillonitic
soils and a similar trend is also apparent in
Sherard's boundary lines (Fig. 15). Heinzen and
Arulanandan (1977, figs. 1 and 2) plot the relationships between sodium adsorption ratio (SAR), ionic
concentration, pH and clay mineralogy, indicating,
as would be expected from the above diagram, deflocculation at generally higher SAR values for a
given ionic concentration in the order montmorillonite, illite through to kaolinite. In addition,
they noted an inverse relationship between pH and
the SAR associated with dispersion at a given
concentration in the particular soil.

Zaslavsky and Zeitlen, 1965; Kassiff and Henkin, 1967).
Susceptibility to piping was found to be particularly associated with increased dispersion due to 1) increased cation
exchange capacity (CEC), 2) decreased cation valency, e.g.
Ca++ to Na+, in the soil, 3) a decrease in the ionic concentration of the pore fluid and 4) increase in water content
(Kassiff and Henkin, 1967). Three years of observations
led Kassiff and Henkin to conclude that engineering properties

are not very relevant. But chemical analysis of 10 failed
and sound dams paired in terms of similar soil texture and
construction techniques showed that the failed dams were
built of slightly saline soils with total cation contents
(TCC) greater than 150 me l^{-1} in the pore water, sodium
availability ratios (SAR) greater than 25 and estimated ESP
greater than 26%. The loess is described as sandy silt to
clayey silt with low plasticity index, and montmorillonite
was the main clay species accompanied by a little illite.
However, as pointed out by Perry (1975), Kassiff and Henkin
should have measured the TCC of the seepage water or rainwater rather than of the pore water. In the case history of
Flagstaff Gully Dam, Tasmania, Ingles and Wood (1965) found
that a fall in electolyte concentration by one order of
magnitude between pore water and reservoir seepage water
was critical to the performance of the dam (Ingles, 1973).

Work on dam failures in solonetzic and solodic soils
has been conducted at the Scone Research Station of the Soil
Conservation Service of New South Wales by Ritchie (1963;
1965) and Rosewell (1970; 1977) with similar results.
Ritchie developed the Dispersal Index to identify dispersible soils. This index is still used by the Service in
the slightly modified form produced by Rosewell (Charman,
pers. comm., 1978). Rosewell (1970) found that extreme
cases of piping were associated with banks of low density,
high ESP and low soluble salts, plus rapid filling with low
salinity water. Low susceptibility was found with high
density, low ESP and high soluble salts content, plus slow
filling with high salinity water. Cases of low density, low
ESP and high SAR, filled with low salinity water have a high
potential susceptibility depending on permeability, because
this influences the rate at which deflocculation occurs.

Ingles et al. (1969) pointed out that an alternative to
deflocculation might occur when clays such as montmorillonite
or illite adsorb sufficient cations from the seepage water
to cause significant swelling. If swelling of the clays
occurs quickly enough in relationship to attrition by seepage,
macropore channels may be severely constricted. Since discharge is proportional to the fourth power of the macropore
diameter (Quirk and Schofield, 1955), this results in an
exponential reduction in flow, rapidly reducing attrition on
the tunnel walls and re-establishing stability. Ingles and
Wood (1965) observed changes in the volume and salinity of
seepage water from the Flagstaff Gully Dam, Hobart, which
they attributed to irreversible swelling due to calcium saturation in calcium montmorillonite clays. The seepage water
was found to be higher in sodium but depleted in calcium and
magnesium after the exchange and the dam remained stable.
This has led to suggested lime treatment to stabilise earth
dams with high montmorillonite clay content (Ingles, 1964).
However, stability may also be achived by purely physical
means, either by natural collapse causing compaction or by
artificial compaction during dam-building.

Sherard, Decker and Ryker (1971) summarised the

Australian advances of the 1960's as follows:-

1. Some clays, termed 'dispersive clays', are highly
    erodible.

2. Dispersion occurs when repulsive forces exceed
    attractive van der Waals forces. If water is
    flowing, the clay particles are carried away.

3. The main property governing susceptibility to
    piping is the level of dissolved sodium cations
    in the pore water relative to calcium or magnesium.
    Sodium increased the thickness of the double water
    layer around each particle so decreasing the
    attractive forces.

4. A second factor is the total content of dissolved
    salts in the seepage water, lower levels giving
    greater susceptibility. In fact, Australian,
    American and Israeli work shows the likelihood of
    failure is much greater when the water is very pure,
    especially with TDS below 5 me 1^{-1}.

5. Clays with ESP 7-10 are moderately dispersive and
    may be associated with piping if seepage water is
    relatively pure. Clays with ESP 15+ have serious
    piping potential.

6. When a concentrated leak starts either (i) clays
    swell to close it or (ii) the velocity is
    sufficient to carry away dispersed particles.

7. There is no clear correlation between ESP (and
    therefore piping potential) and the ordinary soil
    mechanics tests.

8. High ESP and piping potential are linked with
    montmorillonite, and some illites but rarely with
    kaolin.

Nevertheless, the factors controlling dispersion are
clearly multivariate. For example, to say simply that
clays with ESP > 15 are highly dispersive and therefore
have a high piping potential is an oversimplification of
Fig.16. This is illustrated by Stocking's (1976, 222) data
from natural pipes in Rhodesia which suggested that sufficient deflocculation occurred with ESP > 15 (i.e. in natric
horizons according to the 7th Approximation (USDA, 1960)) to
drastically reduce permeability and that no horizontal pipes
occur in situations with ESP between 15 and 20%. Instead,
vertical pipes can form if there is a pre-existing horizontal
pipe at a lower level; since only vertical pipes have
sufficient hydraulic head to get through the impermeable,
sodic layer. At ESP levels of 40 to 70% spontaneous dispersion occurs on wetting. Stocking found dense micropiping
in such cases with a preponderance of vertical pipes. He
points out that these thresholds are tentative, since the
data on which they are based are limited to his own soils

(cf. section 4.5(i)).

Landau and Altschaeffl (1977) of Purdue University concentrated on the complex interrelationships between controlling variables by conducting laboratory experiments according to a fractional, factorial design with 8 variables (at
2 levels) and their interactions. Their conclusions both
confirm and extend previous results:

1. The lower cationic concentration (1 me 1^{-1})
was more erosive than the higher (8 me 1^{-1}).

2. The interaction between water content, amount of
compaction and cationic concentration of the eroding water was complex. The most highly erosive
combination was high water content (during compaction), low compaction (density) and low cationic
concentration. The lowest erosion occurred under
the same levels of water content and compaction
but with high cationic concentration.

3. Varying the eroding water anions between SO_4^{--}
and Cl^- showed the SO_4 cases to be much more
erosive in cases of compaction with high water
content but Cl slightly more at low water contents.

4. The eroding water anion is only significant if
concentrations are high.

Some of these results are difficult to relate to the
natural situation, but in stressing that significant interactions were evident and that erosion potential can only be
evaluated in the context of these interactions Landau and
Altschaeffl (1977, 250) seem to be making a point of considerable relevance to future research on processes in the
natural landscape.

Recent work has also advanced beyond the stage of
statistical associations to the point of relating the onset
of piping to physical processes deduced from the now considerable body of theory in soil physics and chemistry. According to Lewis and Schmidt (1977) recent literature suggests
that interparticle forces control the erosion or dispersion
of 'dispersive clays'. Essentially, the interparticle forces
consist of (i) the attractive van der Waals force between
the atoms of adjacent particles of clay minerals and (ii)
a coulombic repulsion between the double layer surrounding
each particle. Dispersion depends on the balance between
these forces. The 'double layer' consists of the imbalance
in electrical charge on the surface of the clay particle
and the resulting ionic charge in the immediately adjacent
fluid, which causes a layer of sorbed, oriented water molecules surrounding the particles. Van der Waals attraction
is virtually independent of environment, and is inversely
proportional to a high power of distance away from the
particle. However, the repulsive forces are very sensitive
to environment. The repulsive forces decline exponentially

from the surface and are only significant when the double layers of two adjacent clay particles interact, but the strength of the repulsion also varies inversely with the concentration of ions in the pore water. Furthermore, the type of ion affects repulsion, which is increased as the ions are changed in the sequence of calcium, magnesium, ammonium, potassium, sodium, and lithium. In addition to these forces, Grim (1962) proposed that in unsaturated clays another attractive force occurs due to a 'water bond'. Grim believes the transition between oriented and normal water molecules is generally abrupt and that the water bond occurs when there is insufficient water available to complete the double layer so that the oriented layers of adjacent clay particles come into direct contact. The bond reduces dramatically when sufficient water becomes available, just below the plastic limit, to enable (relatively) unoriented water to fill the interparticle spaces, hence rapid slaking.

Elliott (1977) has calculated the resultant force between van der Waals attraction and the net free energy of interacting clay double layers for a typical piping soil (yellow solodic) in New South Wales, and concluded that double layer theory adequately accounts for the observed instability and response on wetting.

Reséndiz (1977) points out that the minimum distance between clay platelets, rather than the average spacing which is directly related to void ratio, is the parameter deter-mining the magnitude of interparticle forces. This distance is controlled by ionic concentration. Hence for a given dry unit weight of soil (i.e. a fixed voids ratio) lower water content is associated with less repulsion and greater stab-ibility, since the total amount of electrolytes in the pore water will not change much with water content (unless re-moved by throughflow). If percolation occurs as well as saturation, then leaching and ion exchange will increase interparticle spacing if the water has lower dissolved salts or the predominant ion is nearer lithium than calcium in the above series. The net increase in interparticle dist-ances upon saturation is partly due to these effects and partly to the reduction of effective compressive stress resulting from the elimination of pore water tension. Yong and Sethi (1977) likewise point out that dispersivity is generally thought of as a function of a) swelling due to double-layer repulsion and b) desorption or removal of adsorbed ions from clay surfaces, which increases double-layer thickness and assists particle detachment.

Lewis and Schmidt (1977) made the interesting observat-ion that whilst higher density soil eroded less than lower density soil at water contents below the plastic limit, both reached a minimum erodibility around the plastic limit. The decline in erodibility with increased water content below that limit could be explained by less destruction of aggre-gates by slaking processes, principally pressure from en-trapped air and swelling of clays, as water content increases. Above that limit, however, they proposed that the positive

relationship between erosion and water content is related
to the reduction in interparticle forces: double-layer
repulsive forces should increase because of dilution of the
salt concentration of the pore water and van der Waals
attraction should decrease because of increased spacing of
particles. Furthermore, above the plastic limit the 'water-
bond' between the oriented water layers of abutting particles
will be destroyed (ibid.; Grim , 1962).

Nickel (1977) postulated three conditions that are
necessary for dispersive erosion to occur :

1. a passageway filled with water under a head and
    wide enough not to clog as clay particles are
    transported along it.

2. a zone of expansion must form along the walls of
    this conduit, which comprises clay particles
    available for removal in a zone of extremely low
    shear strength.

3. shear stress on these walls greater than the
    shear strength of the zone of expansion.

Nickel looked at this dispersive erosion in terms of a
rheological system. Fig. 17 illustrates the shear stress/
shear velocity gradient relationships of an ideal Bingham
system and of a typical clay slurry, as derived by van Olphen
(1963). Flow in the clay begins at stress τ_0. At applied
shear stresses between τ_0 and the Bingham yield stress τ_B,
the gradient of quasi-plastic flow in the clay increases
slowly with increased shear stress. Above τ_B a rapid increase in velocity gradient occurs, indicative of the breaking down of some type of particle association or bonding.
The strength of this bonding, the type of particle association (face-to-face, edge-to-face or edge-to-edge) and the
equilibrium particle spacing are functions of the attractive
and repulsive forces on the edges and faces of the clay
particles, which depend in turn on clay mineral type, valence,
hydrated radius of exchangeable cations and the concentration of electrolyte in the system.

Nickel attempted to define the shear stresses necessary
to induce piping in pinhole tests of 6 soils found to be
dispersive, intermediate (slowly eroding) and nondispersive.
He concluded that clays are dispersive if the Bingham yield
stress (τ_B) in the zone of expansion is less than 30 dynes
cm^{-2} and nondispersive if greater than c.125 dynes cm^{-2}.
Comparison with data from van Olphen suggested that sodium
clays tend to be nondispersive at very low electrolyte concentrations, most dispersive (i.e. lowest Bingham yield
stresses) between electrolyte concentrations of 1-30 me 1^{-1}
(Na-montmorillonite) or 1-10 me 1^{-1} (Na-illite), with gradually increased resistance above c. 5 me 1^{-1}. This happens
to be the same range in which the lowest sodium percentages
also produce a dispersive condition in many clays (Fig.16).
Decker and Dunnigan (1977) state that Fig. 16 is accurate

102

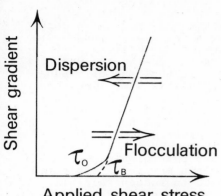

Fig. 17
Relationship between shear
stress and rate of shear for a
Bingham system and a typical
clay slurry (after Nickel,1977).

τ_B = critical shear stress for
a Bingham system.

τ_0 = critical shear stress for
a typical clay slurry.

in about 85% of cases, i.e. this is not the complete answer,
and Nickel (1977, 310) believes that 'high sodium percentage may be only one of the more common symptoms of the actual
cause of dispersiveness'. This presents an as yet unexplained
problem. However, Nickel invokes the evidence of Arulanandan
and others, which has shown that erodibility can vary with
gradients in electrolyte concentration between pore fluid
(PF) and eroding fluid (EF) without sodium necessarily being
important: Statton and Mitchell (1977) confirmed this in terms
of pH and salt concentration. We can summarise thus:-

Gradient in electrolyte concentration	Result
PF < EF	no erosion
PF > EF	swelling caused by osmotic migration of water into clay mass.

Perry (1975) summarises similar results from a doctoral
thesis by Alizadeh (1974) at the University of California,
showing an increase in erosion rate from zero when electrolyte concentration in the 'eroding fluid' is equal to or
greater than that in the pore water to higher rates as the
relative concentrations in the seepage water are reduced.
Thus the critical shear stress is reduced with increased
electrolyte gradients. Conversely, the critical shear stress
is increased by increases in clay mineral content of the
soils, no matter whether kaolinite, illite or montmorillonite.
Critical shear stress may also be increased by increased
moisture content in soils with high SAR.

Nickel (1977) suggests that in some cases osmotically
induced expansion and expansion due to double layer repulsion
might combine; for example, if a calcium-montmorillonite
swelled at the surface to a clay concentration of less than
9.5% by weight, the Bingham yield stress would be less than

30 dynes cm^{-2}. He also proposed that this mechanism could be extended to mixtures of sand, silt and clay. Sand particles may plug passageways reducing flow velocities, or else provide a key between the zone of expansion and the unexpanded mass. But set against this, sand in the zone of expansion could inhibit expansion, though at the same time reducing τ_B. Silt is potentially detrimental to resistance because it would not provide as much 'keying or plugging' as sand. On balance sand would probably not have much effect, but silt might reduce resistance.

Francq and Post (1977, 168) have put forward an hypothesis which they believe might explain why soils subjected to long periods of drying and exposed to sudden rainfall are often more erodible than soils that are constantly damp, albeit subjected to heavier rainstorms. Comparing results between air-dried and moist soil samples, air-drying made some nondispersive soils dispersive and it also made some clays release less sodium and more calcium and magnesium to the saturation extract. They postulate that sodium 'sticks' more to the clay particles in these cases than the calcium and magnesium and that it is this sodium rather than the sodium ions released to the saturation extract which is in the end responsible for a dispersive reaction.

Despite considerable advances in our understanding of the processes of dispersion particularly over the last decade, it is clear that the ESP/clay minerals/cationic concentration relationships which have attracted most attention do not explain all cases. Indeed, it is possible that the often expressed view that these interactions hold the key to the development of piping erosion may well become just as outmoded in the next few years as the once widely held view that piping was a preserve of arid and semi-arid environments. Even within sodium-rich clays, Sherard and Decker (1977, 469) have noted that some resist piping by rainfall even though laboratory tests show them to be dispersive. They observed that 'the panel discussion did not provide significant insight into this question' at their symposium, except possibly that the laboratory tests did not adequately simulate the natural conditions. They themselves hazard the suggestion, along the lines of Ingles *et al.* (1969), that it may be a matter of relative rates of expansion, closing cracks, and erosion, but no laboratory tests yet take this into account.

In soils of neutral and acid reaction hydrogen ion activity is important to maintaining a flocculated state. In strongly acid conditions, hydrogen tends to displace aluminium in the soil exchange complex and create good flocculation. Ingles and Aitchison (1970, 346) believe that piping only occurs in weakly acid podzols of pH 5-7 which are more susceptible to dispersion, or in the solodic soils of pH 7-9, rather than in the strongly acid cases, and point out that 'some doubt still remains as to the relative contributions of the acidity and the sodium ion content in such soils'. Landau and Altschaeffl (1977, 254-255) note

that the hydrogen ion has been shown to be a potential determining ion for the edge charge of clays and that edge
effects are very important to clay minerals such as kaolinite
which obtain a large proportion of their exchange capacity
from broken bonds. Edge charge is usually positive (attractive) but can become neutral or repulsive above pH 7 and could
augment the negative face charge in weakly alkaline clays.

Soil granulation may be destroyed by physical means
irrespective of soil reaction, i.e. by stresses generated
by the compression of entrapped air during wetting (Payne,
1954; Kemper and Chepil, 1965, 503) and cohesion may be
reduced by wetting beyond the plastic limit and is virtually
non-existent beyond the sticky limit (cp. Black *et al.,*
1965, 391-399). In alkaline soils, the latter is largely
the result of dilution of salt concentration in the electrolyte at the points of particle contact (Ingles and Aitchison, 1970, 345-346). On the other hand, in soils bonded by
insoluble oxides, silicates or carbonates, e.g. krasnozems
and laterites, the structure is effectively stable (*op. cit.*
344). Krasnozems were really the only major soil group
analysed by Charman (1969, 1970 a and b) in New South Wales
that were found to be 100% non-susceptible to piping (cf.
Fig. 7). However, piping processes are fairly common in the
weaker horizons beneath laterite duricrusts (cf. section
8.6.).

Lessivage may also be a significant predisposing factor
particularly in base-deficient soils. Clay enrichment of
the B horizon may create the necessary impermeable underlayer *in situ* (cp. Jones, 1971; 1975). Moreover , deposition of slimy colloidal gels or cutans (argillans) typically
8-60 μm thick on ped surfaces or aggregates during lowering
of the phreatic surface after a storm or a wet season may
also lubricate the soil and aid settlement or failure.
Sleeman (1965) noted these films of oriented, illuviated
clays also encourage cracks to re-open in the same place
time after time, calcium saturated clays encouraging wider
and more continuous cracking than sodium or hydrogen saturated clays. These cutans tend to form only in the voids which
are not swelled closed during wetting and are most common
in soils with low clay contents. An alternative form of
cutan, the 'stress clay mineral cutan' is caused by the
build-up of overburden pressure in soils with >50% clay
content and high salt content, and these may also aid slumping.

As we approach the end of this section, we are in fact
reaching a bridging point to the next. In so far as lessivage involves the washing of fine silt and smaller particles
down the vertical profile through voids in the soil, it is
clearly germane to entrainment piping, which is the process
that occurs when horizontal, rather than vertical, hydraulic
gradients generate sufficient velocities of drainage to wash
out whole layers of soil. Moreover, settlement and subsidence may result from and in local destruction of soil
aggregation, but perhaps more importantly subsidence may
locally increase hydraulic conductivity within the profile

permitting higher and more erosive velocities, whether or not soil aggregates have been destroyed.

In other words, there seem to be good theoretical grounds for doubting the universal necessity for the destruction of soil aggregation which has received so much emphasis in the literature, and perhaps at no time in the past as much as now. The British experience in particular tends to bear this out.

Thus, any process which results in internal subsidence may begin the concentration of seepage water necessary to develop soil pipes, as indicated by the early observations of Jennings and Knight (1956; 1957) on earth dams in South Africa. Knight (1959) reported volume reductions of 13% caused by subsidence due to wetting on sandy soils and Hardy (1950) reported subsidence of 3.5 m due to setting in silty soils.

Volk (1937) in America and Burton (1964) in Australia have suggested that piping may be concentrated at the phreatic surface, where air voids have developed as a result of subsidence in the saturated soil below. Parker and Jenne (1967) pointed out that such collapse may lead to either 'desiccation-stress' or 'variable permeability subsidence' forms of piping. Subsidence may occur both after saturation and after drainage. In the latter case, drainage may cause numerous stresses to develop, particularly due to the lowering of pore water pressure associated with drawdown (Terzaghi and Peck, 1948; Carroll 1949). Drawdown following the development of piping and accelerated drainage is likely to cause local collapse of pipe roofs which may either enlarge or block the pipes. N.O. Jones (1968) found this blocking of pipes a major factor in inhibiting pipe development in his laboratory simulation. More general subsidence due to percoline drainage may, however, be partly responsible for the common occurrence of piping under slight surface hollows in Britain (Jones, 1975, 1978a; Newson, 1976), New Zealand (Ward, 1966) and elsewhere, which sometimes show signs of landslippage around the edges. On a smaller scale, minute collapse events may encourage the development of piping during drainage of a soil body. Shiobara (1970), for example, observed the development of a 'honey-comb structure' of 'pipe-like paths' due to lowering of pore water pressure in fine sands and silts, and Nogushi, Takahashi and Tokumitsu's (1970) experiment suggested that sinkholes of the Chikuho coalfield in Japan might be due to collapse induced by pipe development and hydraulic drawdown in loosely packed soils even without total saturation.

The suggested concentration of piping at the phreatic surface due to subsidence is interesting since a number of workers have noticed this association. Rubey (1928) noted the common occurrence of pipes just below the wet season water table in the Great Plains, and Horton (1936) noted a similar location near the water table peak. However, subsidence may not be the only factor; the long continued concentration of flowlines at the phreatic surface suggested by many speleologists may also destroy homogeneity of permeability and concentrate piping (cp. Davis and De Wiest, 1966, 365; section 6.3).

5.2(ii). Permeability and soil displacement.

In non-cohesive materials or in cohesive materials that have locally developed metastable states, the strength of the displacing hydrodynamic force is largely controlled by permeability and hydraulic gradient.

In hydraulic terms, the onset of piping is a clear indication of the breakdown of Darcy's Law:

$$Q \ = \ kiA$$

where Q is the flux, i the hyraulic gradient, k a coefficient of permeability and A is cross-sectional area. In the tortuous channels within the soil matrix turbulent flow sets in at much lower Reynolds Numbers than for straight channels and once turbulence begins the discharge is no longer linearly related to the hydraulic gradient, but falls off due to wastage in eddying (Hillel, 1971, 94). The inertial forces thus generated tend to cause erosion and enlargement of the micro-channels (Zaslavsky and Kassiff, 1965), which may lead to the growth of a pipe network and once again raise the level of discharge. In the presence of significant amounts of swelling clays, the rate of infiltration may be an important factor in determining whether entrainment occurs, because swelling of montmorillonites and illites may close the fine channels before seepage water can establish the necessary rate of surface attrition (Ingles, Lang and Richards, 1968).

The classic theory of piping evolved by Terzaghi and often termed 'boiling' or 'heaving' by civil engineers (Fig. 18) was considered to be the result of excessive build-up in seepage pressure below the phreatic line in flowlines converging on an exit point reaching the stage in which cohesionless material becomes weightless (Terzaghi and Peck, 1948, 510; and references on page 28). In its simplest form, the hydrodynamic pressure acts directly against the submerged weight of the soil, thus:

$$\gamma_{effective} \ \equiv \ \gamma_{submerged} \ - \ D$$

where the γs are the respective unit weights and D is the hydrodynamic pressure of the water emerging at the surface of the soil (Jumikis, 1962, 332). The right-hand variables are

$$D \ = \ \gamma_{water} \ . \ i$$

where i is the hydraulic gradient, and

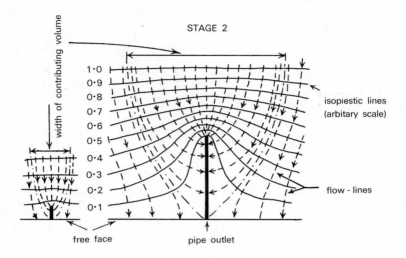

Fig. 18   The process of pipe development by soil 'boiling'
          as outlined by Terzaghi and Peck (1948, fig. 207).

          Stage 1 : incipient indentation in free face
                    starts concentration of flow-lines.

          Stage 2 : after a period of headward erosion the
                    flow-lines have become more concentrated
                    towards the head, increasing boiling
                    at an accelerating rate.

$$\gamma_{submerged} = \frac{\rho_{soil} - 1}{1 + e} \cdot \gamma_{water} = \gamma_{soil} - \gamma_{water}$$

e being the void ratio and ρ_{soil} the specific gravity of the
soil. When

$$D = \gamma_{submerged}$$

the soil appears to be weightless and instability may set in
if the effective unit weight of the soil becomes negative.
In other words, the critical hydraulic gradient is

$$i_{crit} = \frac{\rho_{soil} - 1}{1 + e}$$

beyond which there is a sudden increase in the coefficient
of permeability, k, and in seepage velocity and the soil
becomes 'quick'. Glossop (1945, 77-80) took representative
values for the specific gravity (2.65) and void ratio (0.6)

and estimated that $i_{crit} \approx 1$.

This suggests that the tendency to 'boil' depends directly on the specific gravity and void ratio of the soil and that lower porosity reduces the susceptibility to boiling. But Glossop noted an apparent contradiction: although a few dams founded on gravel have been known to fail by piping, in general finer materials are more susceptible to piping. He attributed this to the facts that fine materials are often more uniformly sorted so that there is little relative movement until the whole mass moves at i_{crit} and that finer materials are often loosely packed.

However, it is now clear that the correct factors had not been identified. Terzaghi and Peck (1948, 502-514) introduced the 'creep ratio' to account for some effects of vertical inhomogeneity and the anisotropic hydraulic behaviour of 'alluvial subsoils' beneath masonry dams, reducing $i_{crit} << 1$. But cohesion, flocculation and dispersion and resulting changes in the stability and permeability of finer grained materials are now seen as the main problems. This is not to say that the boiling process is not important, simply that the engineers' safety measures based solely on the boiling theory are inadequate (cp. Ingles and Aitchison, 1970; Sherard and Decker, 1977). 'Boiling' has been directly observed in the natural landscape. Scheidegger (1961, 86) refers to it on alluvial toeslopes. Morawetz (1969) and Morgan (1979) mention water coming out of bare slopes in Austria and England sapping headwards and creating rilling or small gullies. Perhaps if the roofing materials had been sufficiently cohesive, pipes could have formed. Jones (1975, 75) argued that the critical hydraulic gradient for the onset of boiling would be less than predicted by the standard equations of Terzaghi and Peck (1948) and Jumikis (1962) when it is acting on a free face or at the head of a pipe which is sapping headwards by boiling. With hydrodynamic pressure acting at angles (α) between 90° and 180° to gravity.

$$\gamma_{effective} = \gamma_{submerged} - D \cos \alpha$$

Should the hydrodynamic pressure act at angles of $0-90^\circ$, it will aid soil breakup by loading and

$$\gamma_{effective} = \gamma_{submerged} + D \cos \alpha$$

The process may also operate on the bed, walls or even roof of developed pipes and Terzaghi and Peck (1948, 52) noted that boiling along the walls of a pipe could be aided by suction developed by pipeflow. Martin (1970) observed that the force of effluent seepage at a sand/water interface is lower than within the body of sand, because the interfacial grains have only one adjacent layer of particles, i.e.,

$$D_{bed} \approx \tfrac{1}{2} [\gamma_{water} \cdot \frac{\delta h}{\delta y}]_{bed}$$

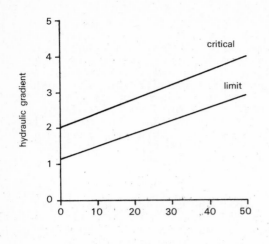

Fig. 19

Susceptibility to piping in relation to hydraulic gradient and texture according to Lehr and Stănescu (1967).

These conditions may apply both to the headward extension of piping and to subsequent enlargement.

In material where cohesion is present, a cohesive force should be added on the side of resistance to boiling. Unfortunately, cohesion is a very difficult parameter to measure or predict in absolute terms. Zaslavsky and Kassiff (1965) and Kassiff, Zaslavsky and Zeitlen (1965) developed their own formula for safety from piping for clays, based on laboratory experiments and employing a measure of soil cohesion,

$$F = \frac{b\sigma_t}{\gamma_{water} \; jd}$$

in which σ_t is the tensile strength of the clay, as estimated from drained triaxial tests, b is a dimensionless soil constant (≈ 1 but varying with d and aggregate shape), j is the actual exit hydraulic gradient and d the mean size of eluviable aggregates. For $F > 15$ there is a marked reduction in collapse during wetting, which they considered important for piping. They concluded that the critical exit gradient, j_c, is given by $j_c = \frac{2q_{crit}}{\pi D^2 k}$ where k is the coefficient of

permeability, D the average diameter of holes in the clay and $q_{crit} = \frac{q_t}{n}$, the total discharge up to failure divided by the number of holes.

The Terzaghi theory and subsequent elaborations have concentrated on failure around an exit point. However,

110

Ingles and Wood (1964) and Aitchison and Wood (1965) have
studied the details of a slightly different process of subsurface erosion, showing that piping in earth dams may begin
as very minor lateral eluviation extending some distance
through the soil and growing until an open pipe develops.
From both laboratory and field evidence they were able to
establish a relationship between permeability (k) and eluviated particle size, confirming and extending the work of
Peterson and Iverson (1953) on Canadian dams. Assuming a
fixed exit gradient of i = 0.5 and a uniform k and using
the relationship between particle size and settling velocity
established by Hazen, particles of 5.2 μm diameter will,
for example, be carried with k = 0.1 mms^{-1} (Aitchison and
Wood, 1965, fig. 3, 444). They proposed that in soil sus-
ceptibile to dispersion an upper limit of k = 1 x 10^{-4}
mms^{-1} (10 ft yr^{-1}) should be set to avoid piping in small
dams (10^{-3} mms^{-1} being sufficient to initiate piping). Above
this value attrition appears to heavily outweigh the effect
of clay expansion. However, in larger dams which tend to
dry out more, piping may be initiated in permeabilities as
low as 10^{-6} mms^{-1} (Ingles, 1968, 42). Such values would
be very low indeed for natural soils. Terzaghi and Peck
(1948, table 15, 331) describe a k of .0001 mms^{-1} as 'very
low', k = 0.1 mms^{-1} as 'medium' and only as 'high' above
1 mms^{-1}. Soils with strong fabric units bonded to act as
silt-size particles will only be eroded if the permeability
equals or exceeds 0.1 mms^{-1} (Fig. 20). On the other hand,
dispersed clay particles of sub-micron size can be eroded
in velocities associated with permeability of 10^{-4} mms^{-1}
(Ingles, 1968). Aitchison and Wood (1965, 445) also proposed
a *lower* limit to safe permeability of 1 x 10^{-6} mms^{-1}. Below
this there is negligible pore-pressure dissipation in dams
and a built-up of pore-water pressure may lead to landsliding.

Independent evidence comes from engineering studies of
suffosion in Polish loess, where Kühn (1963) observed that
seepage rates of 0.01-0.06 mms^{-1} were sufficient to prepare
the ground for liquid bursts : he noted that liquid and
plastic limits were very close, 3.6-7.0%, thus making the
loess very susceptible to water action. It is worth noting
here that piping and 'bog bursts' may be alternative
solutions for natural drainage in peat bogs, and piping or
landslides may be alternatives in soils, perhaps with some
variable zone of stability in between.

Ingles (1964a) used Pouseuille's equation for flow in
a tube to calculate expected diameters for eluviation channels under laminar flow:

$$v = \frac{r^2}{32\eta} \cdot i$$

where v is mean velocity of flow, r is the radius of the
capillary, i the hydraulic gradient and η the coefficient
of viscosity of the fluid. He first calculated that a clay
particle of 2 μm diameter and a specific gravity of 2.65

111

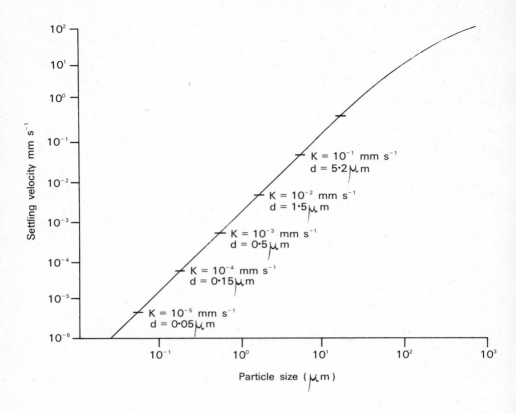

Fig. 20   Relationship between particle size and settling
          velocity, assuming a hydraulic gradient of 0.5
          and a uniform coefficient of permeability (based
          on Wood, Aitchison and Ingles, 1964, fig. 3 and
          Ingles, 1968, fig. 1.17).

requires a minimum velocity of flow of 3.6×10^{-3} mms^{-1} to
transport it. Then, applying the Pouseuille formula with
$i = 0.5$, Ingles obtained a channel diameter of 9.7 μm and,
by Darcy's Law, a permeability $k = 7.2 \times 10^{-3}$ mms^{-1}. Ingles
also observed a tendency for the tortuous capillaries to
clog with clay particles on sharp bends. In these calculations Ingles was assuming laminar flow. However, in the
presence of 'pseudo-piping' a significant degree of turbulence is to be expected, along with anisotropic conductivity,
which strictly invalidate the Pouseuille and Darcy formulae
(cf. Childs, 1969, 159), although up to this point these
laws may be reasonably accurate even in unsaturated conditions.

There has been a strong tendency to associate higher
permeabilities with greater susceptibility to piping in
earth dams. Aitchison, Ingles and Wood (1963) pointed out
that dry-side compaction of dams tended to make them more
susceptible to piping because the aggregated condition of
the clay resulted in more silt-sized aggregates and therefore

more voids and greater porosity. However, greater porosity and permeability does not necessarily result in greater susceptibility to piping. Landau and Altschaeffl (1977, 250) noted that of two samples the one with larger aggregates and voids needed over four times the hydraulic head of the other before piping was initiated. In this case a very high initial permeability was substantially reduced later by swelling or clogging. Glossop's paradox (v.s.) suggests another reason for being cautious over the association, i.e. the need for 'eluviatable' material.

Once a recognisable pipe exists then the problems of 'soil displacement' within that pipe are little different from corrasion and transportation in open channels and we can conveniently refer the reader to the large body of literature on open channel hydraulics. We might note that one possible difference is that the gradient term in the Darcy-Weisbach equation may be hydraulic gradient (for full pipes) rather than slope of the water surface (for open channels or partially-filled pipes).

6. THE MORPHOMETRY OF SOIL PIPES AND SOIL PIPE NETWORKS

Recent interest has been focused on morphometric aspects of pipe networks from the point of view of modelling hydrological response (e.g. Morgan, 1977; Newson and Harrison, 1978; Humphreys, 1978; Gilman and Newson, 1980). Earlier interest in morphometry was also concerned with the evolutionary changes between pipe and channel geometry, the relationships between pipe morphometry and pedological environment or with the general evolution of the pipe networks themselves (e.g. Jones, 1971; 1975).

It is intended in this chapter to build up from the geometry of individual pipes, or 'pipe-links', to the geometry of whole networks. Incidentally, this involves a progression from the crudest, most commonly quoted information to the rarer information on network form only recently available.

6.1. Pipe geometry.

Systematic study of pipe geometry is very recent and limited to work in Britain. However, a wealth of information may be gleaned from the literature describing occurrences around the world. The problem is that by and large we have no means of telling how representative that information is. Much of the subsequent analyses must therefore be taken as indicative rather than definitive.

6.1(i) The size of soil pipes.

The size of pipes is the most frequent measure of pipe geometry recorded in the literature, although as with pipe shape only a small minority of reports are based on statistical analyses of rigorous sampling experiments. Table 5 collects together measurements of pipe diameter from around the world and relates them briefly to soils, geomorphic environment and climate. Qualitatively, pipes appear to be best developed in soils with high silt-clay content and high hydraulic gradients. These sites tend to be on floodplains or river terraces, partly because high silt-clay content can generally only be expected on higher landscape units given suitable parent materials.

Whilst all areas tend to have a wide range of pipe sizes, often in the same networks, in general the largest pipes have developed in arid and semi-arid areas. This must partly explain why piping has frequently been considered as a process specific to these areas and has largely been overlooked in mid-latitude, humid areas until recently. In Britain, New Zealand and parts of S.E. Australia typical diameters are in the 1 - 15 cm range (e.g. Gibbs, 1945; Blong, 1965; Hosking, 1967; Jones 1971; Floyd, 1974; Gilman and Newson, 1980).

There has been little quantitative work on pipe size in relationship to environmental factors. The work of Jones (1971; 1975), already mentioned, analysed size, shape and vertical location and indicated joint relationships between these and soil properties. A similar approach was taken by Morgan (1977) in terms of the size and vertical location of pipes on Plynlimon.

Morgan (1977) took measurements of pipes in 37 sample pits at Cerrig yr Ŵyn and removed undisturbed cores for analysis. Histograms of the measurements showed a general right-skewness, except in soil properties, and square-root transformations were made prior to Pearsonian correlation analysis. The correlation matrix is set out in Table 7 (from Morgan, 1977, tab. 2; Gilman and Newson, 1980, tab. 4b). It was concluded that:(1) the strong correlations between pipe depth and size and the depth to the surface of the Eg horizon indicate the importance of the permeability break at the mineral horizon (perhaps we could say that pipes near the base of thick, peaty Oh horizons have larger catchment volumes and are larger in size). Morgan (1977, 33) commented that pipe beds need not necessarily be on the impermeable layer since water ponded up may act as another impermeable layer. However, she also pointed out that shallower, smaller pipes tended to be at the extreme ends of the network and the clay layer was nearer the surface on steeper slopes, so there may be some confounding of 'pipe order' and environmental factors; (2) the negative relation-ship between contrasts in dry bulk density from roof to bed and depth of the mineral horizon is simply due to the fact that contrasts are less when pipes run wholly within the peat (none run wholly in the mineral layer); (3) the negative correlation between slope and mineral layer depth indicates thinner peat horizons on steeper slopes.

Gilman and Newson (1980) extended this work to the Nant Gerig. They removed 14 soil blocks containing pipes and took samples from the adjacent profile, which they analy-sed for ash content. Thirteen of the pipes were shown to be in the peat or at the peat/mineral interface.

Howells (1980) recorded the dimensions of 12 soil pipes in the Nant Maescadlawr basin near Bridgend, South Wales, for which she also obtained discharge data (cf. Section 7.1). The pipes were developed in gley soils of the Pendoylan Series which contain an impermeable clay horizon that con-trols pipe location. Pipes issue into seeps near the stream and the measured outlets range from 5 to 32.75 cm in mean diameter. Comparing her cross-sectional data with the record-ed discharge data (*ibid.*, tabs. 2.1 and 4.1), one can see a good correlation between the mean diameter of pipe and both mean and maximum recorded pipe discharges (r_s = 0.87 and 0.84 respectively, $\propto$ = 0.05). This supports the assump-tion, at least within a homogeneous area, that size may be taken as a reasonable surrogate for discharge (cp. Jones 1975, 1978a, in section 4.6(iii)).

Table 7. Correlation coefficients for pipe geometry and
          soil variables after Morgan (1977, tab. 2).

	1	2	3	4	5	6	7
1		0.627**	0.424**	0.092	0.122	-0.245	0.006
2			0.348*	0.182	-0.0001	-0.270	0.041
3				-0.428**	-0.185	-0.602**	0.232
4					-0.023	0.233	-0.178
5						0.420+	-0.565**
6							-0.695**

Significance: ** 0.1%, * 0.5%, + 5%

pipe geometry   1.   depth of base of pipe

2. cross-sectional area of pipe

3. depth of Eg horizon

4. angle of slope

soil variables   5.   difference in field wet bulk density
between bottom and top of pipe

6. difference in oven dry bulk density
between bottom and top of pipe

7. % difference in loss on ignition
between bottom and top of pipe

6.1(ii) The shape of soil pipes.

Qualitative references to pipe shape are relatively
numerous. But only recently have sample surveys been conducted to produce quantitative data which have been used to
explore theories of genesis and evolution and to determine
hydraulic geometry (section 6.1(iii)).

There is evidence to suggest that in some cases tunnel
form evolves from small rounded pipes to larger flat-bottomed
or rectangular pipes. Where this occurs it seems reasonable
to assume an origin due to boiling below a storm period
phreatic surface during the circular phase and subsequent deve
lopment   as a phreatic surface pipe with open channel hydraulics, giving a horizontally lenticular or ⌒-shaped cross-
section (cf. Jones, 1975, 170). Guthrie-Smith (1921) seems
to describe a similar process in New Zealand, with initial
downcutting developing vertically lenticular pipes followed
by widening as 'springs' begin to play a more dominant role

than percolating rainwater.   In Australia Downes (1946)
described evolution from circular pipes of c. 12-25 cm
diameter to rectangular pipes which tend to collapse with
diameters exceeding 1.5m.   The reference by Cumberland
(1944, 99) to pipes of circular cross-section 6m down in
South Island, New Zealand, is unusual and probably reflects
the unusual character of the relatively erodible loess soils.
In general, circular or slightly horizontally lenticular
cross-sections are typical of small, shallow pipes particularly in Britain (e.g. Weyman, 1971; Jones, 1975; Morgan,
1977; Humphreys, 1978; Atkinson, 1978).   Berry (1970) reports the same forms in similar sized pipes in the Sudan.
However, frequent mention is also made of a characteristic
⌐-shaped section elsewhere, for example by Blong (1965, 84)
and Ward (1966) in New Zealand.   Both observed the form in
larger pipes, 30cm and over or up to 90 cm in diameter
respectively, but Blong also observed it in small pipes of
1.5 - 5.0 cm.   Conacher and Dalrymple (1977, 26) reported
small circular or ⌐-shaped G areas on unit 2(2) toeslopes
on London Clay, which appear to be non-active soil pipes
blocked by translocated soil materials (cp. section 8.3;
Jones, 1971).   Jones (1975, 167) recorded that 37% of pipes
on Burbage Brook and 12% on Afon Cerist were ⌐-shaped. This
is interesting, since it is the type of bed geometry generally expected in open channels and it therefore suggests that
non-capacity flows control the geometry (cf. section
6.1(iii)). Morgan (1977) also observed smooth beds and partially smoothed walls but rough roofs at Cerrig yr Wyn,
Plynlimon (Gilman and Newson, 1980).

        Jones (1975, 162-170) analysed the distribution of 138
outlets on the Burbage Brook and 17 outlets each on the Afon
Cerist and on Bourn Brook, near Cambridge, in terms of percentage bankfull height of outfall, mean diameter of pipe
and pipe shape (Fig. 21).   The overall breakdown in terms
of shape was: 40% horizontal-lenticular, 34% round and 26%
vertical-lenticular on Burbage, compared with 12%, 18% and
70% respectively on Cerist.   All but four of the outlets on
the lowland stream were round.   However, in the two upland
area streams, Burbage and Cerist, vertically-elongated outlets were dominantly in the lower half of the bank and the
mean bankfull heights for the different shapes on Burbage
were: 34% vertically-elongated, 44% round pipes (which also
happened to be the mean for all pipes), and 48% for horiz-
ontally-lenticular outlets.   84% and 83% of all vertically-
lenticular pipes occurred at below 50% bankfull in Burbage
and Cerist respectively and these were the only group to be
significantly non-randomly distributed between the upper
and lower halves of the bank.   Vertical-lenticular outlets
tended to be larger and a high percentage of large outlets
were of this form.   The same form is suggested by Carroll's
(1949) description of pipes on the San Pedro in Arizona
keeping pace with rapid stream rejuvenation, and in general
the form suggests active bed erosion in the pipes, i.e.
active 'master' pipes.   Heede (1971) reported the same form
in quickly eroding piping in Colorado.   However, once a
static baselevel is reached, the larger perennially-flowing

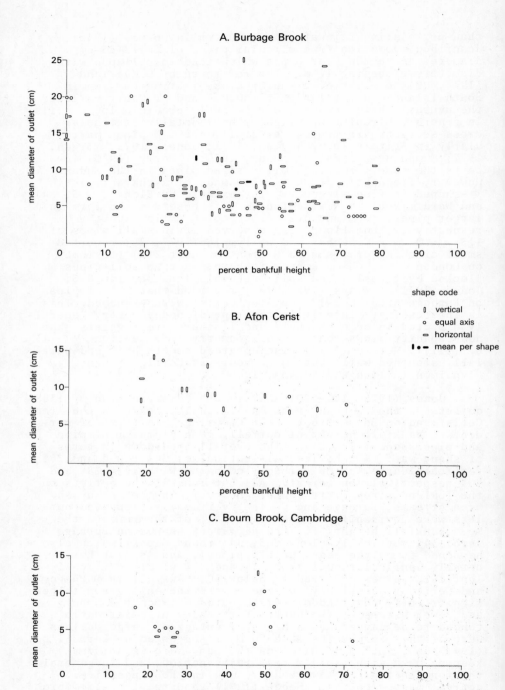

Fig. 21 Shape, size and bank location of pipe outlets at three British sites, after Jones (1975, fig.6.1.2.).

118

pipes become horizontally lenticular (Jones, 1975).

The shape of soil pipes is also controlled by variations in the soil profile which may or may not affect hydrological status. Flat-bottomed forms have been reported above soil pans, for example, above claypans in Wales (Bell, 1972), which are also foci for perched water tables. Despite the insistence of Fletcher *et al.* (1954) that such zones of reduced permeability are necessary for the initial development of piping generally, it is now clear this is not so (e.g. Parker, 1963). Indeed, it is possible that such locations may be due to arrestment of vertical migration by less erodible material as well as to initiation at a frequently saturated interface.

Jones (1975, 178-194) took measurements of pipe geometry and soil texture and erodibility from 20 randomly-selected pipe outlets on Burbage Brook, which he subjected to simple and multivariate correlation analyses (Table 8). He used canonical correlation analysis since, as is commonly the case in morphometric analysis, the concept of 'pipe geometry' is not uniquely described by any single measurement: a major limitation of multiple regression analysis as a basis for hypothesis-forming is that it will only give results in terms of the specific dependent variable used, whereas canonical correlation seeks new variates. The analyses produced variates which suggested that (1) smaller pipes tend to be found in generally less erodible horizons (greater aggregate stability), (2) larger pipes are associated with stronger, less erodible roofs, but with more erodible walls or beds, (3) vertically lenticular pipes tended to have weaker beds, low in silt content and aggregate stability and (4) horizontal development seemed to be controlled more by the relative stabilities of bed and roof than by any consistent property of the walls. The latter finding appears to conflict with the importance qualitatively attached to the properties of the sidewall material by Parker (1963, 104) and possibly with the need for an erodible layer above any impeding layer in Fletcher's list of necessary conditions for piping (Fletcher *et al.*, 1954), although perhaps the latter phrased it sufficiently generally to be acceptable. Fig. 22 summarises the relationships in the form of a map of soil variables in canonical variate space. These results seem to establish the importance of soil erodibility in determining cross-sectional form. It therefore becomes very difficult to disentangle the relative roles of pedological and hydrological factors in creating pipe form.

6.1(iii)   Hydraulic geometry of pipes.

The first surveyed sequence of cross-sections appears to be the series of eight sections produced by N.O. Jones (1968, fig. 11) of a large, 2.5-3.2 m high, partially-collapsed ephemeral pipe in the semi-arid San Pedro valley, Arizona, although he did no quantitative analysis of geometry. Jones (1975, 196ff) analysed N.O. Jones's cross-sections and compared them with 30 sections measured on two perennial pipes in the Burbage Brook basin, in the English Peak District.

Table 8. Simple correlation matrix between soil variables and pipe geometry (after Jones, 1975).

Variable key

outlet geometry	31 % bankfull height
	32 – horizontal axis
	33 – vertical axis

soil variables

sample location variable	top	side	base
gravel %	1	11	21
organic %	2	12	22
sand %	3	13	23
silt %	4	14	24
clay %	5	15	25
aggregate(1)	6	16	26
stability(2)	7	17	27
MWD	8	18	28
conductivity	9	19	29
saturation	10	20	30

Top sample variables / side sample variables (columns 1–20)

var	1	2	3	4	5	6	7	8	9	10	11	12	13	14	15	16	17	18	19	20
1	1.0000	-.1356	-.1562	-.0130	.2936	.2223	-.5403	-.0814	-.1690	.2623	.7206	-.0947	-.2351	.0568	.2734	.1467	-.3691	-.0156	-.2793	.0498
2		1.0000	-.0471	.1768	-.1747	.1255	.3386	.2290	.2978	.0508	.2717	.5192	.1259	-.0424	-.1281	.0593	.2798	.2332	.2714	.3002
3			1.0000	-.8225	-.3600	.0518	.0024	-.2387	-.4467	-.2734	-.3486	-.3817	-.5530	-.5721	-.1635	-.0827	-.0214	-.0890	-.3718	.1020
4				1.0000	-.0105	-.2031	.0024	.0140	-.6547	-.2986	-.3424	-.1987	-.1879	.3313	-.2170	-.1101	-.0629	-.1454	.5064	.1924
5					1.0000	.2048		.3994	.0356	.0458	.1138	.3816	-.6985	.5237	-.2726	.3058	.1292	-.3684	-.0869	-.4596
6						1.0000	.3962	.7569	-.1684	.2205	-.0409	.3608	-.1580	.6390		.6390	.5348	.7064	-.0312	.2433
7							1.0000	.7668	.2381	.1699	-.4519	.4386	-.3990	.3771	.0366	.3105	.6026	-.5733	.2641	.1012
8								1.0000	.2296	.2677	-.1970	.4972	-.1194	.1375	-.0267	.2654	.7335	.8553	.1418	.1046
9									1.0000	.5472	-.0692	.4129	-.3979	.4273	-.0417	.2383	.3314	.2306	.8484	.4685
10										1.0000	.3146	-.0099	-.1934	.1375	-.0267	-.1598	.0058	.2165	.4161	.5085
11											1.0000	-.0462	.2677	-.1194	-.0505	.4150	-.4792	-.2400	-.1188	-.0218
12												1.0000	-.0099	-.3979	-.1777	.1277	-.6046	.5730	.4751	.0762
13													1.0000	-.2115	-.1934	.4485	-.1994	-.4369	-.1335	.0147
14														1.0000	-.0483	.4568	.6046	.4863	.0675	-.0320
15															1.0000	.3974	.1277	.1199	.1047	.0263
16																1.0000	.3240	.9142	.2053	.2236
17																	1.0000	.7791	.4029	.2296
18																		1.0000	.8313	.2067
19																			1.0000	.4402
20																				1.0000

base sample variables (21–30) / geometric variables (31–33)

var	21	22	23	24	25	26	27	28	29	30	31	32	33
1	.8338	-.1459	-.5441	.4395	.1920	-.0764	-.3997	-.3142	.1380	.0478	.0179	-.0612	-.0457
2	-.0826	-.0363	.1396	-.0881	-.0404	.0690	.2512	.2598	.1427	-.2514	.3232	.3974	.5106
3	-.1741	-.0541	-.4594	-.4848	-.0995	.3604	.3588	.0772	-.1824	-.1724	-.4232	-.1583	-.1402
4	.0881	.0737	-.2035	-.0299	-.1244	-.0981	-.2014	-.2396	-.2983	.0658	.3446	-.2930	-.2716
5	.1777	-.0122	-.5112	.5175	-.0063	-.3676	-.2949	-.2132	-.1138	-.3989	.2422	-.1484	-.1492
6	.0664	.1548	.0015	-.2301	.3701	.6314	.4204	.6860	.3217	.3501	-.3405	-.1666	-.1390
7	-.4956	.2715	.2843	-.3015	-.0107	.3943	.4824	.7888	.1421	.1383	-.1705	.0477	.0752
8	-.2005	.1361	.0141	-.1712	.1968	.5538	.4960	.7917	.3431	.1500	-.2957	.0283	.0311
9	-.0525	.1143	.1214	-.1242	-.0113	.0751	.0858	.0658	.5885	.4309	.1327	.2493	.3680
10	-.2540	-.0374	-.1084	-.0467	-.0627	.1651	-.0084	-.0526	.7234	.5831	-.0947	.1073	-.0656
11	-.7434	-.1242	-.4481	-.5738	-.2258	-.3705	.5190	.4188	.1678	.0174	.0112	.1670	.3231
12	-.0336	.1281	-.0699	-.0989	-.2666	.1857	-.0387	-.2469	-.2678	-.0981	-.0841	.0502	.3252
13	-.0550	-.0608	.5665	-.3848	-.1937	-.1145	-.0371	-.1233	.1308	-.1082	-.1363	.3196	.2576
14	-.0547	-.0191	-.3306	-.5266	-.3779	-.0500	-.1233	.1059	-.2678	-.1082	.1609	-.0745	-.1383
15	.1674	-.1222	-.3639	-.2130	.8723	.3631	.1157	.2167	.2108	-.0499	-.0364	-.0366	-.1838
16	-.1425	-.2202	-.1660	-.0087	.2522	.4797	.5190	.5333	.4480	.1770	-.2645	-.0955	-.1025
17	-.5250	-.1796	.0858	-.2444	.2063	.4168	.6897	.6339	.2550	.1515	-.2865	-.0068	-.0064
18	-.2529	-.0787	-.0931	-.1101	.3211	.5097	.5437	.7436	.4833	.2195	-.2568	-.1064	-.0947
19	-.1736	-.0993	.0668	-.1793	.1258	.0588	.0588	-.0375	.3800	.3106	.1508	.3847	.3073
20	-.0753	-.1720	.1835	-.3331	.1828	.0917	.1598	-.0759	.6288	.7089	-.1282	.0444	-.0663
21	1.0000	.1275	-.3158	.2010	.1751	-.0073	-.4386	-.3446	-.0126	-.0002	-.1587	.2018	.2981
22		1.0000	.2020	-.3173	.1751	-.0073	.0141	.4158	.0014	-.0163	-.0385	-.0463	.0088
23			1.0000	-.7515	-.3458	.2928	.2744	.2740	-.1851	.3015	-.2055	-.1260	.3983
24				1.0000	-.3458	-.5095	-.3365	-.3599	-.0304	.1482	.4142	-.2811	-.2523
25					1.0000	.3382	.0932	.1621	.3175	.1575	-.2515	.1201	-.1999
26						1.0000	.7569	.6882	.1480	.1277	-.5476	.0261	.1119
27							1.0000	.6291	.0620	-.4961	-.3522	-.0489	.0751
28								1.0000	.2040	.2542	-.2559	-.0944	-.0944
29									1.0000	.6103	-.0021	-.2559	-.1687
30										1.0000	-.2573	-.0517	-.0759
31											1.0000	-.1994	-.2610
32												1.0000	-.3196 / .7125
33													1.0000

120

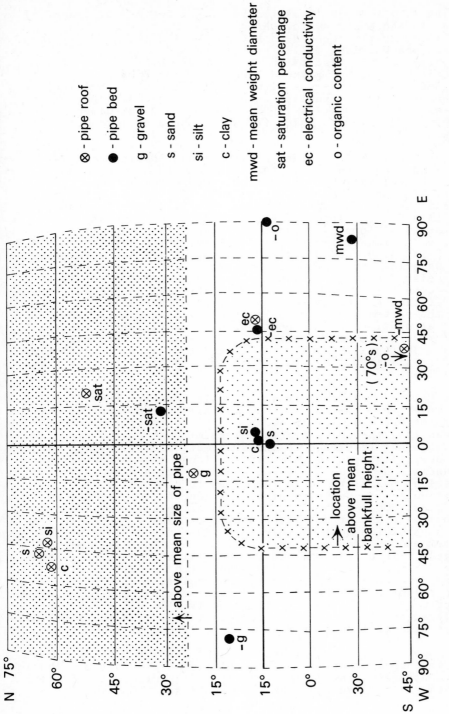

Fig. 22 Spherical map of piping soil properties in canonical variate space according to Jones (1975, fig. 6.1.10).

Morgan (1977, 48-49) measured 84 cross-sections of ephemeral
pipes in the Upper Wye catchment, mid-Wales, and her data
has subsequently been used in pipeflow modelling studies by
Newson and Harrison (1978) and Gilman and Newson (1980).
Humphreys (1978) also tried to calculate hydraulic radius,
friction factor and Reynolds numbers using 7 cross-sections
and bedslope measurements from ephemeral and perennial pipes
on Wansfell in the English Lake District and 3 on Plynlimon,
mid-Wales. And Howells (1980) has recorded the dimensions
of 12 pipes in South Wales for which she also recorded pipeflow.

An obvious difference between measured pipe cross-sections and open channels is that maximum width occurs not at
the 'channel-full' stage, but some way below. Under capacity flow one might expect maximum width at about 50% full
height, but in the San Pedro and Burbage pipes the figure
is 25% and 30% respectively, which probably suggests that
the 'channel-forming' discharge is less than capacity flow,
i.e. open channel in nature. Similarly, the level of maximum
width in the perennial Burbage pipes was very near to observed baseflow level, in marked contrast to surface stream
channels (Jones, 1975, 196). Some corroboration for this
view comes from Morgan (1977, 43), who used length of pipe,
travel time and cross-sectional area to calculate discharge
and the percentage of cross-section filled by high flows.
Her calculations suggested that the ephemeral pipes in the
Upper Wye catchment were only 25-50% full for most high
flows. Morgan (1977, 25) also noted that in some cases pipe
floors exhibited channels-within-channels indicating erosive
low flows.

However, although it is probable that the more frequent
but relatively low stage flows may affect the level of maximum
width or even locally increase depth, Jones (1975) has
argued that the level of 'tunnel-forming discharge' should
be sought by reference to the systematic relationships within
the overall hydraulic geometry, as found in surface stream
channels. He argued that one would expect a well-adjusted,
efficient tunnel to be so formed as to give the most even
distribution of velocity along its length (cp. Leopold et
al, 1964, 308) and he analysed the variability of inferred
velocity levels at different depths of flow in order to find
the depth which most closely approximated a well-adjusted
system and by deduction is therefore the 'tunnel-forming
stage'. Since bedslope did not vary significantly over the
lengths of pipe measured, Jones inferred velocities from
cross-sectional area, assuming continuity, i.e. that the
same volume of discharge passed each section, and scaled them
as percentages of the outfall velocity. In all three pipes
analysed there was markedly less variability at the pipe-full
stage than at the quarter-full stage (Table 9B).

Similarly, Jones argued that the level in the pipe with
the most consistent width-area relationship downstream should
be the stage of the dominant erosive discharge. The results
of his analysis in Table 9B.ii. suggest some improvement in

the consistency of the relationship above the half full
stage, especially in Burbage tunnel 1, although he found
the relatively small size of the Burbage tunnels did not
permit meaningful measurement of the 25% and 75% points. It
is worth noting that Linsley and Franzini (1964, 253) state
that the maximum discharge occurs in artificial drainpipes
when water depth is 0.94 of the pipe diameter because of
frictional losses on the roof at higher levels.

Comparison of the exponents in the width-area equations
led Jones to note that at near-capacity flows both the semi-
arid San Pedro pipe and the humid, temperate Burbage pipes
seem to approach the range associated with the respective
surface channel types to which they are tributary. Jones
(1975, 195) deduced width-area relations from published
relationships between width, depth and discharge, which
suggested higher exponents for ephemeral stream channels.
He also showed that Schumm's (1960) relationship between
silt-clay content and channel width could be applied to the
San Pedro pipe given an empirical correction which may subsume hydraulic differences between the pipe and open channels.
In fact, the San Pedro pipe consisted of alternating segments of tunnel and open channel, collapsed areas in the
same material (95% silt-clay according to N.O. Jones, 1968,
93) and Jones (1975, 208) calculated that the width/depth
ratio in the open channel was 1.8 times that in the pipe.
Empirically correcting for this gave a width/depth ratio
very close to the Schumm prediction, 3.7 against 3.5.

All this is highly speculative, but it points to some
exciting possibilities when we can obtain more data.

A number of observers have noted marked variability in
width, depth and cross-sectional area along soil pipes.
Indeed, Morgan (*ibid*) notes that because of this irregularity
some sections of the pipes may be completely full although
on average the pipes are less than half full during storm
flow. Gilman and Newson (1980) note that these same pipe
cross-sections can vary rapidly over short distances, and
this is illustrated in Fig. 23 by the measurements of percentage of cross-sectional area occupied by baseflow taken
by Jones (1975). Coefficients of variation on the 20 m long
Arizonan tunnel were 87.6% (width), 7.2% (depth) and 30.9%
(area). On the Burbage pipes, the respective coefficients
were 26.9%, 24.1% and 43.6% over 21 m in one tunnel and
41.7%, 22.6% and 64.0% over 17 m in the other (Jones, *ibid*).
There appears to be a tendency for depth to vary least here,
as in stream channels. Unfortunately, Morgan did not tabulate
her width and depth measurements, but cross-sectional area
indicates a much greater uniformity than in the other
examples, c.v. = 8.2%   (cf. Morgan, 1977, App. E; Gilman
and Newson, 1980, tab. 5). Indeed, this is less than is
suggested by these authors' statements. Note that Morgan's
results relate to small, shallow ephemeral pipes, 10 cm
below the surface with mean cross-sections of 67.5 cm^2,
whereas the samples taken by the two Joneses relate to much
larger tunnels (Table 9A).

Table 9 Hydraulic properties of pipes

A. Measurements of hydraulic geometry

site	pipeflow type	number of sections	width x̄ (cm)	width C.V. (%)	depth x̄ (cm)	depth C.V. (%)	cross-sectional area x̄ (cm²)	cross-sectional area C.V. (%)	surface slope x̄ (°)	surface slope C.V. (%)	source
San Pedro	ephemeral	8	402.3	87.6	295.7	7.2	70,000	30.9	10.0	-	Jones (1975)
Burbage 1	perennial	15	41.3	26.9	26.7	24.1	749	43.6	8.0	-	
Burbage 2	perennial	15	52.0	41.7	25.2	22.6	842	64.0	-	-	
Wye	ephemeral	84	3-5	-	3-5	-	67.5	8.2	9.0	0.7	Morgan (1977)

B. Analyses of hydraulic geometry relationships

i) downstream changes in inferred velocity, C.V. (%)

type of pipeflow		stage 25%	Full
San Pedro	ephemeral	47.4	33.4
Burbage 1	perennial	87.3**	59.4

ii) downstream power-law relationships between width or width/depth ratio and area of wetted cross-section.

			stage 25% r*	stage 25% b†	stage 50% r	stage 50% b	stage 75% r	stage 75% b	Full r	Full b
San Pedro	ephemeral	width	0.72	0.50	0.80	0.56	0.87	0.63	0.84	0.76
Burbage 1	perennial	width			0.42	0.28			0.67	0.43

			0.90	0.79	0.93	0.77		0.93	0.62
Burbage 2 | perennial | 97.9** | | | | | width | | |
San Pedro | ephemeral | 54.7 | 0.47 | 0.42‡ | 0.59 | 0.56 | 0.93 | 0.77 | width/depth 0.78 0.85 |

——— significant at 5% level.

═══ significant at 1% level

* correlation coefficient, r.

† exponent, W ∝ A^b.

** actually baseflow stage, approximately a quarter full.

‡ exponents derived from unpublished data for W/D ∝ A.

(tabulated from the results of Jones, 1975, 196–209)

Table 9 continued..

C. (i) Hydraulic parameters (derived by Humphreys, 1978).

site	pipeflow type	hydraulic radius (m)	slope (o)	velocity m s^{-1}	Reynolds Number	D'arcy-Weisbach Friction Factor
Wansfell	ephemeral	0.0071	0.2	0.01	71	1114.0
Wansfell	ephemeral	0.0075	0.2	0.01	75	1177.0
Wansfell	ephemeral	0.0083	0.19	0.01	83	1237.6
Wansfell	ephemeral	0.0070	0.198	0.0134	94	605.8
Wansfell	ephemeral	0.0067	0.198	0.0134	90	592.9
Wansfell	perennial	0.0241	0.2	0.034	819	327.3
Wansfell	perennial	0.2260	0.2	0.017	3842	1227.4
Plynlimon	ephemeral	0.0125	0.275	0.1	1250	26.98
Plynlimon	ephemeral	0.0125	0.275	0.4	5000	1.69
Plynlimon	ephemeral	0.0070*	0.279	0.1	695	15.23

——— = turbulent flow.

* = calculated not measured.

126

Table 9 continued...

C.(ii) Hydraulic parameters of a Plynlimon pipe after
        Gilman and Newson (1980)

Depth of flow (m)	θ rad.	cross-sectional area (m²)	hyd. radius (m)	discharge (m³s⁻¹ x 10⁻³)	veloc-ity m s⁻¹	Rey-nolds Number	Frict-ion Coeff.	Re x f
.020	.970	.001066	.01195	.0750	.0704	58	30.0	1740
.025	1.097	.001462	.01449	.1332	.0911	91	21.7	1975
.030	1.216	.001884	.01684	.2113	.1122	130	16.6	2158
.035	1.329	.002320	.01897	.3085	.1330	174	13.3	2314
.040	1.440	.002773	.02093	.4266	.1538	222	11.0	2442
.045	1.549	.003232	.02268	.5636	.1744	273	9.3	2539
.050	1.658	.003692	.02420	.7179	.1944	324	8.0	2592
.055	1.768	.004148	.02550	.8871	.2139	376	6.9	2594
.060	1.880	.004591	.02654	1.0668	.2324	425	6.1	2593

It is impossible to say as yet how characteristic these differences are.

Humphreys (1978) assumed that pipeflow was steady and uniform and that the hydraulic gradient was the same as the surface slope, in order to apply the Colebrook-White and Darcy-Weisbach equations, obtaining f from the latter and using it in the Colebrook-White equation to obtain a measure of pipe roughness.

$$\frac{1}{\sqrt{f}} = \ln\left(\frac{aR}{D_t}\right) \qquad \text{(Colebrook-White)}$$

where f is the friction factor, a is a constant, assumed to be 13.46, although a lesser value may have been more appropriate, R = area/wetted perimeter and D_t is the roughness height of the soil, and

$$f = \frac{8\,Rg}{v^2} \cdot \frac{\Delta h}{L} \qquad \text{(Darcy-Weisbach)}$$

where $\Delta h/L$ is the hydraulic gradient. He notes, however, that field measurements indicated that the assumption of

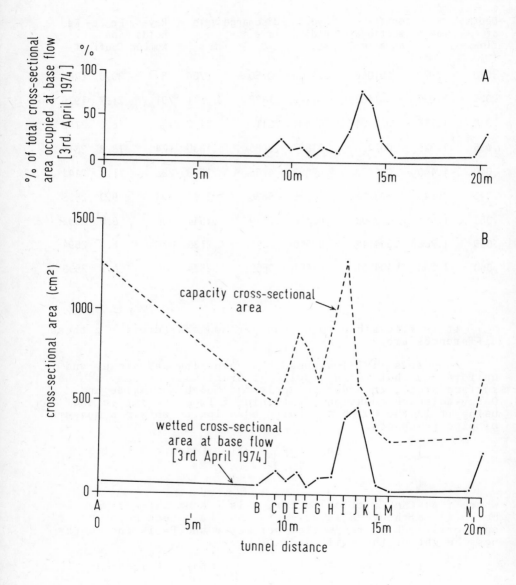

Fig. 23 Variation in cross-sectional area along a meandering pipe on Burbage Brook, Derbyshire, according to Jones (1975, fig. 6.2.3B).

128

steady, uniform flow was only approximately true and that
the assumptions take no account of energy losses on bends.
Calculated Reynolds Numbers suggested generally non-turbulent
flows except in two of the four pipes with larger hydraulic
radii, in which higher velocities and hydraulic radius combined to give turbulent flow (Table 9C). One of these was
a perennial pipe on Wansfell, the other an ephemeral pipe
on Plynlimon.

All the Wansfell pipes had higher friction factors and
roughness than the Plynlimon examples, including the perennial pipes. However, the friction values were abnormally
high and some roughness values physically impossible, so
that the absolute values, if not the relative values also,
should be treated with caution. It seems most likely that
the assumptions were not adequately met.

Gilman and Newson (1980) were able to combine measurements of hydraulic geometry with measurements of actual
discharge for the first time. They obtained a relationship
of $Q = 19.00 \ A^{1.818}$ for discharge $(m^3 s^{-1})$ in an ephemeral
pipe and mean cross-sectional area (m^2), which they generalised to : $Q = 41.2 \ \sqrt{S} \ A^{1.818}$. Their estimates of hydraulic
parameters (Table 9C(ii)) for the different levels of flow
in the pipe suggested a range in friction factor from 30 in
flows of 2 cm depth to 6.1 in depths of 6 cm, and in Reynolds
Number from 58 to 425 over the same range, both of which
were much more realistic than those of Humphreys. Gilman
and Newson concluded that the pipes are extremely rough
channels because flow is dominated by channel form rather
than by skin resistance, a feature also of the surface
channels in the same area of upland. The authors also used
their cross-sectional area-discharge relationship to estimate
rates of movement of a kinematic flood wave for one particul-
are storm and concluded that a lag of c.23 minutes occurred
over a 300 m slope with no appreciable change in hydrograph
shape. Hence a constant kinematic wave velocity would be
sufficient for routing in the Institute of Hydrology model.

A few interesting observations have also been made of
meandering channel plans in soil pipes. Both N.O. Jones
(1968) and J.A.A. Jones (1975) have mapped meadering pipes
in Arizona and England, and Parker (1963) stated that meandering is a common feature in Navajo County, Arizona. Sinuo-
sity in the two mapped pipes was 3.9(San Pedro) and 1.68
(Burbage) according to Brice's definition (Jones, 1975, 219).
Wavelength on the short meandering section of the Burbage
pipe was estimated at 12m, which would give a much higher
ratio of wavelength (L) to wide (W) than reported for open
channels, $L = 30.57W$, again indicating a narrow width in
relation to other geometric variables than in stream channels.
(Leopold and Wolman (1960) found $L = 10.9W^{1.01}$ and Inglis
(1949) $L = 6.6W^{0.99}$.) Spectral analysis of the Arizonan pipe
suggested that the only significant principal wavelength of
30m +, with $L = 22.4W$, also showed a narrower width in
relation to wavelength than observed in open channels (*ibid,*
223). However, the radius-to-width ratio, 2.05, falls within

the observed range for channels. The author comments that the bedslopes (c8°) in these pipes are much higher than might be expected of meandering open channels but that braiding is probably not so readily available as an alternative planform in piping as it is for open channels with steep slopes. It is also possible that lower discharge in the pipes 'permits' meandering on the higher slopes (cf. Leopold and Wolman, 1957).

In general, therefore, analyses of hydraulic geometry to date have thrown up some exciting possibilities, but by and large the evidence is too scant to allow firm conclusions at this stage.

6.1(iv). Soil pipe slopes and long profiles.

Very little information is yet available on long profiles, but what is to hand suggests generally very irregular and often convex upward profiles with not infrequent waterfalls, except in the most mature systems. Surveyed long profiles have been drawn up by Drew (1972), Jones (1975) and Morgan (1977) and are reproduced in Fig. 24. Drew's pipe is excavated in hard, silty sandstone in the Big Muddy Valley, Saskatchewan and is much larger than those of Jones and Morgan, which are from pipes a few centimetres in diameter running generally just above the Oh/Eg interface in peaty podzols on Plynlimon, Wales, and from one at the interface between soil and reworked solifluxion parent material in a lessivé brown earth in the Peak District. The latter pipe appears to be seasonal and its outlet is near the stream bed, yet it contains major waterfalls within a few metres of the streambank. Maesnant pipes show increased upward convexity and waterfalls as they approach outlets on the scarp of the river terrace. Jones (1975, 209) noted that the general pseudokastic nature of these systems means that long profiles are irregular both in pipe source areas and near the streams.

Morgan (1977) observed that both the bed of one of her pipes nearest the base of the peat and the surface of the clay layer showed periodicities, but not exactly in phase, and she produced a correlogram for various lags from measurements taken at 10 cm intervals over a 6.6 m length of pipe. She concluded that the pipe inflexions responded to the inflexions in the clay surface, but with a downslope lag of 2.6 m. This interval compared with the observed spacing of surface spouts during storms and Morgan argued that the upward inflexions of the pipe had weakened the roof and eventually broken surface (Gilman and Newson, 1980). Wilson (1977) also commented upon upward arching in pipes in the Brecon Beacons, which would require a hydraulic head to operate them. A series of such arches in the long profile could cause the pulsating flow which Jones (1975, 248; 1978a, 16) recorded on Maesnant and suggested could be due to 'periodically syphoning reservoirs or overflowing ponds'.

In contrast, N.O. Jones (1968, 151) noted in Arizona

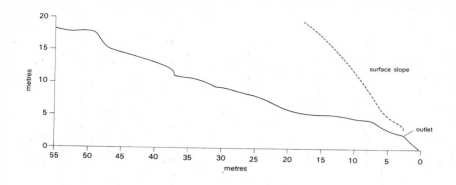

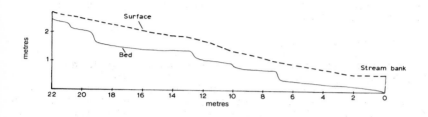

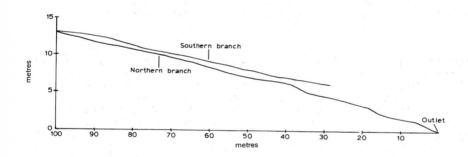

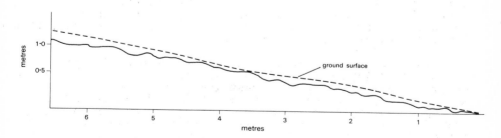

Fig. 24 Long profiles of soil pipes. Adapted from
original diagrams by Drew (1972, fig. 3), Jones
(1975, figs. 6.2.6 and 8.1.3b) and Morgan (1977,
fig. 7)

that 'the long profile of major pipes rapidly assumes the characteristic exponential curve of graded stream channels' and referred to the 'collapse of all pipes to form an integrated arroyo system, with long profiles indistinguishable from arroyos formed by headcut advance' (*ibid.*, 147). Jones (1975, 214) measured bed-slopes off N.O. Jones's 1:300 scale plan and concluded that they obeyed neither Horton's nor Broscoe's cumulative slope laws. A marked break occurred between second and third order pipes, which Jones considered could be due to (i) higher order pipes only having adjusted to a new baselevel or (ii) marked increase in discharge due to seepage influx in higher order pipes. In the light of the known history of rejuvenation in the area and of no break in channel width along the pipes which might suggest the latter cause, it was concluded that the systems were adjusting to a new baselevel.

6.i(v). The length of soil pipes.

At present, we have very little comparative data on the lengths of soil pipes. There are obvious problems here with tracing pipes with anastomosing and partially discontinuous networks. However, an interesting observation is that length appears to be reasonably independent of cross-sectional area. Long traces have been made in pipes of c. 10cm diameter in Wales: Wilson (1977) performed dye tracing on a 363 m long pipe, Newson and Harrison (1978, tab. 10) recorded a 300 m stretch, and Jones and Crane (1979) have followed a pipe over $\frac{1}{3}$ km by dye tracing.

Elsewhere, Fuller reported a few tunnels extending over 150 ft. (46 m) back from plateau rims in the loess of China, where pipes were typically 3 ft. (1 m) wide and 10 ft. (3m) high. Subakov *et al.* (1968) state that maximum lengths in loess plainlands in Uzbekistan were 100 m, but more usually they extend no more than 3-5 m back from gully walls closely approximating the curved surface of the water table, with occasional extensions up to 10 m. There is an interesting parallel here with the author's own unpublished observations of piping on the Ribble Marshes near Southport, where lines of piezometers set along the course of pipes discharging into the banks of creeks indicated that the beds closely approximate the convex curve of the water table adjacent to the creek. These salt marsh pipes are typically no more than 5 m long and 5-10 cm in diameter.

In South Africa, Beckedahl (1977) has mapped a pipe system 140 m long with maximum diameter of 1-5m. On a loam-covered slope in the East Carpathians, Galarowski (1976) monitored a pipe system extending 150 m and consisting of 11 pipes varying from 'small ones' 3 m in diameter to big ones 3 x 9 m. Presumably drawing largely on American experience, Butzer (1978) refers to pipes up to 90 m long and one to several feet in diameter; very similar to the pipes 30 cm wide and 90m long reported by Crozier (1969) in Eastern Otago. N.O. Jones (1968) reported pipes of 6 ft. (1.8 m) diameter and 200 ft. (60 m) long in sandy material in Arizona, and pipe systems were found up to 1200 ft. (366m) from arroyo banks.

At the other end of the spectrum, Conacher and Dalrymple (1977, 21) report that small pipes of 1-3 cm diameter in discrete seasonally active G horizons within gleyed lessivé soils in the London Basin are essentially discontinuous and have rarely been traced for more than 25 cm. Many of these 'linear holes' do not 'exhibit the traditional shape and dimensions of classical soil pipes', but are clearly modified root or animal holes.

6.2. Spacing and distribution.

6.2(i). Vertical distribution of pipe outlets in the streambank.

Whilst there is a profusion of data on the vertical distribution of pipes in the soil profile, there is only limited information on the vertical location of soil pipes in relation to open channels. Buckham and Cockfield (1950) reported that outlets 'often occur' at $\frac{1}{2}$ to $\frac{2}{3}$ the way down gully walls near Kamloops, British Columbia. Parker (1963) noted that 'many outlets' occurred at the base of gully walls in Arizona-New Mexico. N.O. Jones (1968) observed that the deepest pipes in the San Pedro Valley, Arizona, were generally near the gully bed, except in one case where a new, deeper system had developed beneath an old one in response to a rapid lowering of base level, and Carroll (1949) also noted that pipe beds had generally kept pace with stream downcutting in the area. Heede (1971, 5) quotes a typical outlet 1m above the gully floor and says that none were close to the channel bed near Silt, Colorado. Newman and Phillips (1957) illustrate outlets half way down a stream bank and mention shallow pipes developed where the B horizon is relatively impermeable irrespective of gully depth. Jones (1971; 1975, 162) concluded from random sample surveys of bank pipes in basins in England and Wales that outlets were almost normally distributed around a mean height of c.40% bankfull or slightly skewed towards the channel bed, the larger pipes tending more towards the bed. He suggested that this general distribution might be due to a balancing between greater volume of seepage available with greater depth and greater hydrualic drawdown higher in the bank. He also suggested that the baselevel for streambank piping is determined by the flood recession curve, although Bryan (pers. comm., 1978) has suggested that grading could be to the rising rather than the falling stage.

Clearly two separate groups of factors are in conflict, the properties of the soil profile and the hydraulic environment, together on occasions with an historical factor causing relict pipe outlets. Moreover, different factors may have the same effect. Superimposed levels of piping may reflect either two impeding layers or else rapid rejuvenation. It is also possible that in some cases a vertical hierarchy of pipes or pipes and channels may exist in current equilibrium, each activated by storms of differing magnitude, rather than the single level of phreatic surface 'bournes' envisaged by Horton (1936). Both Jones (1975, 211; 1978a, fig. 6) and Gilman and Newson (1980, fig. 14) illustrate vertical profiles which show overflows from pipes to surface channels or mires and return inlets. And it has been proposed that a vertical hierarchy can exist in terms of frequency of flow within pipe systems at multiple levels, with ephemeral pipes generally near the top of the profile (Fig. 25). However, in the absence of marked aquicludes or impermeable layers the distribution of outlets might be expected to be continuous through the bank and marked discontinities of distribution suggest strong control from soil properties.

6.2(ii). Horizontal spacing of pipes

Data on the lateral spacing of pipes have been collated in Table 10. There are still very few quantitative measurements available to enable a comparison between sites. N.O. Jones (1968, 152-153) pointed out that the irregular spacing of pipes in the San Pedro Valley, Arizona, was controlled by inhomogeneities in the soil and sediments, particularly subsidence fractures, the counterpart of the traditional view of stream network development on consolidated bedrock, and suggested that as the drainage system develops (mainly after collapse) fluid hydraulics exert increasing control.

Analyses by Jones (1971; 1975) showed that distances between pipe outlets in the banks of Burbage Brook and the Afon Cerist, N. Wales, were positively skewed with modes in the range 0.1 to 1.0m. Cell count analyses on Burbage also revealed a clustered distribution of outlets along the bank (Jones, 1971, tab. 1), whilst histograms of cluster size showed that the distribution of cluster sizes deviated significantly from a Poisson random growth model (α = 0.05) primarily because of four extra large clusters.

A contingency test showed a significant association between pipe outlets and percoline/seepage line outfalls along the streambank, and the spacing of the percolines was found to be almost random. Jones observed that the extra large clusters of pipe outlets occurred on percoline outfalls and found that the large clusters generally contained the larger pipes, as indicated by the shift in the mode of the histograms of mean pipe size in clusters of increasing size (Fig. 26). 77% of above average size outlets occurred in clusters, each cluster tending to have one or two 'master' pipes much larger than the rest. However, looking at the most advanced, semi-collapsed perennial pipes, Jones found no evidence of any cluster of pipes around them. Jones concluded that:

(i) pipes develop to larger sizes in more favourable locations, particularly locations with prior concentration of drainage (percolines). Redox measurements support this view (section 4.6(iii)).

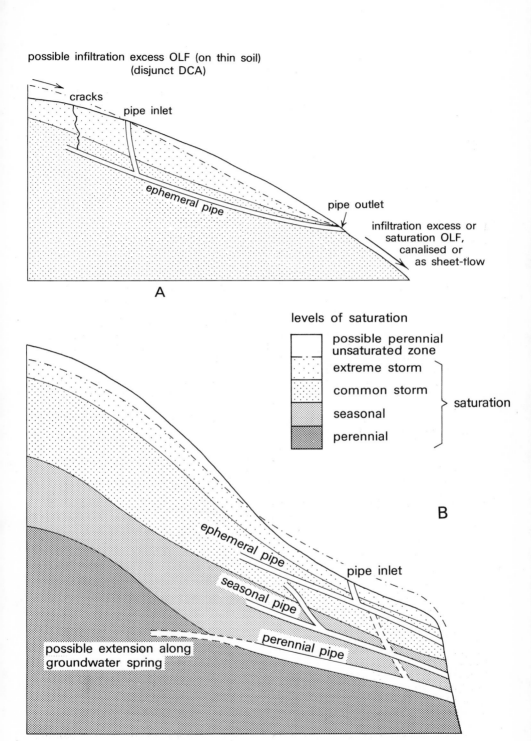

Fig. 25   Frequency of hydrologic response in relation to
          pipe levels and connectivity, after Jones (1978c,
          fig. 5; 1979, fig. 3).

Table 10.   Spacing of pipes on sideslopes(based on
            Humphreys, 1978, tab. 1).

site and source	mean spacing m	width of measured section of slope m	closed spacing over 20m section m
Nant Gerig, Plynlimon			
(Pond, 1971)	28.24	380	3.33
(Humphreys, 1978)	9	20	5
Cerrig yr Ŵyn, Plylimon			
(Newson, 1976)	17.8	100	2.86
Wansfell			
(Humphreys, 1978)	5.98 (ephemeral) 50 (perennial)	60	1.54

(ii) there is some evidence for cyclical development
at the pipe-channel interface, beginning with a prolif-
eration of pipes in particularly wet areas followed by
preferential growth of master pipes which eventually
take over, starving smaller competitors and developing
to a size where roofs collapse. The low percentage
of pipes activated by major storms in the area supports
the inference that many are relict features with their
catchment volumes captured by more successful competitors; cp. drain spacing laws (Kirkham and Powers, 1972).

Conacher and Dalrymple (1977, 21-22, 26) also observed
spatial clustering in small (1-3 cm) pipes in G horizons in
gleyed lessivé soils in the London Basin ending in field
drains or unit 8, and at the base of the A horizon of lessivé
soils in New Zealand above a fragipan (Btg horizon). Humphreys (1978, tab. 1) collated measurements of mean and
closest spacing of pipes on the side-slopes of Wansfell and
the Wye experimental catchment. Unfortunatley, few conclusions can be drawn from this limited data. The marked difference in estimates of spacing on Nant Gerig indicates
inadequate sample sizes and the observations that the pipes
are spatially clustered (50% reflexive neighbours in 12
streambank outlets at Wansfell; *ibid.*, 14) indicates that
the 'mean spacing' is not the best measure.

6.3.   Soil pipe networks: form and evolution.

The pattern of soil pipe networks is clearly of interest
both from the point of view of inferring the factors responsible for creation of the network and from the hydrological
viewpoint, since, as in stream networks, the spatial and
topological features of the networks must in large measure
control their hydrological response. Both hydrological and
genetic interest also focuses on evolutionary changes in the
networks. Maps of pipe networks have been produced in
Poland by Czeppe (1960) and Galarowski (1976), for parts of
the San Pedro valley, Arizona by N.O. Jones (1968), in Canada

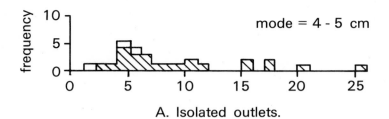

A. Isolated outlets.

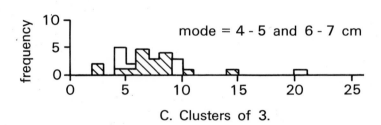

B. Clusters of 2.

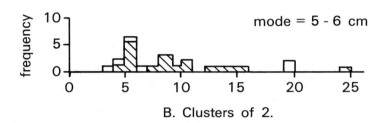

C. Clusters of 3.

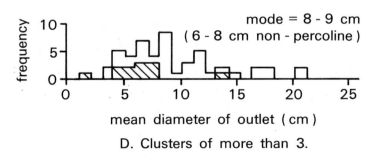

mean diameter of outlet (cm)

D. Clusters of more than 3.

☐ = percoline pipes   ◩ = non - percoline pipes

Fig. 26   Distribution of mean pipe diameter for different
sized clusters of pipes on Burbage Brook, Derbyshire (Jones, 1975, fig. 6.1.7).

by Drew (1972) and in Britain networks have been mapped in
the East Twins basin, Mendip (Weyman, 1971; Stagg, 1973), on
the Nant Gerig and Upper Wye, Plynlimon (Gilman, 1971; Pond,
1971; Davis, 1972; Newson, 1976a and b; Morgan, 1977;
Atkinson, 1978, fig. 3.9; Humphreys, 1978; Gilman and Newson,
1980), on Burbage Brook, Peak District (Jones, 1975), on
Maesnant, Plynlimon (Lewin, Cryer and Harrison, 1974; Jones,
1975, 1978a and c; Jones and Crane, 1979), on Nant Cwm-llwch
Brecon Beacons (Wilson, 1977; Atkinson, 1978, fig. 3.13),
on Wansfell, Lake District (Humphreys, 1978), and at Slitheroe Clough, Pennines (McCaig, 1979a and b). Fig 27 collects
these together in a similar form to help comparison.

One feature that is immediately obvious is the differing nature of the network in the same area when mapped on
differing scales. This is particularly clear in the series
of three maps covering the Nant Gerig. At the most detailed
level, Pond (1971) produced a map at 1:2000 which showed
the Nant Gerig side-slope network to be anastomosing and
very discontinuous. In contrast, the maps of the whole Nant
Gerig subcatchment by Newson (1976b) at 1:2500 and of the
whole Upper Wye basin at 1:17000 by Newson and Harrison
(1978) show little anastomosis and give more the impression
of a discontinuous but vaguely dendritic network. The
dendritic appearance of the network is increased when the
surface stormflow routes mapped by Newson (1976b) are added,
highlighting the close relationship between pipeflow and
linear elements of overland flow (cf. section 7.1(i)).
Measurement of pipe density on these maps also shows marked
differences, from 43.65 km km^{-2} measured from Pond (1971),
to 7.44 on Newson (1976b) and 18.55 (Newson and Harrison,
1978). Differences are also apparent though to a lesser
degree, between the maps of Maesnant produced by Lewin, Cryer
and Harrison (1974) and Cryer (1978), Jones (1975; 1978a)
and Jones and Crane (1979). The map by Lewin *et al.* (1974)
at 1:5700 (as first presented at conference by J. Lewin, Dept.
of Geography, Aberystwyth) is purely dendritic and continuous, whereas those of Jones and of Jones and Crane show
varying degrees of anastomosis and discontinuity, at scales
of 1:625 and 1:3750 respectively. The main dendritic arteries of the Maesnant system are seasonal or perennial,
whereas on the Nant Gerig the arterial system seems to be
less well developed. In fact, there seems to be more of a
generic division between the ephemeral pipe net and the
'subsurface streams' of Newson (1976b) or 'flushes' of Newson
and Harrison (1978) (cf. section 7.1(i)). The pipes of
Slitheroe Clough show a similar pattern to those of Maesnant,
with ephemeral pipes tending to feed perennial arteries which
follow seepage lines : McCaig's map omits the ephemeral pipes,
recording only collapse features on the dendritic arteries.
The semi-collapsed system of ephemeral pipes in the San
Pedro valley are also generally dendritic though with the
occasional triple junction. Note that these are semi-arid
ephemeral pipes and much more advanced than similar features
in humid areas (cf. section 6.1(iii)). Jones (1978a, 9)
also reported that the larger pipes on Burbage Brook tended
to follow dendritic percoline routes. Although he was unable

Fig. 27   Collated maps of soil pipe networks.

These maps are shown on the following five pages (140-144) and show the following areas:

A.   Networks in the San Pedro Valley, Arizona, mapped by N. O. Jones (1968, plate 3).

B.   Networks in the upper basin of East Twins catchment, Blackdown, Mendips, U.K., mapped by Stagg and Weyman in 1970 (Stagg, 1974, fig. 11).

C.   Pipe distribution over the whole Nant Gerig catchment, after Davis (1972) and Newson and Harrison (1978).

D.   Networks on a side slope of the Nant Gerig, Upper Wye catchment, Plynlimon, Wales, based on Pond (1971);   parts of the network also appeared in Gilman (1971) and Davis (1972).

E.   Networks in the Cerrig-yr-Wŷn subcatchment, Plynlimon, Wales, after Newson (1976b), Morgan (1976) and Gilman and Newson (1980).

F.   Networks on Wansfell, Lake District, U.K., after Humphreys (1978).

G.   Distribution of pipe collapse sites on Slitheroe Clough, Yorkshire, U.K., after McCaig (1979b).

H.   Networks in the Maesnant catchment, Plynlimon, Wales, after Jones and Crane (1979; 1980); a detail of the network appears in Jones (1978).

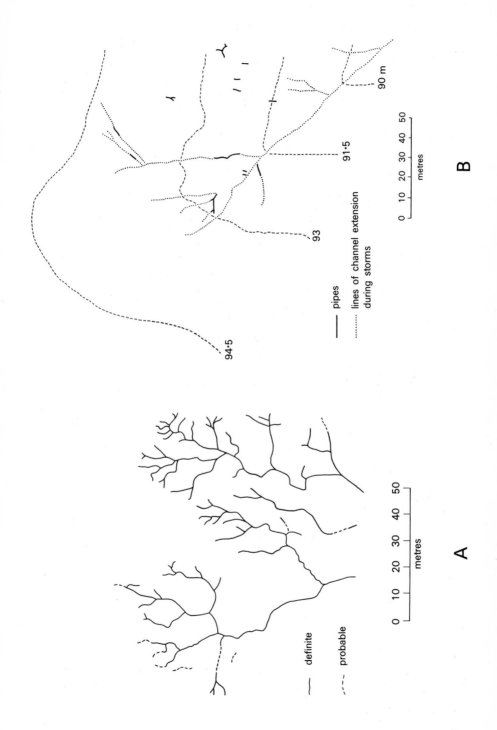

A

definite

probable

metres

0 10 20 30 40 50

B

pipes

lines of channel extension
during storms

94·5

93

91·5

90 m

metres

0 10 20 30 40 50

140

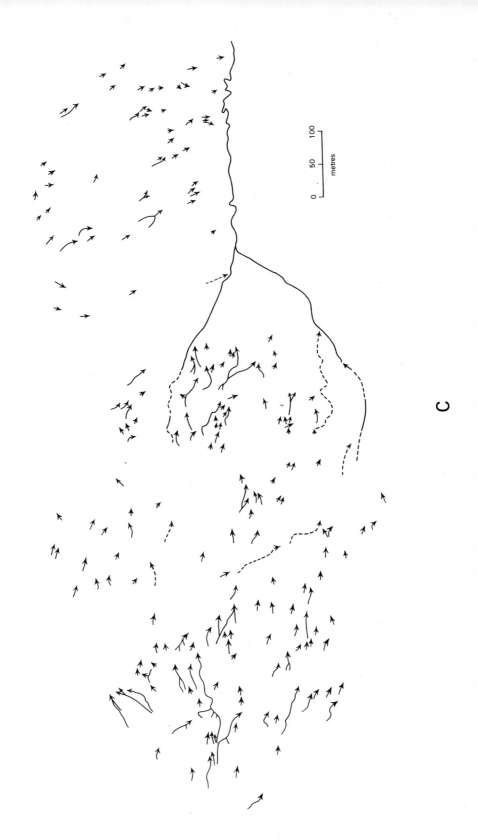

C

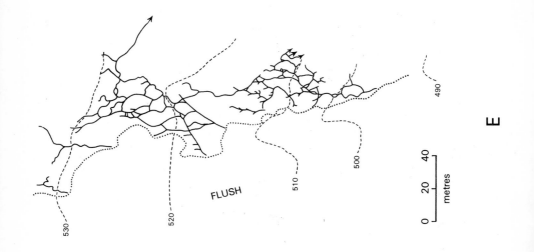

FLUSH

530

520

510

500

490

0   20   40

metres

E

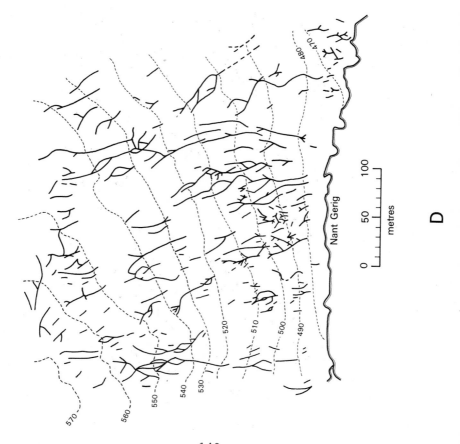

570

560

550

540

530

520

510

500

490

480

470

Nant Gerig

0   50   100

metres

D

142

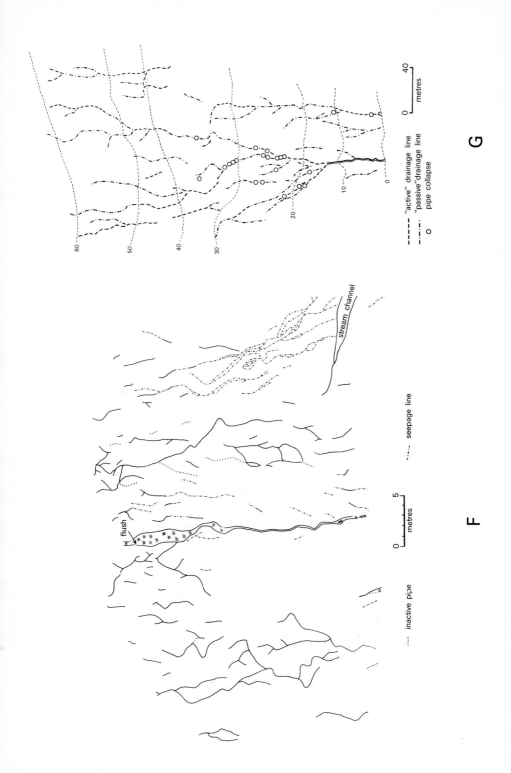

"active" drainage line

"passive"drainage line

O   pipe collapse

G

seepage line

inactive pipe

flush

stream channel

F

143

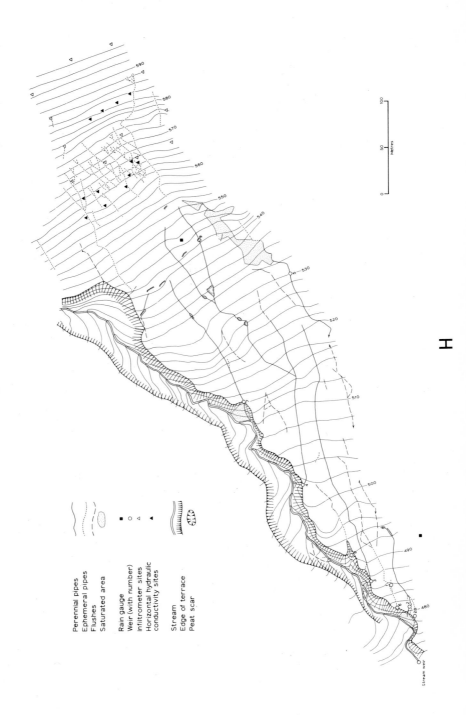

Perennial pipes
Ephemeral pipes
Flushes
Saturated area

Rain gauge
Weir (with number)
Infiltrometer sites
Horizontal hydraulic
conductivity sites

Stream
Edge of terrace
Peat scar

Stream weir

590
580
570
560
550
540
530
520
510
500
490
480

100

50
Metres

0

N

144

to follow the ephemeral pipes there because of their small size (c. 9 cm diameter), depth (c. 50 cm) and general lack of flow (only 13% flowed in a storm of 1% probability), Jones (1975) presumed that the majority of ephemeral pipes on Burbage Brook did not extend much further back from the streambank than the edge of the small floodplain. These ephemeral pipes therefore fall into a different hydrological, and probably also genetic, group than those of Gilman and Newson (1980), i.e. pipes adjacent to and linked with the main stream that may owe their formation more to 'boiling' compared with disconnected, discontinuous pipes that are basically caused by desiccation cracking (Jones, 1979).

It has frequently been stated that pipe networks are discontinuous and anastomosing (e.g. Atkinson, 1978; Humphreys, 1978), but observations to date suggest that there are wide variations between areas and that different impressions can be gained of the same area depending on the scale of study. Gilman and Newson (1980) note that the connectivity of the Cerrig yr Ŵyn network appears greater than in the Gerig, 'but it must be remembered that it was mapped on a smaller scale'. However, the Gerig network is also on a larger, steeper slope which would affect connectivity. Extensive distributary systems have also been noted on the Burbage Brook (Jones, 1971; 1975) and at Cerrig yr Ŵyn (Morgan, 1977; Gilman and Newson, 1980).

In general, however, it seems quite feasible to suppose that continuity and dendritic form are indicative of more advanced development. At the microscale, Morgan (1977, fig. 10) demonstrated by her excavation of part of the Cerrig yr Ŵyn network how piping had used, expanded and integrated anastomosing desiccation cracks into one main downslope route, and this process seems likely to continue at larger scales. In many cases, anastomosing branches are not always activated at the same time, one branch possibly forming a high flow alternative. Anastomosing branches and discontinuous networks probably in general reflect the location of initial weaknesses exploited by the flowing water, e.g. desiccation cracks in peat (Newson, 1976a) or subsidence cracking associated with percolines (Jones, 1975; 1978a). Interestingly, N.O. Jones (1978, 122-124) considered the possible influence of polygonal cracking on the plan of his Arizonan networks, but Jones (1975, 227; 1978a, 9) concluded that there is no evidence for this in his relatively mature, partially collapsed networks. Moreover , Horton analyses of N.O. Jones's plans by Jones (1975, 292) suggested a streamlike structure, in so far as these can be regarded as useful indicators (Fig. 28). Fig 28 also shows the results of analyses of the Maesnant pipe network based on the map by Lewin et al.(1974), which gave similar results. Although further ground survey has revealed a more extensive network, Jones (1975, 289) argued that the cutoff point in the size of feature mapped was sufficiently consistent over the area to permit an exploratory analysis.

Bifurcation ratios in the two dendritic Arizonan

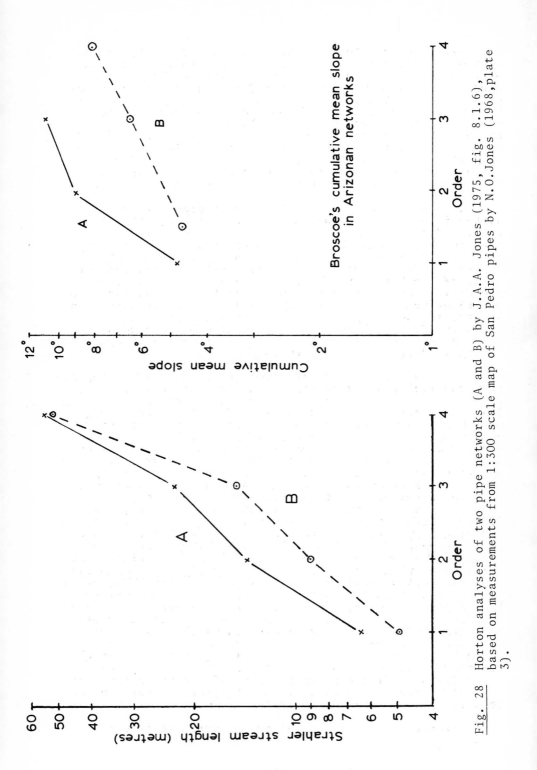

Fig. 28 Horton analyses of two pipe networks (A and B) by J.A.A. Jones (1975, fig. 8.1.6), based on measurements from 1:300 scale map of San Pedro pipes by N.O.Jones (1968, plate 3).

networks were 3.2 and 3.3, compared with the range of 2 to
4 recorded for stream networks and 4 for a random-walk
model. Interior link length distributions showed the positive skewness and dearth of shorter links recorded in channel
networks (Fig. 29). The latter feature would not be expected
under a purely random process, which is likely to
produce an exponential distribution, and suggests the presence of some rationalisation, such as the elimination of
close, same-sided tributaries proposed by James and Krumbein
(1969) for stream networks. Histograms of cis- and trans-
link distributions and Markov chain analyses confirm the
lack of short links and a particular lack of short-to-mod-
erate cis-links. In fact, only 39% of the links were cis-
links, i.e. approximately two-thirds the probability of
trans-links just as was observed by James and Krumbein in
stream networks. Moreover, those authors found that this
trend was less marked in streams of magnitude 4 than of
magnitude 10 and over, whereas Jones has used all links of
magnitude 2 and above in the pipe networks (because of lack
of data). James and Krumbein suggested that the rationalisation trend marked the onset of 'maturity'. However, Jones
(1975, 296) noted that measurements taken on badland rill
networks show similar features and he speculates that
either 1) these pipes are similarly developed in highly
erosive material, for which there is much evidence, or 2)
the feature is due to a compression of scale, i.e. the pipes
are a semi-independent system performing the same function
as stream channels but on a smaller scale and reaching their
own 'maturity'.

Analyses based on the spacing of junctions in the
Arizonan networks suggested greater spatial clustering than
found on a few sample stream networks by Jones (1975, 381-
389; 1978b). This he took as indicative of more primitive
drainage systems, with the sources showing far greater concentrations in preferred localities within the drainage area
than do surface streamheads. The internal junctions tended
to be less clustered than sources in the larger of the two
pipe networks analysed, and this was in keeping with the
general trend found from analyses of stream and percoline
patterns. This was a reversal of the trend indicated by
computer simulations based on a random-walk model of network
development and suggested a measure of increased spacing due
to internal competition between drainage links.

Random-walk simulation experiments were also performed
on pipe network development by Humphreys (1978). He tested
three models in terms of the frequency of pipes per square
in a grid of 1000 squares compared with observed frequencies
on Wansfell. Each model permitted movement down and across
slope only, according to throws of a dice. Model 1 began
with equispaced sources with a spacing equal to the mean
lateral spacing of pipes at Wansfell and allowed crossing
and anastomosing. In model 2 starting points were randomly
chosen and the number of sources increased to allow for
reduction in the number of pipes caused by combining downslope. Model 3 allowed for random source locations and

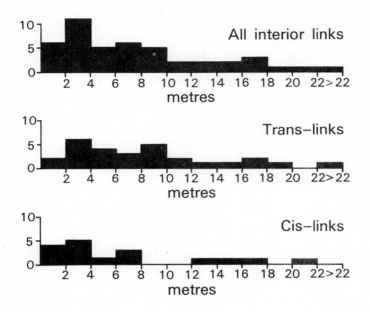

Fig. 29   Interior link length distributions derived from
          maps of two Arizonan pipe networks.   Based on J.A.
          A. Jones (1975, fig. 8.1.7), derived from original
          map by N.O. Jones (1968, plate 3).

discontinuities : 5 successive movements in the same direction caused termination for 5 further moves, then the pipe
was automatically revived.   Despite some artificiality,
especially by following Leopold and Langbein's (1962) procedure of all sources beginning in a line, the cell counts
of model 3 proved to be a significantly good fit to the
observed Wansfell pattern by χ^2.

        Humphreys (1978, tab. 4) also abstracted similar data
from the Institute of Hydrology map of Nant Gerig (Pond,
1971), but did not use it in a simulation comparison.
Humphrey's data for Wansfell and Nant Gerig have been
compared with a Poisson model in Table 11.   The comparison
indicates a very different distribution in the two areas,
with a tendency to clustering on Wansfell and a tendency to
a more regular than random distribution of Nant Gerig.   Part
of the difference may be due to more developed competition
for drainage amongst the Nant Gerig pipes.   Although Humphreys (1978) measured a pipe density of 175 km km^{-2} on Wansfell (and a constant of pipe maintenance of 5.71 m), the
pipes were small with a mean of 2.98 cm diameter against
10 cm on Nant Gerig.   Another plausible factor is that piping
on Nant Gerig is developed along desiccation cracks in the
peaty A horizon of the podzols which are themselves more
regularly spaced.   Morgan (1977) stripped an area on the
Upper Wye (on Cerrig yr Ŵyn, cf section 7.1) of its grass

Table 11. Pipe frequencies and densities.

A. Quadrat analyses of pipe distribution at Wansfell and Nant Gerig, using data from Humphreys (1978).

number of pipes per quadrat

	0	1	2	3	4	5	6	6	$\bar{x}$	variance/ mean ratio (Poisson ratio = 1)
Wansfell frequen-cies.	715	222	57	5	1	0	0	0	0.355	1.75
Nant Gerig frequen-cies.	39	33	32	29	19	13	7	8	~2.35	~ 0.84

Quadrat counts derived by Humphreys (1978) from personal survey of Wansfell and Institute of Hydrology map of Nant Gerig (Pond, 1971).

B. Pipe network densities.

	Drainage density (km km^{-2})	mean diameter of pipe (cm)	constant of pipe maintenance
Cerrig yr Ŵyn, Plynlimon (Gilman and Newson 1980)	180	5	
Wansfell, Lake District (Humphreys, 1978)	175	3	5.71
Nant Gerig, Plynlimon (measured from Pond, 1971)	43.7	c.4	
Measnant, Plynlimon (measured from Jones and Crane, 1979)	27.3	c.9	

sod after the 1976 drought and found that the pipes were, indeed, following these desiccation cracks (*ibid*, fig. 10; Institute of Hydrology, 1978, fig. 21; Newson and Harrison, 1978, plate 5; Gilman and Newson, 1980, fig. 23).

Morgan (1977) studied the variability of pipe drainage density in 10 m^2 quadrats across the Cerrig yr Ŵyn slope. Pipe density was 42.3 km km^{-1} over the whole area or 180 over the actual piped area (Gilman and Newson, 1980).   There appeared to be a greater density in areas of uniform or con-cave slope, but no significant correlation was found.   Gilman and Newson (*ibid*.) note that as well as the high density the network is distinguished by a large number of interior nodes.

Although they noted that the relationship between frequency and density is similar to dendritic stream networks, many of the nodes are distributary or anastomosing so that less than 50% of the interior nodes are caused by exterior links. They also noted that part of the network dominated by exterior links does not have a source area of blanket peat and they speculate that exterior links result from subsurface runoff generated on the slope itself rather than from peat drainage (cf. section 7.1(i)).

There is very little direct evidence on which to build a general model for the evolution of pipe networks. Indeed, from what has been said, it is clear that no one single model could totally encompass all possible developments, largely because of differing initiating processes. Direct observations of network growth in areas of rapid development have at best produced maps of linear extension of roof collapse (Czeppe, 1960; Galarowski, 1976; N.O. Jones, 1968, 49) (cf. section 8.1). In Britain, development seems to be slow or non-existent. Jones (1975) observed no reportable change in the bank outlets on the Burbage Brook over a 7 year period, and speculated that piping could attain a quasi-equilibrium state (Jones, 1971). Gilman and Newson (1980) also regard the Wye networks as quasi-stable.

On the other hand, Shreve (1972) has provided a mathematical explanation of competitive forces in the development of pseudokarst networks in glaciers. He suggests that tunnels replace shifting, unstable sheetflow at the base of glaciers because of disproportionate growth in the larger voids increasing the concentration of erosion and eventually leading to stable networks 'like a three-dimensional river network' (*ibid.*, 206). Jones (1975, 79) suggests an analogy with diffuse seepage at the phreatic surface exploiting larger voids in soils. Taking the radii, a and b, of two passages of the same length and connecting the same areas, with M as the rate of erosion of the walls and discharge $Q_a + Q_b = Q$, Shreve showed that:

$$\frac{\dot{a} - M_a}{a} = \frac{\dot{b} - M_b}{b}$$

Differentiating
$$\dot{b} = -\dot{a} \left(\frac{\partial Q_a / \partial a}{\partial Q_b / \partial b} \right)$$

and
$$\dot{a} = \left(\frac{M_a}{a} - \frac{M_b}{b} \right) \bigg/ \left(\frac{1}{a} + \frac{1}{b} \left[\frac{\partial Q_a / \partial a}{\partial Q_b / \partial b} \right] \right)$$

Hence, should a > b, $\dot{a}$, the rate of change of radius a, is positive, and the larger passage increases in size quicker, capturing ever more of the total discharge. Eventually growth may be halted in pipe b by lack of discharge and $\dot{b}$

150

may become negative owing to factors such as soil creep, and an anastomosing branch is lost.

The analogy with karst cave networks has been made by a number of workers (e.g. Drew, 1972; Jones, 1975, 1978a) and is implicit in the term 'pseudokarst'. Describing piping in the bedrock walls of the semi-arid Big Muddy Valley, Saskatchewan, Drew (1972) suggests that there is strong evidence that development is according to the master conduit sequence of Rhoades and Sinacori, who proposed that underground flow concentrates near to resurgences at the level of the watertable: as the movement of water just below the watertable becomes more efficient due to des- truction of the homogeneity of permeability by flowlines converging towards the outfall, less water will move along deeper-seated routes and more is concentrated in master conduits near the surface of saturation (cf. Davis, 1930; Davis and De Wiest, 1966, 365; Sweeting, 1972, 238). Drew therefore suggests an evolutionary sequence beginning with initial throughflow in many channels and eventually leading to turbulent flow in an integrated, dendritic pipe network. Although the pipes are initiated on perched water tables, latter-stage breaching of the underlying aquiclude leads to the development of a further, lower level of piping. It is in- teresting that Rubey (1928) described piping on the Great Plains as commonly occurring just below the wet season water table. This is also the location of Horton's 'bournes' (Horton, 1936). It has also been suggested that soil voids might be created near the surface of saturation by the collapsing of saturated soil (Jennings and Knight, 1957). However, Woodward (1961, 40) has suggested that velocities are generally too low at the phreatic surface to cause app- reciable cavern development without a 'sudden' event, such as stream rejuvenation, increasing the hydraulic gradient. Rejuvenation has also been frequently invoked for both the initial development of pipes and the development of multiple levels (e.g. N.O. Jones, 1968; Barendregt and Ongley, 1977). Indeed, Carroll (1949) describes the development of pipes with high, vaulted roofs as pipes beds have kept pace with the rejuvenation of the San Pedro River, Arizona. Barendregt (1977) and Barendregt and Ongley (1977) describe two levels of piping in the Milk River Canyon, Alberta, caused by a process similar to that described by Drew (1972). Both Heede (1971) and Stocking (1976b) plotted the development of vertical profiles over periods of time, illustrating the growth of new piping at lower levels as the adjacent gullies incise their beds. These lower levels were linked by vert- ical shafts to the older, higher levels, capturing their drainage. In other words, vertical link pipes may result either from downward breaching of a retarding layer or from upward drainage capture from lower levels.

The frequent existence of piping at more than one level is a factor which complicates any attempt to map the hori- zontal structure of pipe networks. At present there appears to be no map which shows networks at two or more levels, although a handful of vertical cross-sections exist (v.s.).

Research workers have either taken the 'horizontal view' or the 'vertical view' and side-stepped the problems of representing the three-dimensional reality. This can intro-duce problems of interpretation. It is known, for example, that pipes exist at at least two levels in parts of the Maesn-ant percoline network. Where they happen to overlie each other, in general only the uppermost flowing system will have been mapped, because as a rule pipes have been traced by listening for the sound of flowing water (Jones, 1975, 277; 1978a, 8) or by injecting dye through surface collapse features (Jones and Crane, 1979). Where they do not overlie each other it is possible that systems at two different levels may be mapped as though they were one. Sometimes they may, in fact, be linked and sometimes not. Perhaps we can never hope to find all the linkages.

Jones (1975, 70-73; 1978a, 9-12) considered the karst analogue in terms of an analogy between the controversy over phreatic or vadose origins for caves and that between hyd-raulic boiling and desiccation cracking for pipes. He considered that, although the division between these two environments may fluctuate with individual storms, the dist-inction could be meaningful both in terms of process of formation and of sources of pipeflow. Tubular cross-section and occasional reverse gradients may indicate a phreatic origin (Davis, 1930) suggesting pipe initiation by the clas-sic process of boiling, whereas meandering, irregular long profiles and poor integration of levels between adjacent pipes may indicate a vadose origin using cracks and biotic courses. It is possible that networks due to both origins may overlie each other in the soil profile. Observations on the Maesnant seepage line suggest that arterial perennial/seasonal pipes are probably due to phreatic water using weak-nesses created by subsidence in the seepage line at depths of 50 cm or so, whereas shallow, ephemeral pipes at 15 cm depth follow desiccation cracking. It has also been sugg-ested that pipe networks on the upper slopes may more commonly be due to desiccation, whilst hydraulic boiling is more important nearer the stream, with mass movement and desic-cation sharing dominance on mid-slope (Jones, 1979).

Finally, mention must be made of the suggestion that pipe networks may be nascent stream channel networks. There is abundant evidence that gully development follows lines of piping in many parts of the world (cf. section 8.1), e.g. Downes (1946), Heede (1971), Stocking (1978a), Morgan (1979). It is also apparent that some perennial pipes and features described as 'subsurface streams', for example by Newson (1976a), may easily convert into surface channels as a result of roof collapse. Jones (1971; 1975) saw an evolutionary sequence in which pipes in the most favoured positions de-velop by competition into master pipes or arterial pipes, capturing drainage from others and eroding their beds to place themselves in ever more favourable positions. Evid-ence adduced from the spacing of these arterial pipes tends to support the view that they may form foci for open channel extension on Burbage Brook, although at present little development seems to be taking place.

In the San Pedro Valley, Arizona, where there is active development (N.O. Jones, 1968, 152), the evidence suggests that there is little change in the plan of the network prior to roof collapse, but that roof-collapse initially produces complex, angular gully patterns reflecting the development of pipes along fracture patterns. Only later do gullies derived from pipes acquire the normal branching form. Heede (1974) produced other evidence of similar import in Colorado. He analysed the network structure and hydraulic geometry of the gully network at Alkali Creek, which is at least partly developed from piping, and he noted an excessive, 67% of the area of a fourth-order basin drained by first-order gullies, as opposed to an average of only 1% quoted by Leopold *et al*. (1964) for the United States as a whole. He also noted that the ratio of maximum to mean depth of gullies of 2.0 ('the shape factor') is was relatively high, indicating hydraulically inefficient channels. These features, he said, suggest that the network is at a 'youthful' stage, perhaps (we might add) still exhibiting some of the characteristics of the tributary pipe network.

Arnett and Conacher (1973) have also inferred that throughflow is important in *valley* initiation in the coastal area of Queensland and describe collapsed tunnels in seepage lines in valley bottoms. However, they conclude that in general subsurface removal is not sufficient to cause pipes in the area (*ibid*, 242). The situation seems to be different in the Sudan, where Berry (1970) observed piping in linear subsurface flow zones uphill of gully heads and suggests that 'nearly every river valley in the area' results from this form of erosion. This topic will be continued under the 'relationships between piping, rilling and gullying' in section 8.1.

There is therefore considerable variation in network form and relationships to stream channels. The largest and most highly developed networks studied closely resemble stream channel networks, whilst the smallest tend to be discontinuous and more clearly reflect the influence of varied initiating factors. So far, the information on network evolution is meagre. It is tempting to envisage gradual integration of active networks, conceivably by both upslope and dowslope extension, but little more than simple headward extension of roof collapse in integrated networks has so far been observed. Moreover, Drew (1972, 208) has expressed the opinion that the large scale networks of semi-arid regions are not likely to be genetically similar to the 'micropipe' networks of humid regions. This may be an overstatement, but it underlines the danger of basing a theory of evolution on the scraps of information from different environments as yet available.

7.   THE HYDROLOGICAL ROLE OF SOIL PIPING.

Most research into the hydrological function of soil pipes has to date been performed in Britain, with the exception of Morgan (1972) in Malaysia, Bryan *et al*. (1978) in Canada and Yair *et al*.   (1980) in Israel.   From the very limited point sampling of pipe discharge by Jones in 1968 and Weyman in 1969 (Weyman, 1971; Jones, 1975) in the Burbage and East Twins catchments respectively, fuller sampling schemes were set up by Stagg in 1970   in the latter catchment (Stagg, 1974) and by the Institute of Hydrology in 1971 on the Nant Gerig research catchment (Gilman, 1971 a and b; Pond, 1971; Newson, 1976b, Newson and Harrison, 1978; Gilman and Newson, 1980).   The first autographic records were obtained for 20 storms on a single pipe on the Maesnant catchment, Plynlimon, over a six month period in 1975 (Jones, 1975; 1978a) and the most extensive autographic data set available to date was obtained for 9 storms on a number of pipes (27 hydrographs) in the Nant Gerig basin during 1976/77 (Gilman and Newson, 1980).   A current research project on the Maesnant is designed to obtain the first comprehensive autographic records of pipeflow, overland flow, seepage and stream flow (Jones and Crane, 1979).

As a result of this and other British research, the hydrological role of soil pipes has been considered in a number of recent hydrological textbooks.   Gregory and Walling (1973, 286) supported the view of Jones (1971) that pipes must contribute significantly to water flow in the soil and hence to stream discharge and that the conceptual model of a drainage basin system should be revised to include pipe and pseudopipe networks.   Writing on stream networks in 1976, Gregory emphasised the dynamic nature of drainage networks and the complication introduced by the identification of subsurface flow networks and pipes, which he sees as 'a further manifestation of the scale problem'.   Weyman (1975, 12) says that it is now generally recognised that these routes may be a very widespread phenomenon and that they respond very rapidly to rainfall, probably carrying the greater part of storm runoff.   He also noted that pipeflow velocities of around $0.1 ms^{-1}$ are nearer to overland flow than throughflow and that the hydrological importance of piping will therefore depend more on the frequency and network characteristics of the pipe systems.   Ward (1975, 255-256) noted that evidence of substantial pipeflow adds support to the proposition of Hewlett that water passing through the soil is the major source of streamflow.   And Rodda, Downing and Law (1976) state that permanent and intermittent pipes are the obvious source of stream baseflow on Plynlimon and may cause minor floods after moderate rain, when the ephemeral pipes are not operating.   Again, clearly based on the Plynlimon experience, they believe that most pipes are ephemeral and suffer from large losses to storage, so that the water may not reach the base of the slope and therefore fail to contribute to stream flood water.   It must be added here that this idea of large losses to storage may be due in part to the use of artificial

flows generated by pump to study pipeflow when the surrounding soil was unsaturated (cf. Newson and Harrison, 1978),
which is probably atypical for natural storm events.

In summarising a British Geomorphological Research Group
Symposium on current and future problems in geomorphology,
Embleton, Brunsden and Jones (1978) put piping briefly in
hydrological context, particularly in respect of dynamic
network size and sources of solutes (*ibid,* 47 and 55).

In the first text specifically on 'hillslope hydrology'
(Kirkby, 1978), the role of soil piping is considered by a
number of authors. Chorley includes pipeflow and pipe storage components tapping saturated throughflow in his illustration of the hillslope hydrological cycle (Fig. 30) (*ibid,*
5) and quotes the work of the Institute of Hydrology to
support the view that pipeflow can provide an important element in stream runoff (*ibid,* 24). However, the absence of
any explicit reference to pipeflow in Chorley's section on
'models of hillslope hydrology' highlights a continuing gap
in the range of current models. The first steps towards
filling this gap have now been taken by Gilman and Newson
(1980) (*v.i.*). Atkinson contrasts matrix throughflow and
pipeflow and suggests that pipeflow is turbulent whereas
matrix flow is laminar (*ibid.,* 73-77) (cp. section 6.1(iii)).
He says that there are almost no measurements of pipeflow
velocity and none at all of pipe roughness, so that the
Darcy-Weisbach equation cannot be applied and direct measurements are needed, and he discusses methods of measuring pipeflow, based primarily on the work of the Institute and
observations in the East Twins basin (*ibid.,* 88-92) (v.i.).
Finally, Whipkey and Kirkby (*ibid.,* 132-133) note that pipeflow can result in a quick response from otherwise slowly
permeable material. They rate the components of hillslope
discharge in order of rapidity of flow as Hortonian overland
flow, saturation overland flow, saturated subsurface flow,
return flow, unsaturated subsurface flow and groundwater
flow (*ibid.,* 131). They consider return flow to be less
important than saturation overland flow derived mainly from
rainfall on a wet surface. Saturated subsurface flow is
usually slow unless large pipes are present, but it may
contribute to hydrograph peaks especially if flow in the
other categories is small. Indeed, they note that there is
'a bewildering number of possibilities for the dominant
processes': at moderate storm intensities saturated subsurface flow is often important whereas in extreme events infiltration-excess overland flow may dominate, whence many of
the difficulties of creating an all-embracing model. It
should be added that piped return flow, as described by
Newson (1976b) and others, might alter the relative rating
they give to return flow, and that if piping is present,
then it would appear that saturated subsurface flow in the
form of pipeflow has every likelihood of contributing to
peaks in the stream hydrograph in a positive way rather than
depending upon the demise of other sources.

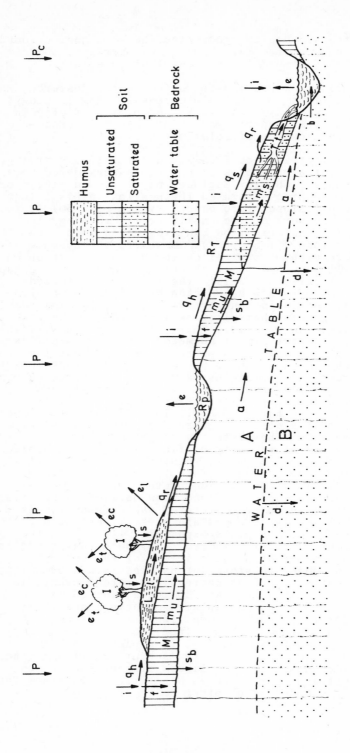

Key to symbols (on page 243)

A

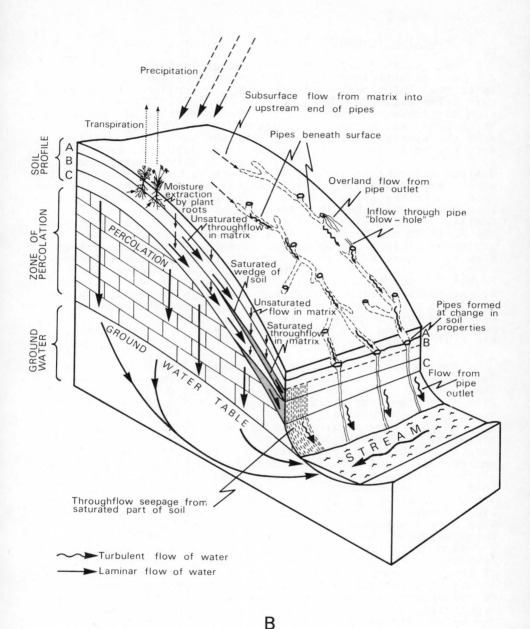

Precipitation

Subsurface flow from matrix into upstream end of pipes

Transpiration

Pipes beneath surface

SOIL PROFILE
A
B
C

Overland flow from pipe outlet

Moisture extraction by plant roots

Inflow through pipe "blow – hole"

Unsaturated throughflow in matrix

ZONE OF PERCOLATION

PERCOLATION

Saturated wedge of soil

Unsaturated flow in matrix

Pipes formed at change in soil properties

A
B

GROUND WATER

GROUND

Saturated throughflow in matrix

C

Flow from pipe outlet

WATER TABLE

STREAM

Throughflow seepage from saturated part of soil

~~➤ Turbulent flow of water
——➤ Laminar flow of water

B

Fig. 30   Two views of the hydrological function of soil
          pipes: Chorley (1978a, fig. 1.2) and Atkinson
          (1978, fig. 3.1).

7.1. Pipeflow in humid regions

Amongst the first published suggestions that pipeflow may be a significant source of stream runoff, Weyman's (1970) inference that pipeflow was responsible for flood peaks in the East Twins basin was criticised by Freeze (1974, 632) as 'far from convincing'. Weyman had isolated a section of the stream by an upper and a lower weir and monitored throughflow input from the side-slopes. He found that the measured contributions were too small to account for the flood peak and, in the words of R.A. Freeze 'the author had to plead more favorable conditions external to his measured reach to obtain a water balance'. Weyman actually suggested that pipeflow, which has been observed in the upper catchment, and channel extension in hollows may be the cause, but Freeze does not appear to have been familiar with the implications. Indeed, Freeze argued that on the basis of simulations with a mathematical model 'there are stringent limitations on the occurrence of subsurface storm flow as a quantitatively significant runoff component'. However, it would appear that the limitations lie more with Freeze's model than in reality. Quite apart from the limitations of the model (Freeze, (1972, a and b) highlighted by Hewlett (1974), pipeflow was not included. Oddly enough, Weyman's (1970) reference to the probable role of piping in his basin was also recently ignored in a summary of his work by Beven, Iredale and Kew (1977, 2-3).

Weyman actually discovered pipeflow in the upper East Twins basin in collaboration with Stagg (Weyman, 1971, 175; Stagg, 1974). They measured velocities of 0.3 ms^{-1} compared with 1 cm min^{-1} for matrix throughflow and 0.05 ms^{-1} for overland flow. Discharges were over 1 l s^{-1} for the largest pipe at peak flow. Weyman (1971) decided that 'subsurface stormflow is probably only important in non-capillary systems (pipes). Otherwise, throughflow is likely to be an important recession and baseflow process'. In some cases surface water was found upslope of active pipes and they inferred that the pipes conducted water from the 'plateau' slopes across the mid-slopes to the hollows, which form extensive linear zones of saturation overland flow during storms. From direct observations taken over a three week period in November, 1969, Weyman (1971, 183) concluded that although a good correlation existed between discharge and extended 'channel length' along hollows (r = 0.93), the substitution of length of surface water for the dynamic contributing area should be made cautiously in view of the probable importance of subsurface routes. However, although he could see the possibility that soil pipes might generate runoff by collecting water, Weyman (1971, 217) saw the East Twins pipes as a 'runoff maintenance system' rather than a 'runoff generating system' in so far as they merely redirected and transmitted water already concentrated but not linked to the channel system.

Stagg (1974) obtained discharge data by bucket and stop-watch for 5 storms on a single pipe about 4cm in diameter in the A horizon of a peaty podzol, although without access to

autographic equipment it is not certain that peak discharge
was actually measured and no complete pipe hydrograph was
obtained (*ibid.*, table 8). Pipeflow appeared to occur later
than matrix throughflow and was particularly high when saturation overland flow was prevalent. This suggested to Stagg
that the pipes were transmitting overland flow and probably
some water from the surface layers of the saturated peat.
Using data kindly supplied by Stagg, Jones (1975, 263) was
able to show that the hydraulic conductivity of the peat was
insufficient to create even the lowest discharge measured
in one of Stagg's pipes, given its length and size, and
that the source of pipe discharge must be other than by diffuse effluent seepage, as, indeed, was indicated by Weyman's
and Stagg's conclusions. A composite recession curve based
on the data from Stagg's storms suggested a steeper decay
than the familiar exponential time relationship (Jones, 1975,
263; 265). Since Stagg did not find the pipeflow confined
by pipe capacity, the convex exponential recession curve may
be related to the low storage capacity in the pipeflow supply
system.

Stagg also used the Chezy formula to calculate maximum
possible pipeflow for 9 surveyed pipes (*ibid.*, App. H). This
suggested capacity flows of 0.62 to 1.39 $1 s^{-1}$ compared with
a maximum observed discharge of 0.75 $1 s^{-1}$ in one pipe. The
calculated capacity of the observed pipe was 1.24 $1 s^{-1}$
(*ibid.*, table 8) and the discharge measurement was taken 2h
after peak rainfall intensity in a storm of 6.75 mm in moderately wet antecedent conditions (36 mm 5-day antecedent
precipitation). Stagg's theoretical calculations therefore
appear to give quite a reasonable estimate for these pipes.

Finlayson (1976; 1977) sampled four pipes developed in
brown earths in the lower half of the East Twins basin, in
the reach where Weyman (1970) sampled diffuse throughflow,
two of which were perennial-flowing, the others ephemeral,
together with a couple of ephemeral pipes developed in the
podzols of the upper basin. Frequent point samples taken
during 1972-73 produced discharge patterns which, according
to Finlayson (1977, 13), were 'not consistent in the long
term because of changes in the pipe network initiated presumably by animal activity and collapse'. However, Finlayson's evidence rests on no more than 57 point estimates of
discharge for the longest of his pipe records covering the
whole of the water year 1972-73 (*ibid.*, fig. 4); his conclusions are therefore limited by the lack of a continuous
record for any single storm or of a study of the pipe network. Reading from Finlayson's graphs suggests that the
perennial pipes reached high flow levels of c.110 and 25 ml
s^{-1}, whereas the ephemeral pipes reached 10 and 35 ml s^{-1}
in the lower basin. His throughflow trough nearby recorded
no more than 5 ml s^{-1}. In contrast, discharges of over 200
ml s^{-1} were recorded for the shallow ephemeral pipes in the
podzols of the upper basin (*ibid.*, fig.8). These he regarded
as part of the direct runoff network, whereas the pipes of
the lower basin he regarded as indicating 'the behaviour of
baseflow'. No details are given of the size and vertical

location of the pipes. However, Finlayson's measurements
of dissolved solids load in these pipes are discussed in
section 8.4.

Waylen (1976) studied the hydrochemistry of the same
four pipes in the East Twins brown earths, together with
three in the podzols, as part of a general study of solute
sources and throughput in the basin (cf. section 8.4). Weekly
measurements of discharge were taken by stopwatch and cylinder during 1972-73. These revealed values (as read from
his graphs of solute concentrations) in the ranges 0.0025-
0.120 1 s^{-1} and 0.004-0.050 1 s^{-1} for the two perennial,
brown earth pipes labelled 4A and 4D respectively, but sample
sizes were too small for plotting the data for the ephemeral
pipes (cf. section 8.4). The perennial pipes flowed at
depths of 10 cm in 'vegetation stripes', where throughflow
and overland flow are intercepted by slump cracks, and, he
suggested, directed into animal burrows. The response from
these pipes was very similar to the general response in
his throughflow troughs and together pipeflow and throughflow are responsible for a less steep falling limb at the
lower weir compared with the upper weir (*ibid*, 107). Of
course, such statements must relate to *weekly* response only,
but it is interesting that the two responses may be similar
and to compare the pipe discharges with a maximum of 0.003
1 s^{-1} measured for matrix throughflow by Weyman (1970, 27)
from a 1 m broad strip of brown earth B horizon.

Knapp (1970a and b; 1974) discovered piping in the Upper
Wye catchment, Plynlimon, whilst studying throughflow processes in one of the Institute of Hydrology's experimental
basins. Although his work was not primarily directed towards
the study of pipeflow, spot checks on pipe discharge during
stormflow indicated flows of 40-50 1 min^{-1} (0.67 - 0.83 1
s^{-1}), in 10 cm diameter trunk pipes (Knapp, 1970a, 158). He
observed that these larger pipes had few tributaries as they
crossed the middle slopes and concluded that their discharge
was derived from higher up the slope, because the Hiraethog
soils were deep and rarely saturated. There were no pipe
outlets in the streambanks and outflow occurred through the
surface of valley bottom bogs to the stream (*ibid.*, 170).
In discussing problems of observing and modelling hillslope
throughflow, Knapp (1974, 27) noted that the trunk pipes
tended to occur in slight depressions across the middle
slopes and that overland flow and pipeflow occurred together
in these broad concavities, carrying water to the streams
much faster than matrix throughflow or interpedal void flow.
Drainage hydrographs were therefore shorter in piped areas
because of the quicker drainage. He suggested that pipes
and associated overland flow must therefore be accounted
for in any model of hillslope hydrology, although the irregul-
are distribution of pipes makes this difficult.

The first instrument capable of producing an instantaneous hydrograph of pipeflow was produced for the initial
experiment at the Institute of Hydrology in 1971-72. As
described in the internal reports of the Institute (Gilman,

1971a and b) and more recently by Gregory and Walling (1973, 118), Gilman (1977), Atkinson (1978, 89-92) and Gilman and Newson (1980, fig. 33), the instrument consisted of a siphoning tank for low flows and a Braystoke propeller meter for flows above 0.3 1 s^{-1}. The propeller was set upstream of a U-bend in the flexible ducting to maintain a full tube as pipeflow was directed past the meter (Fig. 31). Gilman (1971a, fig. A5) presented a calibration curve up to 3.5 1 s^{-1} for this instrument. Unfortunately, attempts to link this hybrid instrument with an event recorder failed, and it was only operated when observers were present to produce 15 minute counts (Gilman, 1971a; Gilman and Newson, 1980). Gilman monitored 5-6 ephemeral pipe outflows together with overland flow and channel flow in 3 experiments in 1971 on a side slope of the Nant Gerig, Plynlimon. He observed that the pipeflows tended to rise swiftly from zero to peak flow in 2h despite the gradual onset of rain, and he concluded that flow only occurs once a certain storage has been filled. Recession to zero flow took 8-10h in side-slope pipes, but took longer in the valley or the plateau peat. Gilman (1971, 13) drew attention to the very quick and sensitive response to rainfall and suggested that this may indicate that secondary sheet runoff is draining directly into pipes or via a 'displaced storage' system with water entering the storage directly and in quantity elsewhere and displacing water from that storage into the pipe system (cp. the conclusions of Weyman and Stagg).

Gilman also tabulated total pipeflow as a percentage of rainfall upslope during the three storms and obtained values varying from 4% to 72% (*ibid.*, table 1). In one case, he noted that a low percentage could be related to large amounts of surface flow in that storm. The mean of his 12 observations was 34%, and he pointed out that a highly significant portion of the available rainfall was carried by the pipes.

Apparently despite the results of Gilman's pilot experiments, the Institute of Hydrology's (1973, 20) official report of research suggested that the drainage of the Upper Wye catchment (including Nant Gerig) was 'a classic example of Hortonian overland flow, the water entering the soil profile being removed only by evaporation and plant transpiration'. However, the report goes on to say that overland flow feeds networks of ephemeral pipes and ephemeral channels and that this flow 'probably produces most of the flood peak discharge of the main channels'.

One of the few references to the hydrological function of soil pipes in humid areas outside Britian yet available was made by Morgan (1972), whose study of the factors affecting the behaviour of a first-order stream in Malaysia suggested that throughflow from perched water tables and soil piping control the drainage system. He estimated that only 5% of the catchment area was naturally susceptible to overland flow. And he used the poor correlation between antecedent rainfall and streamflow to argue that streamflow was not generated by a simple rise and fall of the water table.

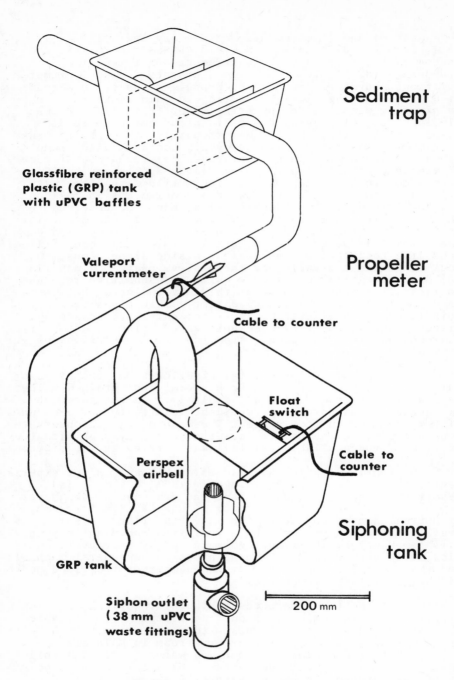

Sediment
trap

Glassfibre reinforced
plastic (GRP) tank
with uPVC baffles

Propeller
meter

Valeport
currentmeter

Cable to counter

Float
switch

Cable to
counter

Perspex
airbell

Siphoning
tank

GRP tank

Siphon outlet
(38 mm uPVC
waste fittings)

200 mm

Fig. 31   Instruments for measuring pipeflow: I. Institute of
          Hydrology combined siphon and flowmeter (Gilman,
          1972, 1977; Gilman and Newson, 1980).

162

Unfortunately, no direct measurements of pipeflow were reported.

Jones (1975; 1978a and c) obtained 20 storm hydrographs
for a single seasonal pipe in the Maesnant valley, Plynlimon,
using a 10-way flow splitter, 80ml tipping bucket and Rustrak multi-channel event recorder (Fig. 32). Rainfall was
recorded on the same time base from an Artech tipping-bucket
gauge. Like Gilman, Jones also observed a very quick and
sensitive response to rainfall, although in his pipe baseflow continued until the summer dry period (Fig. 33). The
mean delays of one and three-quarter hours between first
rain and first rise in the hydrograph and just under 4 hours
peak to peak were far too short to have been generated by
diffuse seepage through the peaty A horizon, which had a
mean saturated hydraulic conductivity of 0.0094 mms^{-1},
without the peat being saturated to allow piston flow, which
was generally not the case. Jones (1975, table 7.1.1.)
obtained a correlation matrix comprising antecedent precipitation for various periods between half a day and one
week, measures of storm rainfall (total rainfall, total
prior to peak pipe discharge and mean rainfall intensity),
baseflow discharge, and parameters of the pipe and stream
hydrographs (peak discharge, total storm discharge, and lag
times between the beginning of the storm and first rise in
the hydrograph and between hyetograph and hydrograph peaks).

Essentially, the statistical analyses indicated that
there was a close correlation between certain pipe response
characteristics and stream response (Table 12), for example,
$r = 0.80$ ($\alpha = 0.01$) between peak lag times, and $r = 0.69$
($\alpha = 0.01$) between magnitudes of peak discharge. Missing
data caused by instrument failures reduced the number of
cases for comparison of response times and total discharges
below the minimum needed for meaningful parametric analysis.
However, nonparametric correlation indicates a significant
relationship between total storm discharges ($r_S = 0.86$,
$\alpha = 0.05$), but not in terms of response time, i.e. the lag
between the beginning of the rainstorm and the first rise
in the hydrograph.

From an analysis of the correlation matrix, Jones
concluded that :- (1) Baseflow levels in the pipe immediately
prior to a storm were the most significantly correlated
factor explaining variations in the speed of initial pipeflow
response ($r = -0.61$, $\alpha = 0.05$). This suggested that pipe
response is largely due to a rise in the phreatic surface,
and this seems to be corroborated by the pattern of correlation with antecedent rainfall which shows a rise in the level
of correlation to reach a 5% level of significance only at
4 days ($r = -0.55$), suggesting that the importance of antecedent rainfall to pipeflow response time may lie more
generally in determining phreatic levels than in determining
the moisture content of the surface and overlayer.
(2) Lag times between rainfall and pipeflow peaks were not
significantly correlated with any single factor, although
the closest correlation was with rainfall intensity

II.

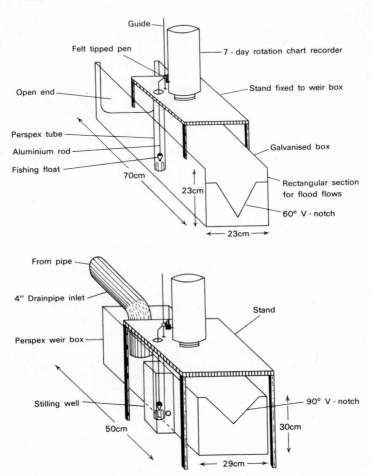

Guide

Felt tipped pen

7 - day rotation chart recorder

Open end

Stand fixed to weir box

Perspex tube

Aluminium rod

Fishing float

Galvanised box

70cm

23cm

Rectangular section for flood flows

60° V - notch

23cm

From pipe

4" Drainpipe inlet

Stand

Perspex weir box

Stilling well

90° V - notch

50cm

30cm

29cm

Fig. 32   Instruments for measuring pipeflow (see also Fig. 31):   II.   Pipeflow weirs used by Jones and Crane (1979). Jones and Crane have since used sheet metal weir plates alone.   III. (top of opposite page) Tipping bucket system used by Jones (1975; 1978a and c).

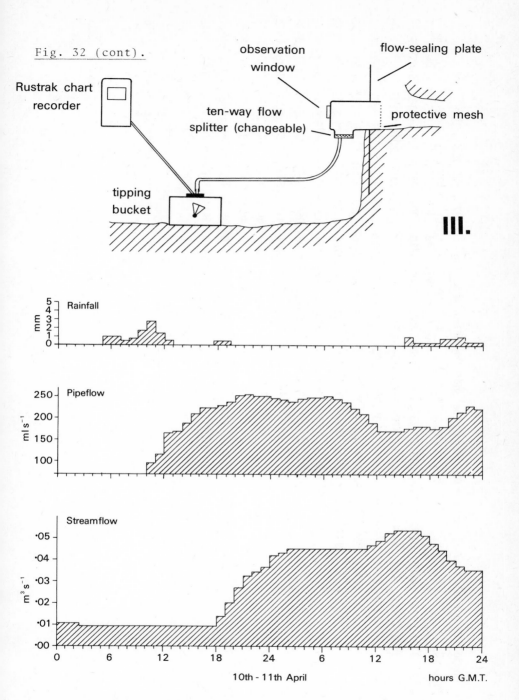

Fig. 32 (cont).

Rustrak chart recorder

observation window

flow-sealing plate

ten-way flow splitter (changeable)

protective mesh

tipping bucket

III.

Fig. 33 Hyetograph and hydrographs for pipe and stream on Maesnant (Jones, 1978c, fig. 8).

165

Table 12. Composite correlation matrix for Maesnant pipe and streamflow data. (Jones, 1975, tab 7.1.1.)

	Q_{pk}	Q_{tot}	Q_b	P_i	P_{tot}	P_t	$A_{\frac{1}{2}}$	A_1	A_2	A_3	A_4	A_7	R_{lag}	P_{lag}	R'_{lag}	P'_{lag}	D	A_{i7}	Q'_{pk}	Q'_{tot}
Q_{pk}		.5737	.4121	.6248	.4685	.5446	.2351	.2045	.2073	.2963	.3826	.4047	.2513	-.0770	-.0315	-.1163	.3132	-	.6905	.7958
Q_{tot}			-.1151	.3435	.4541	.5424	-.0992	-.0862	-.0478	-.0306	.1364	.3154	-	-	-.1309	.2016	.0567	-	-	-
Q_b				-.3320	-.0838	-.0740	.4532	.3281	.3113	.3713	.3989	.4840	-.6081	-.2372	-.6585	-.9585	.5485	.3831	.0009	-.3753
P_i					.7182	.7357	.1188	-.0026	.0821	.1991	.2985	.0135	-.3363	-.3900	-.4880	-.4250	.0147	.0162	.4869	.4271
P_{tot}						.9849	-.2928	-.4218	-.4099	-.2785	-.2864	-.4509	-.2520	-.1186	-.5453	-.3434	.4432	-.5132	.1939	.4632
P_t							-.3002	-.4000	-.3892	-.2575	-.2567	-.3731	.2313	-.0434	-.4855	-.3100	.3587	-.4652	.1667	.4447
$A_{\frac{1}{2}}$								.8604	.6935	.6778	.6513	.3226	-.4223	-.2706	-.0643	.0180	-.3766	.6955	-.3739	-.0948
A_1									.8535	.7667	.7096	.3809	-.3035	-.0401	.4103	.5084	-.4557	.8030	.3438	.0644
A_2										.9490	.8965	.5457	-.3606	-.0869	.6089	.4276	-.2711	.9205	.5990	.2688
A_3											.9311	.5068	-.4021	-.2413	.4665	.2419	-.1697	.8721	.4870	.2369
A_4												.6758	-.5498	-.3318	.3529	.1725	-.1665	.8981	.6137	.0049
A_7													-.3312	.1740	.3414	.1020	-.3224	.7781	.4641	-
R_{lag}														.7077	-	-.6066	.4879	-.3804	-	-
P_{lag}															-	-.8735	-.1344	-	-.0210	-
R'_{lag}																.7989	.0486	.6165	.1616	.1716
P'_{lag}																	-	-	-	-
D																		-.3590	.2469	.3163
A_{i7}																			.6108	.1833
Q'_{pk}																				.7444
Q'_{tot}																				

(r = -0.39). This and the previous results for initial
response times suggested that transmission of the storm peak
may be due to a combination of seepage supply with 'crack-
age supply', i.e. relatively quick transmission of fresh
rainwater via cracks and pipe inlets; a view corroborated
by analyses of pipeflow chemistry by Cryer (1980) (cf.
section 8.4). The relative dominance of these sources may
vary between storm events.
(3) Total storm discharge in the pipe was best correlated
with total rainfall prior to peak discharge (r = 0.54,
α = 0.05) and antecedent rainfall was not significant
(although correlation improved monotonically up to α = 0.32
at one week).
(4) Peak discharge was best related to mean rainfall intensity (r = 0.68, α = 0.05) and the rainfall amount prior to
peak discharge (r = 0.61, α = 0.05). Baseflow prior to
the storm and weekly antecedent rainfall had the next highest
correlations, though not significant at α = 0.05 (r = 0.4).
Correlation with shorter term antecedent rainfall showed
some improvement if the effect of antecedent moisture was
regarded as exponential, but this was still not statistically
significant although theoretically likely (e.g. r = 0.19
raised to r = 0.35 for log one-day antecedent rainfall in
Jones, 1978a, 16).

Finally, comparison of the most significant factors
in multiple regression equations for pipe and streamflow
parameters revealed close similarity in terms of the quantities of both total 'direct runoff' and maximum discharges.
For both pipe and stream the two most important factors
were total precipitation and weekly antecedent rainfall for
total storm discharge, and precipitation intensity and weekly
antecedent rainfall for peak discharge. However, different
factors appeared most important in terms of timing, initial
response in the pipe being best explained by a combination
of prior baseflow levels and two-day antecedent rainfall,
whereas weekly antecedent rainfall and precipitation intensity took marked precedence in the stream. Similarly,
prior baseflow and one-day antecedent rainfall dominated
peak lag times in the pipe, whereas three and seven-day
antecedent rainfall dominated in the stream (Jones, 1978c,
tab. III). The author emphasised the tentative nature of
these results, not only in terms of significance levels,
but also because of the likely variability of pipe response
on a catchment-wide basis. Nevertheless, some exciting and
cogent possibilities seem to have emerged.

Jones's results were very similar to those of Weyman,
Stagg and Gilman, despite the fact that his pipe had a
seasonal regime whereas the others studied pipes that receded
to zero flow after storms. The basic difference was that
the response of the seasonal pipe was superimposed on a background baseflow. An important similarity was that stormflow
was clearly generated by a quick influx of rainwater into
the pipe system by some means other than diffuse percolation
through the upper soil layers.

Cryer (1980) believes that Jones would have found even greater similarity between pipeflow and streamflow in the larger pipes on Maesnant, and in fact this seems to be substantiated by recent monitoring of these perennial pipes (Jones and Crane, 1979). Cryer also professed to be 'unconvinced' by Jones's use of 'advanced statistics' on this relatively small sample. However, there seems to be reasonable justification for the analyses bearing in mind i) that most of Jones's conclusions were based on a simple correlation analysis and that multiple regression was used only as an exploratory tool, and ii) that these were the only data available on which to base a quantitative exploration of the factors behind pipeflow. The limitations of Jones's experiment should perhaps be sought elsewhere, namely in the obvious need for more significant investment in instrumentation to study spatial variations in pipe response, as, indeed, was noted by Jones (1978a, 18) and is now being rectified (Jones and Crane, 1979).

Dovey (1976) undertook an independent analysis using Jones's data. He attempted to estimate the total contribution of pipeflow to stream discharge by multiplying storm discharge in the monitored pipe by a factor expressing the ratio of the likely piped area in the catchment (derived from soil maps, field observations and air photography) to the supposed drainage area of the gauged pipe. Estimates made on this basis for 7 complete storm records suggest that between 19% and 54% of the total stream response, including delayed flow, could be attributed to pipeflow, with a mean of 35% (Table 13). *These estimates must, however, be regarded as very approximate, since current research in this catchment has shown that, on the one hand, the catchment area of this pipe is much more extensive than previously supposed, and on the other, to counter-balance this, there is not one but three or more outlets draining this area interconnected by anastomosing links, the exact number depending on the severity of the storm (Jones and Crane, 1979). Probably, Jones's (1979) crude estimate that pipeflow could account for 25% of *direct* stream runoff, based on the observation that flow in the monitored pipe accounted for 1% of stream discharge and that there were about 25 trunk pipes feeding the stream, is as good as any using the data.

Nevertheless, speculating on the basis of Dovey's estimating method, Jones (1978a, 17) observed that the percentage contribution from pipeflow was generally greater in lighter storms (Fig. 34) and tentatively suggested that alternative routes, such as overland flow, may become relatively more important in heavier storms. Carrying the speculation even further, Jones (1978c) drew up drainage pathway

---

* Dovey calculated percentages on the basis of hydrographs separated by extrapolating an exponential decay from the preceding peak, whereas Jones considered only the area of the hydrograph above a line delimiting the rise in the hydrograph (cf. Linsley, Kohler and Paulhus, 1949, 401, fig. 15.7, method 'AB'). 'Delayed flow' in the diagram is the difference between these two definitions.

Table 13. Pipe and stream response on Maesnant after Dovey (1976, tabs. 5 and 6).

Rainfall $\div 10^3$ l	Stream discharge $\div 10^3$ l	Percent runoff	Pipe discharge $\div 10^3$ l	Percent contribution to stream	Pipeflow as percentage of rainfall (i)*	(ii)*
13068.0	761.9	5.83	144.65	18.99	1.70	1.10
9612.0	2917.0	30.35	584.53	20.04	9.35	6.08
5076.0	3876.0	76.36	1297.43	33.47	39.29	25.56
1080.0	910.0	84.26	417.12	45.84	59.37	38.62
3628.8	1740.0	47.95	947.45	54.45	40.13	26.11
3780.0	1943.0	51.40	805.80	41.47	32.77	21.32
7506.0	3259.0	43.42	1219.58	37.42	24.97	16.25

* (i) percentage of estimated rainfall within catchment of measured pipe

  (ii) percentage of estimated total basin rainfall accounted for by estimated total pipe discharge (added by this reviewer).

diagrams to illustrate the volume of flow via different routes as a percentage of storm rainfall. These showed a considerable variety (Fig. 35). On average 19% of rainfall passed through the pipes, but this fell to 2% in dry weather even under heavy rain. A maximum of 26% of the rainfall was estimated to have passed through the pipe system, occurring when the catchment was relatively wet prior to the storm, although the rainstorms themselves were not very heavy (cp. Fig. 35). Maximum pipeflow contributions to streamflow also occurred under the same circumstances. In the examples available, the maximum contribution from pipeflow (54%) to overall streamflow appears to have resulted from a heavy contribution from pipeflow to delayed flow (84%), but in the case of maximum contribution to the stream floodflow (40%) the amount of delayed flow was much smaller (Fig. 35). Perhaps significantly, the case of maximum pipe contribution to flood peak also happens to have the least rainfall unaccounted for, i.e. presumed lost to baseflow, leakages and deeper stores.

Wilson (1977) studied pipeflow and contributing areas in the Nant Cwmllwch basin, Brecon Beacons, using a trench pump to generate artificial flow in ephemeral pipes and dye tracing with pyranine and rhodamine WT. He delimited the probable area of pipeflow contribution by interpretation of air photos. The pipes begin 130 m from the streambank in a seepage area which appears to act as a 'collecting area', perhaps a subsurface counterpart of the surface collecting areas described by Downes (1946) in Australia. The pipes were about 6 cm in diameter and maximum measured discharge

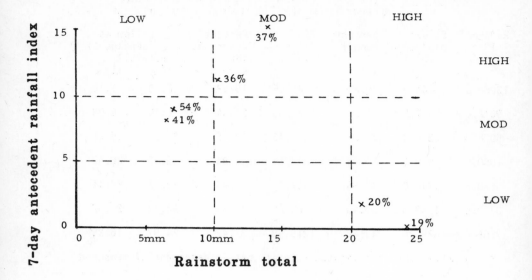

<u>Fig. 34</u> Percentage total pipeflow contribution in terms
of rainstorm severity and weekly antecedent rainfall on Maesnant (Jones, 1978c, fig. 12).

was 1.5 ls^{-1} in one pipe, giving an estimated 4.5 ls^{-1} maximum from the 3 outlets on his study slope. Mean velocity
was 0.8 ms^{-1} in the main pipe system and 0.2 ms^{-1} in an
auxilliary system. The storage capacity of the pipe system
was small, 1.03 m^3 (pump estimate) or 0.664 m^3 (dye estimate)
in a 363 m long pipe, and hence concentration times were
short (Table 14). Wilson observed that the upper part of
the network was very efficient at choosing alternative routes
when full and overland flow occurred when pipe capacity was
exceeded, re-entering the system via pipe inlets lower downslope.

     However, Wilson concluded that although the pipes do
extend the contributing area for the stream, they are only
very localised in his catchment and that the measured flows
were not a significant contribution to streamflow, even if
he had undermeasured by an order of magnitude. It is perhaps
worth adding that Wilson also observed covered channels in
peat areas in the catchment but regarded these as distinct
features, and he concentrated on the ephemeral pipe system.

     Meanwhile, continued research in the Institute of
Hydrology's Upper Wye experimental catchment has tended to
show that even purely ephemeral networks are capable of
significant contributions (Morgan, 1977; Newson and Harrison,
1978; Institute of Hydrology, 1978; Gilman and Newson, 1980).

170

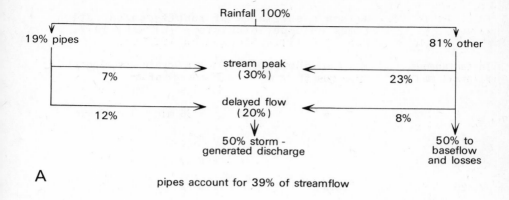

A

pipes account for 39% of streamflow

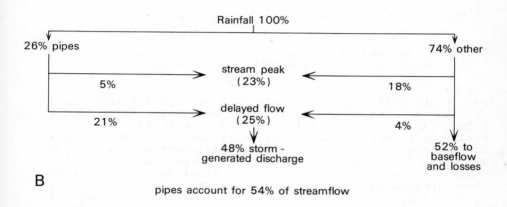

B

pipes account for 54% of streamflow

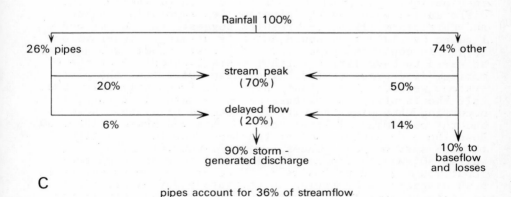

C

pipes account for 36% of streamflow

Fig. 35   Drainage pathways on Maesnant.   (Jones, 1978c, figs.11 and 14).

Table 14. Wilson's estimates of runoff concentration times on upper pipe slope (Wilson, 1977, tab. 4.6.II). Slope area 1925 m^2.

Instantaneous rainfall mm	Maximum volume of pipe runoff in m^3	Time to drain assuming a discharge of *	
		4.5 l s^{-1}	1.5 l s^{-1}
5	9.6	36 min	1 h 47 min
10	19.2	1 h 11	3 h 34
15	28.9	1 h 47	5 h 21
20	38.5	2 h 22	7 h 8
25	48.1	2 h 58	8 h 54
30	57.8	3 h 34	10 h 42

* 1.5 l s^{-1} was maximum measured pipe discharge, 4.5 l s^{-1} was estimated maximum including possibly pipe-fed seeps.

Newson and Harrison (1978) used salt and dye dilution and artificial watering by gravity feed or pump, similar to Wilson (1977). Artificial pipeflows were in the range 0.08 to 1.0 ls^{-1} compared with previously measured natural flows in the area of 0-2 ls^{-1}, but velocities were much lower than Wilson's, c. 0.1 ms^{-1}. Comparison between runoff from flush areas (seepages), overland flow and pipeflow clearly showed that pipeflow was the fastest within the range of discharges studied (Fig. 36), but the results also showed that velocity increased almost linearly with discharge for overland flow, whereas it increased at almost the square root of discharge in pipeflow. Thus the higher the discharge the less the differential advantage of pipeflow: extrapolating Newson and Harrison's (1978, 56) results suggests that at around 1 cumec overland flow would actually attain a higher velocity ! In contrast, pipes had the lowest dynamic storage and appeared to have little further storage potential at the maximum discharges tested. They also noted that the pipe systems were leaky and that output was frequently only about half the input. Some of this may be a natural feature, leaks occurring when pipe capacity is exceeded or along cracks and the peat/mineral interface. Gilman and Newson (1980, 80) describe a case where 68% of the input was lost, partly through surface resurgences, with an input rate of 3 ls^{-1}, whilst 50% was lost purely by internal leakages when the input was reduced to 1.5 ls^{-1}. However, leakiness may have been exaggerated by trying to activate pipeflow artificially in soils at less than saturation level: indeed, conditions at the time were described as 'dry' (Gilman and Newson, 1980).

As noted by Newson (1976a), these pipes tend to link

172

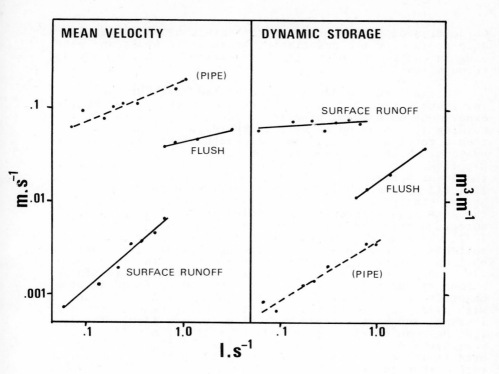

<u>Fig. 36</u> Mean velocity and dynamic storage for pipe,
flush and surface runoff. After Newson and
Harrison (1978, fig. 20) and Gilman and Newson
(1980, fig. 57).

blanket peat source areas (seepage slopes) across valley
midslopes to valley bottom flushes. According to Newson and
Harrison (1978), the fact that the pipes rarely discharge
directly into the stream reduces their hydrological significance, although they still control the partial contributing
area, and some networks do connect to perennial pipes
(Gilman and Newson, 1980). Ward (1966, 68) appears to have
recorded a similar situation in North Auckland, N.Z., in
that many pipes are developed upslope of bog zones and only
a limited number actually discharge into gullies.

In the adjacent Institute of Hydrology experimental
catchment of the Upper Severn at Plynlimon, Newman (1976, 43)
observed many pipes draining from the base of the peat, often
with collapsed caverns upslope which direct surface water
into the system. Most of these pipes feed into valley
flushes or bogs, but dye tracing indicated that some discharged directly into the stream. Several of these pipes
continued to flow throughout the 1976 drought. She estimated
that the pipes reached peak flow 10h 'after the rain has
fallen' and therefore contributed only to recession flow in
the stream. However, her project concentrated on measuring

velocities of surface flow and certainly provided little data to justify the association of specific peaks and plateaux in the recession of the Upper Severn with pipeflow and throughflow (*ibid*, 36).

Gilman and Newson (1980) collected together the Institute's records since 1971 and performed a further field experiment on natural pipeflow using a new autographic recording system based on weir tanks (Fig. 37). This gave them a total of 66 pipeflow hydrographs from ephemeral pipe systems, which were used to test techniques for modelling pipeflow. The data consisted of 1) the side slope area on the lower Nant Gerig studied by Gilman in March and June 1971 (12 hydrographs), 2) a headwater site (upper Nant Gerig) studied in July-August 1972 (29 hydrographs) and 3) the same side slope site as studied in 1971 equipped with autographic recorders between February 1976 and February 1977, which gave 25 hydrographs from 8 storms which were suitable for use in the optimisation program. The most obvious features of the hydrographs were a) the jagged outline showing an immediate response to changes in rainfall intensity and a lack of the normal smoothing found in stream channels due to combining water from different locations with differing travel times and the diffusive effects of routing, and b) the rapid recession to zero flow. The lag time between onset of rain and the start of pipeflow was variable and, like Jones (1975; 1978a and c), they considered it to be closely related to antecedent soil moisture, rather than to 'time of concentration' in the pipes catchment area. In general, flow began after 10 mm, though it could be as high as 50 mm. High rainfall intensity also elicited quicker response, presumably by bypassing 'the usual routes by which water enters a pipe' (*ibid*, 62).

Recessions were not exponential but showed a steeper decline, probably due to increasing influent seepage as flow levels fall. Gilman and Newson suggest that such losses probably occur at all times when the pipes are flowing, but acquire more significance as the quantity of water in the pipe decreases. On the other hand, if pipeflow is generated by a rising phreatic surface (cp. Jones, 1975), then influent seepage is likely to increase in absolute terms during recession.

Gilman and Newson (1980) noted that, nevertheless, recessions continued for longer than could be explained simply by the lag in transmission of water from the furthest points of the contributing area, and investigated the possible sources of storage in a pipe surveyed in detail by Morgan (1977) at Cerrig yr Ŵyn (Fig. 38). They found that the storage area (mean cross-sectional area of flow) A was related to discharge Q, in $m^3 s^{-1}$ as:

$$A = 0.20 \ Q^{0.55}$$

And they concluded that flow levels would reduce to about one-third in just 2.5 min from measured peak flows, compared

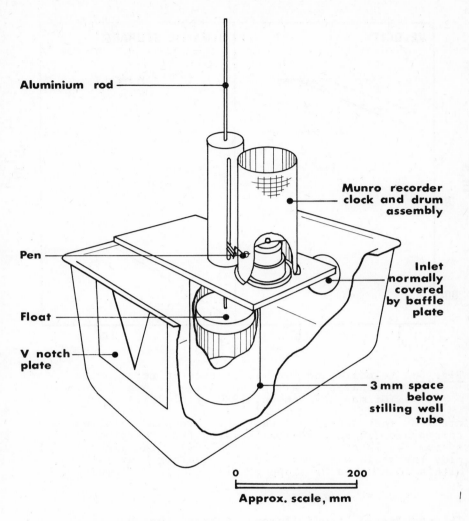

Aluminium rod

Munro recorder clock and drum assembly

Pen

Inlet normally covered by baffle plate

Float

V notch plate

3 mm space below stilling well tube

0   200
Approx. scale, mm

<u>Fig. 37</u> Instruments for measuring pipeflow: IV Weir boxes
          used by Gilman and Newson (1980), designed by
          Truesdale and Howe (1977).

with total measured recession times of 1.4 to 5 h. The
storage sustaining recession was clearly not to be found in
the pipe itself, and this led them to consider pipe discharge
in their modelling studies as virtually ready-made before
it enters the pipe system, with routing along the pipe having
only a minor effect.

The pipe hydrographs suggested a linear reservoir model
with a threshold below which no flow occurs. Possible physical interpretations of the reservoir considered were: 1)
depression storage (likely to be small), 2) unsaturated soil
moisture (not fast enough), and 3) storage in pipes and

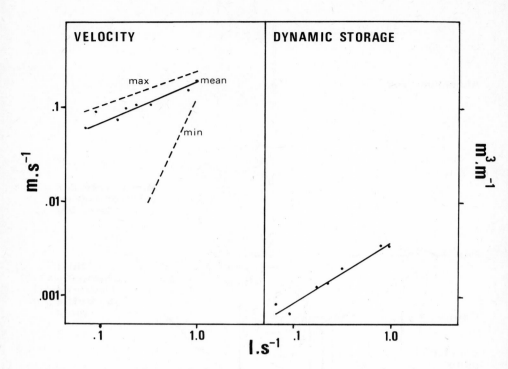

Fig. 38   Velocity and dynamic storage in relation to
          discharge in a pipe on Cerrig yr Ŵyn, Plynlimon
          (Gilman and Newson, 1980, fig. 49).

cracks. They considered that the threshold value was due
to storage 'in some region isolated from the pipe runoff
system' by interception or in the soil. The linear model
plus threshold was simulated on a quarter-hour interval
with a 10 m² area of slope generating flow according to

$$Q = k (z - z_t)$$

'k' and 'z-z_t', the difference between the conceptual moisture content of the soil (z) and saturation level when flow
begins (z_t), were treated as two parameters (k and z_o
respectively) to be optimised for each pipe hydrograph,
thus obviating the need for soil moisture data.

The authors extended the two parameter model to take
account of certain hydrographs which showed a rapid response
by adding a parameter α for the proportion of rainfall input
to the storage reservoir, with (1 - α) going directly into
pipeflow, and a parameter d for influent drainage losses
from the pipes to the unsaturated zone, assumed constant
during pipeflow (Fig. 39). In fact, 45 of the 66 simulated
hydrograph were not significantly improved by adding the last
two parameters, 5 required three parameters and 16 four,
but goodness of fit varied widely (Fig. 40). The simulations

176

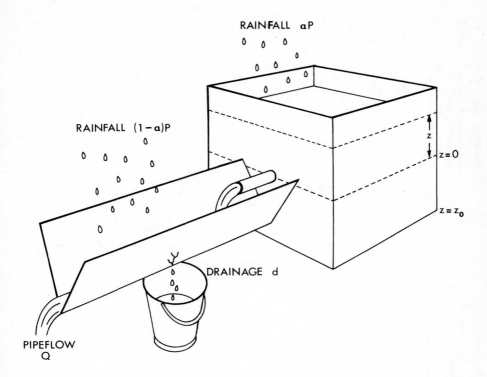

Fig. 39   Model of pipeflow generation used by Gilman and
          Newson (1980, fig. 42).

indicated a number of significant features:

    1) the parameter z_0 varied inversely with antecedent
    stream runoff (which is taken as indicative of moist-
    ure status in the basin) in both summer and winter,
    but in summer the relationship is weaker and higher
    values of z_0 occur for a given stream flow. This is
    probably due to summer desiccation and it is interest-
    ing that the data from winter 1976, following the
    great drought, show effects reminiscent of the summer
    norm.

    2) the constant k indicated recession times for 1/e
    - fold reductions of half to 5 hours, with higher
    values for the upper Nant Gerig ($k = 1.01$ h^{-1}) than
    for the lower Nant Gerig (0.68 h^{-1}).

    3) neither the proportion of delayed runoff α nor the
    loss d were very important parameters.

    4) introducing lag times generally confused the results.

    The second part of their study concerned hydraulic
routing of pipeflow downslope in the Cerrig yr Wyn ephemeral

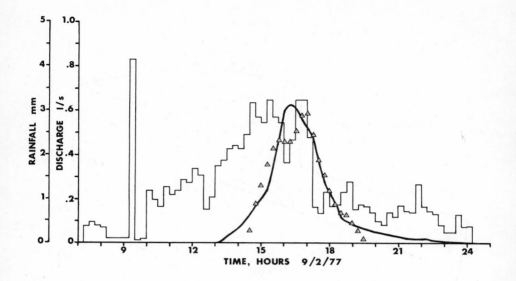

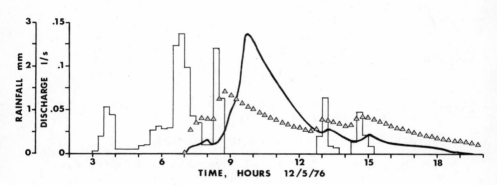

Fig. 40   Comparison of pipe discharge and Institute of
            Hydrology simulations, as indicated by Gilman and
            Newson (1980, figs. 44 and 46).

pipe. Calculation of the Darcy-Weisbach friction factor and
Reynolds Number suggested that pipes behave as extremely
rough channels; their results are compared with those of
Humphreys (1978) in Table 9. The high roughness is the
result of abrupt bends and changes of slope, and is also
present in the surface streams of the area (Beven, Gilman
and Newson, 1980). Using the empirical relationship between
cross-sectional area of flow and discharge in the pipe in
the kinematic wave formula, Gilman and Newson obtained values
for wave velocity. The average lag was 23 min and change
in shape of the hydrograph was minimal. They concluded that
typical values for cross-sectional area of flow plus a constant kinematic wave velocity would be sufficient to give
a fine time resolution in a runoff model.

178

An additional observational network was set up on the Cerrig yr Ŵyn slope consisting of a raingauge, three bore-holes to monitor saturation levels and instantaneous pipe-flow measurements, which were used to calculate the necess-ary parameters to test the linear reservoir model (Fig. 41)*. In the model the pipes were given an average depth of 0.15 m and allowed to flow after saturation of the 0.4 m thick B horizon, equivalent to 75 mm of rainfall. Surplus water was also added in from the blanket bog store above, but leakages were ignored. Somewhat enigmatically, they suggest that this omission may be misleading for catchment scale modelling since Knapp (1974) did report throughflow in the B horizon; how do significant losses to B horizon seepage square with a model that requires the B horizon to be 'filled' before pipeflow begins? The 'threshold' figure of 75 mm also seems rather high. Unfortunately, storm rainfall data are not given, but reading off the six hyetographs/hydrographs illustrating data used in the parameter optimisations suggests thresholds of less than 1 mm up to about 25 mm on different pipes. Indeed, 75 mm of rain would constitute a major storm event with a return period of over a year (cf. page 40), yet emphemeral pipeflow is clearly much more frequent.

Although the authors say the 'experiment proved a feasible approach to lumped piped-slope modelling' (ibid, 92), they omitted to give results to substantiate successful modelling. During the modelling experiments, 'available water capacity' values for the B horizon, taken from the work of the Soil Survey, had to be reduced from 19% to 10% in order for any pipeflow to be generated, although they offer no physical justification for this. Hydrographs taken from the flush area at the base of the piped slope were qualitatively annotated in cases where observed pipeflow con-tributions occurred (ibid., fig. 55), but they state that 'unfortunately, without continuous pipeflow records it was impossible to discover whether pipeflow was a major contri-butor to each event analysed'.

Before reviewing the more general conclusions on the hydrological role of piping in humid areas reached by Gilman and Newson and others, other recent field investigations of pipeflow deserve a mention. Anderson and Burt (1978) placed v-notch weirs and autographic stage recorders on what they describe as a 'throughflow-fed spring', a headwater marsh, a topographic hollow containing a spring and the stream in a small, 1.5 km² basin. Their paper was primarily concerned with the instrumentation used, but they illustrate results from a storm in December, 1976, which they describe as supporting the dominance of hillslope hollows in generat-ing throughflow peaks. It is, however, noticeable (ibid, fig. 7 and tab. I) that the measured hollow did not reach a peak during the period of record available here, whereas overland flow in the headwater marsh and the piped through-flow both reached peaks shortly before the lower stream weir. Admittedly, neither were volumetrically large.

---

* See section 7.1(i) for discussion of the model.

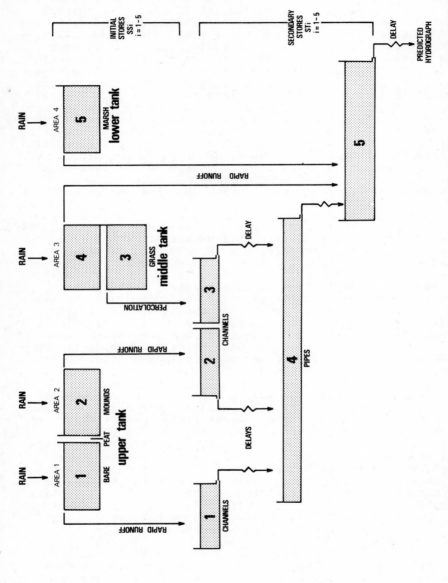

Fig. 41   Institute of Hydrology 'tank' model of Plynlimon hydrological system (IOH, 1978, fig.3; Gilman and Newson, 1980, fig. 53).

180

McCaig (1979a and b) measured pipeflow in the Slitheroe Clough subcatchments, Yorkshire, in a study of stream head development processes. He noted particularly that some isolated areas of surface saturation and overland flow were caused by pipeflow resurgences occurring at discontinuities in the pipe network. The onset of pipeflow was found to be closely correlated with surface saturation at points throughout the catchment, supporting Gilman's (1971) suggestions of a storage threshold before pipeflow. The primary source areas for pipeflow were the flatter, heather moorland areas near the summit of the catchment which act as reservoirs which are quickly filled during storms. Only overland flow or pipeflow appeared to be capable of generating the quick response and single peak of the main stream hydrograph. Measurement of pipeflow velocities suggested an average of c. 0.1 ms^{-1} compared with 0.0025 ms^{-1} for throughflow. Hence average travel time from the saturated areas would be 30 min and 20 h respectively. Peak to peak lag times of a few hours in the two sample stream hydrographs illustrated by McCaig suggested to him that a combination of processes was present with throughflow over short distances feeding the pipes, and perhaps with slower velocities in the smaller headwater tributary pipes.

Most recently, Howells (1980) has undertaken a study of pipeflow on farmland at an altitude of 150 m on a tributary of the River Llynfi in South Wales (cf. section 6.1(i)). The stream actually begins as a partially collapsed pipe feature, but Howells used upper and lower weirs to define a reach c 70 m long, some 300 m downstream from the final resurgence of the stream, and studied 12 pipes feeding two major stream-side seeps. Rainfall was measured by standard and tipping-bucket raingauges and streamflow by observing stage boards also equipped with crest stage recorders. The pipes appeared to be seasonal, flowing throughout an extended period of qualitative observation from April to September with the exception of a six week period around mid-summer. Pipeflow was measured volumetrically by hand as frequently as possible over a 32-day period during the late summer of 1979.

There was considerable variation both in terms of discharge and response times between pipes and also in terms of percentage contributions to streamflow over time. Mean discharges for individual pipes ranged from 0.02 l s^{-1} to 0.21 l s^{-1}, with maximum recorded discharges of 0.04 l s^{-1} to 0.96 l s^{-1} (the extreme means and maxima being for the same individual pipes). Delay times between the start of rain and the rise in pipeflow averaged 12 h, compared with only 7 h for the stream, and the average time from start of rain to peak pipeflow was 21.5 h against only 9 h for the stream. However, occasionally pipes in one of the seeps responded before the stream, perhaps, she suggests, because the pipe system was already relatively full.

In general, she regarded the pipeflow as sustaining but not initiating streamflow, and she related the relatively

slow pipe response to the general lack of recognisable inlets or surface collecting areas for the pipes. Nevertheless, total measured pipeflow accounted for 20.3% of the
streamflow increment between the weirs over the period of
measurement. Daily average contributions ranged from 10%
to 81% on a stream with maximum discharge of 56 1 s^{-1} and
an increment of up to 14.5 1 s^{-1} between weirs. Percentage
contributions were lower at higher stream discharges and in
wetter antecedent moisture conditions, tending to confirm
the observations of Jones (1975; 1978a and c) in mid-Wales.

7.1(i)   Conclusions on the hydrological role of soil pipes
in humid regions

The existence of pipes is bound to increase the trans-
missivity of the soil and result in a quicker stream response
(Jones, 1971). Abundant evidence has been accumulated in
recent years to support this view and, indeed, to indicate
that whilst the contentions of the 'anti-subsurface flow'
lobby that throughflow or shallow interflow is too slow to
be a major source of floodwater (e.g. Dunne and Black,
1970a and b; Freeze, 1972a and b, 1974) may be true of
matrix seepage, it is not true of pipeflow (cf. Table 15).
Jones (1979) argues that piping significantly extends and
reinforces the arguments for subsurface flow championed by
Hewlett and others (v.i.),and Ward (1975, 256) and Cryer
(1980) point out that pipes provide an alternative explanation for rapid supply of rainwater to the stream to
Hewlett and Hibbert's hypothesis of 'translatory flow',
for which there is as yet no good field evidence. The main
factors affecting the hydrological importance of these pipes
to the overall response of a catchment are not rate of
transmission but their frequency, capacity, connectivity,
including their relationship to perennial streams, and flow
regime. Considerably more information is needed on these
aspects.

Jones (1975, 396) suggested that the pipe network must
be considered as another subsystem of the conceptual model
of the drainage system, which therefore consists of 1)
channel net, 2) surface rill net, 3) pipe net, 4) pseudopipe
net, 5) percoline/seepage line net, 6) groundwater spring
net, 7) zone of overland flow, 8) zone of diffuse stormwater
seepage, and 9) zone of diffuse groundwater flow. This
extended an earlier list which did not recognise the rill
and percoline nets as separate subsystems (Jones, 1971, 604;
Gregory and Walling, 1973, 289). Subsystems 6 and 9 are
more stable counterparts of 3 and 8, and are generally not
found in the soil mantle. The other subsystems show marked
expansion and contraction with mutual interference in response to rainfall or snowmelt.

Newson and Harrison (1978, 10) noted that channelised
flow occurs far beyond the boundary of the permanent network shown on topographic maps, and they produced a classification of channels based on the Plynlimon experiments (Table
16). The most obvious subdivision is between open and closed

Table 15. Comparison of measured rates of transmission by matrix throughflow, overland flow and pipeflow.

Flow process	Rate $m\ s^{-1}$	Source of data
matrix throughflow	6×10^{-5}	Hewlett and Nutter (1970)
	1×10^{-4}	Weyman (1975)
	2.5×10^{-3}	McCaig (1979b)
overland flow	$8 \times 10^{-4} - 8 \times 10^{-3}$	Newson and Harrison (1978), Gilman and Newson (1980)
	7.5×10^{-2}	Hewlett and Nutter (1970)
pipeflow	1×10^{-1}	Weyman (1975), Newson and Harrison (1978), Gilman and Newson (1980)
	1.5×10^{-1}	Jones (1975; 1978a)
to	8×10^{-1}	Wilson (1977)

channels. Natural closed channels are divided into inter-mittent and perennial. Only the former are referred to as 'soil pipes' and Gilman's (1971) 'perennial pipes' are termed 'flushes'; a term which is unfortunately well-estab-lished in the literature as referring to features which do not necessarily contain any closed channels (cp. Pearsall, 1950). Jones (1971), Gilman (1971), Newson (1976) and Wilson (1977) all recognised a similar dichotomy of types, and in fact Newson and Wilson (pers. comms.) have expressed the opinion that the perennial pipes are not true soil pipes. In contrast, Jones (1978a) recognised three categories, which he termed ephemeral, seasonal and perennial, and re-garded them as forming a typological continuum related to the cyclical and random variations in the depth of the phreatic surface (Fig. 25). Cryer (1980) has gone so far as to say that he sees no grounds for distinguishing per-manent or perennially-flowing pipes from ephemeral or inter-mittent ones since they clearly form an integrated network on Maesnant. However, are there not good grounds for dist-inguishing ephemeral and permanent stream channels even though they may form an integrated network? Perhaps the point Cryer intends to make is that there is in a sense only a difference in degree, not in kind, between the end-members of the network.

Jones (1979) also distinguished between pipes connected to the channel network and disjunct pipes, some of which may connect to the channel net via sheets of overland flow during extreme storms (Table 17 and Fig. 25). In general, lack of a direct connection will tend to reduce the significance of the pipes to the response pattern of the basin, but to a

Table 16. Classification of channels, after Newson and
           Harrison (1978, tab. 1).

CLASS	REGIME	TYPE	COMMENT
	Artificial	Tile drains	
closed			
channels	Intermittent	Soil pipes	
	Perennial	Flushes	The 'perennial pipes' of Gilman (1971)
	Artificial	Drainage ditches	
Open			
channels	Intermittent	Ephemeral channels	
		Peat-lined channels	
		Bedrock channels with falls-and-pools	
	Perennial	Boulder channels with alluvial banks	
		Alluvial channels	

greater or lesser degree depending on the circumstances.

    Gilman and Newson emphasise the fact that side-slope
piping is often not integrated with the main system and that
headwater areas and footslopes are therefore the main con-
tributors. Where side-slope piping ends, flow velocities
are one order of magnitude slower across flushes and two
orders slower in surface runoff. Even so, the writer's
experience from dye tracing tests on Maesnant is that even
though pipes may issue onto valley-bottom mires (albeit
smaller than most on Nant Gerig), the travel time across the
bog to the stream is scarcely slower than if the pipes had
indeed been connected to the stream. Cryer (1980) also
points out that Newson's argument does not apply on Maesnant,
where pipes feed the stream 'with no substantial delay in-
terposed', and Jones's (1971; 1975) observations on Burbage
Brook suggest that sideslope piping there is invariably
connected directly to the channel, although it is perhaps
less frequent than floodplain piping. Jones (1978c, 1979)
has argued that piping can provide a 'dynamic contributing
volume' which extends beyond the limits of any contributing
area of saturation overland flow. Indeed, it may link

Table 17   Hydro-geomorphic classification of piping.

A. Disjunct pipe networks: not directly connected to channel
    network.

    1. Hillcrest or interfluve piping: on 'seepage slope', draining
        from or into hollows or from upland peats.

            Either: concentrates quickflow and directs it towards
                     percolines, thalwegs or diffuse valley side-
                     slope seepage. Ephemeral.

            Or :   provides more permanent flow from high level
                     'sponge' areas to side-slopes or percolines.

    2. Valley side-slope piping: similar function to (1) above;
        may extend to flood plain or extension thereof and contribute
        to saturation OLF (usually on upper or 'transportational
        mid-slope).

B. Pipe networks linked to channel net.

    3. Valley side-slope piping: concentrates quickflow both within
        DCA and beyond (DCV), and supplies it directly to stream.
        (Usually lower or 'colluvial footslope/alluvial toe-slope').

    4. Floodplain piping: typically limited to floodplain, often to
        just a few metres from bank, though (3) may coexist across
        floodplain. in average storms : may speed up OLF, percoline
        or diffuse seepage.   after overbank flooding: may produce
        steeper recession curve by supplementing overbank drainage
        and diffuse seepage.

                                        (Jones, 1979, tab. 1).

otherwise disconnected areas of overland flow or disconnected
elements of the contributing volume into the main drainage
system, which he termed the disjunct DCA or DCV (Fig. 42).
McCaig (1979a and b) has noted how pipes can link apparently
disconnected areas of surface saturation and that some of
these areas are caused by pipe resurgences along discontinuous pipe networks. He called the latter 'secondary source
areas'.   In fact, McCaig noticed pipes feeding a large mid-
slope bog which then drained to the stream via saturation
lines, some of which contained perennial pipes.   Both large
and small secondary source areas have been observed in Wales
(Jones and Crane, 1979, fig. 2) and they clearly represent
a 'half-way house' between the discrete 'pipe sink' and
'pipe source' bogs, observed for example by Ward (1966) in
New Zealand and Gilman and Newson (1980) in Britain.

      Weyman (1975, 12) also made a distinction between
streambank and hillslope piping. He said that the small
pipes underlying extensive areas of hillslope in mid-Wales
seem to be connected to the surface by open routes and respond very rapidly to rainfall. 'This type of pipe, which
functions as an alternative to overland flow, should not be

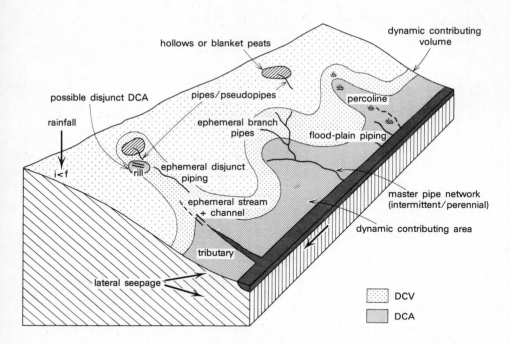

Labels in figure:
- hollows or blanket peats
- dynamic contributing volume
- possible disjunct DCA
- pipes/pseudopipes
- percoline
- rainfall
- ephemeral branch pipes
- flood-plain piping
- i<f
- rill
- ephemeral disjunct piping
- master pipe network (intermittent/perennial)
- ephemeral stream + channel
- dynamic contributing area
- tributary
- lateral seepage
- DCV
- DCA

<u>Fig. 42</u> Post-Hewlett framework for runoff generation model
proposed by Jones (1978, Fig. 3; 1979, fig. 1).

confused with the pipes often seen in stream banks which
represent the concentration of throughflow on the lower part
of the slope and are fed directly from the soil matrix' (*ibid*.).
Whilst there is undoubtedly an element of truth in his statements, they are misleading in so far as we have observational
proof that (i) pipes issuing into streambanks (whether
ephemeral, seasonal or perennial) can react swiftly to rainfall and produce large volumes of discharge, (ii) these
streambank pipes can extend up the hillslope for considerable
distances, some clearly fed at least in part by surface inlets, others by collecting areas which may be largely collecting seepage from the soil matrix, (iii) even short, floodplain pipes may have surface inlets, as described by Leopold, Wolman and Miller (1964, 446-447), and (iv) the extent
to which disjunct pipes on the hillslope are fed by seepage
or surface inflow clearly varies and is still a matter for
research.

Bell (1972a and b; Institute of Hydrology, 1973) suggested 7 processes of drainage on the slopes in the Upper
Wye catchment:

1) laminar flow on the surface,

2) surface ephemeral streams,

3) laminar flow on peat/E horizon interface (mainly in

186

creep colluvium podzols),

4) piping on same as 3), form may interchange,

5) laminar flow on boulder clay/colluvium interface,

6) piping on same as 5),

7) laminar flow on bedrock surface.

In the light of evidence of turbulent flow in soil channels
a few millimetres in diameter (Childs, 1969, 159; Hillel,
1971, 94), perhaps we should replace Bell's 'laminar flow'
with the less specific terms 'sheetflow' or 'diffuse flow',
which are presumably what he meant.

In comparing the character of pipeflow and overland flow,
Gilman and Newson (1980, 97) noted that surface runoff has
substantial storage capacity, although it is weakly related
to flow, whereas pipes have very limited capacity, with
storage limited by efflux and other leakages. Hence pipeflow and surface velocities tend to converge at very high
flows. Because of leakage, pipeflow is, according to them,
unlikely to be surcharged. However, if this were wholly
true then the fountain resurgences observed by Weyman (1971,
175), Gilman and Newson (*ibid*.), the writer and others are
unlikely to occur.

Pond (1971) and Gilman (1971) put forward a provisional
conceptual model of a piped hillslope consisting of a series
of storage and flow elements, which has continued to act as
a successful framework for research in the Upper Wye (Newson,
1976b; Gilman and Newson, 1980). This consisted of 1) a
long-term storage unit, namely the hilltop and plateau peat,
which sustains baseflow in the stream, 2) a mid-term storage
unit, the hillslope peat, which dries out between storms
and contributes to throughflow but not significantly to
pipeflow and 3) short-term storage units consisting of pipes,
cracks, surface depressions and mosses which contribute
directly to pipeflow and are depleted by pipeflow. The
suggested sequence of events during a storm was:

i)   rainfall makes up long and mid-term storage
     deficits.

ii) once these deficits are filled at the top of the
     slope, rain then enters short term storages with
     direct drainage to pipes on the upper slopes.

iii) mid-term storage on side slopes is filled
     by rainfall and pipeflow from higher up, then
     pipes flow right down the slope.

iv) pipes flowing at capacity break surface and flow
     overland.

v)   when rainfall ceases, pipes empty the short-term

storage and stop flowing. Mid-term storage
takes days or weeks to dry out.

The proposed sequence was concerned only with ephemeral
pipes and has been incorporated in a modified form in recent
research aimed at producing a modified, lumped model of the
Cerrig yr Ŵyn subcatchment of the Wye (Institute of Hydrology, 1978, 8-10; Gilman and Newson, 1980). Fig.41 illustrates
the Institute's 'tank' model, which comprises a three-fold
cascade:

a) Upper tank: peat hags with bare channels between.

b) Middle tank: grassy slopes drained by pipes.

c) Lower tank: valley bottom marsh.

The model parameters are :

i) the initial depth of water in each of the initial
   stores or 'tanks', SS_i (i = 1....5).

ii) the initial depth of water in each of the secondary
    stores ST_i (i = 1....5).

iii) the coefficients RK_i, in the non-linear storage
     relationship in which discharge, RO_i is given by

$$RO_i = RK_i (ST_i)^{RK_i} \quad (i = 1....5)$$

iv) the exponents RK_i in that relationship

v) RC the percentage of rainfall on grass slopes
   (Area 3 of Figure 41) which percolates into
   Secondary store 3.

vi) the delays DEL_i in runoff from the secondary stores
    (i = 1....5)

Initial experiments proved encouraging using measured
hydrographs and parameters SS , ST_i (i = 1....4) and RK_i
(i = 1....4) for two storms during the winter 1974-75. The
remaining parameters were determined by averaging values
obtained by minimising a least squares objective function
calculated for each storm separately (Institute of Hydrology,
1978, 10). However, it was admitted that further work is
required before the model can be satisfactorily applied
to other storms. It was to calibrate this model that Gilman
and Newson (1980) conducted the Cerrig yr Ŵyn experiment.

Gilman and Newson (1980) concluded that 'hydrologically,
piping is a spectacular example of soil inhomogeneity leading
to an unexpectedly high permeability and a rapid response
to rainfall'. The response of individual pipes is therefore
'jagged'. It depends considerably on antecedent moisture
conditions for low intensity rainfall, but less so under
high intensities. Routing through the pipes is relatively

188

unimportant because of their low storage capacity and the
high velocity of flow. This, they say, throws emphasis
for response on flow generation in the blanket peat which
feeds most of these pipes: there is a close similarity
here with the view of Weyman (1971, 256) that pipes in the
East Twins basin are a runoff-maintenance system rather than
a runoff-generating system. However, it would be wrong in
our present state of knowledge to assume that this is the
only or, indeed, the general pattern of pipe response.
Indeed, Gilman and Newson (*ibid*) describe a case of a 'dog-
leg' relationship between saturation levels and pipeflow
which they suggest is due to the combining of inflows from
the blanket bog with inflow from the slope giving two
thresholds for flow. An interesting illustration of modification of a flood wave passing through a pipe system comes
from recent work on the Maesnant (Fig.43), in which pipe
weir 5 is c.350 m upstream of weirs 2 and 3 where the peak
discharge is over twice as great, and delayed 4-5 hours,
and the hydrograph somewhat altered in shape. Dye tracing
has proved the connection between the sites and no other
pipe input has so far been found in the intervening distance
(Jones and Crane, 1979). There is clearly room for both
forms of response (cf. Weyman, 1971).

It has been argued that the existence of soil pipes
necessitates some fundamental modifications of the currently-
popular 'Hewlett' model of streamflow generation (Jones,
1978c; 1979). Modifications are needed not only because of
the one or two thousand-fold increase in the speed of soil
water interflow, from c. 5m dy^{-1} (Ward, 1975) or 8.6 m dy^{-1}
(derived from Weyman, 1975) for diffuse seepage to over
9km dy^{-1} for pipeflow (Jones, 1978c; 1979), but also because
it alters the established concept of the 'dynamic contributing area'. Gilman and Newson (1980) note that the general
termination of pipe networks at the edge of the contributing
area in the valley-bottom mires in the Upper Wye mean

'that over the major part of the catchment pipeflow
will not lead to a spectacular extension of the
contributing area, except in very high rainfall
events.

However, the pipes still have an important role in
the uppermost parts of the catchment, where the
valleys are narrow and it is more likely that pipeflow will contribute directly to the perennial
streams or flushes....where direct connection occurs
it is possible that pipeflow, considered as a form of
channel extension, can bring large areas of hillslope
into hydraulic connection with the streams, leading
to the generation of flood peaks in the headwaters.
Elsewhere discontinuities in pipes will lead to a
dissipation of potential flood water, and a less
dramatic rise in the hydrograph than might otherwise
be expected' (*ibid.*, 96).

It is interesting that Day (1978) has recently drawn

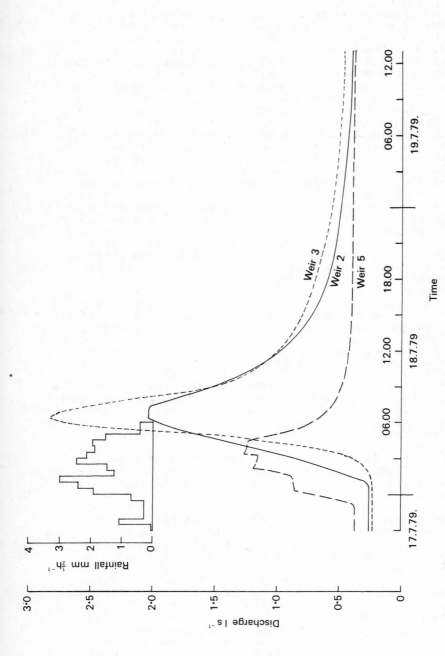

Fig. 43 Comparison of pipe hydrographs at upslope and downslope sites on Maesnant, mid-Wales, based on Jones and Crane (1979, fig.8). At the outlet of a saturated area (weir 5), the flow response closely matches the variation of rainfall intensity. At the pipe outlet weirs about 350m downslope (2 and 3) the peak is delayed by 4-5 h and is higher.

attention to the discontinuous nature of expansion of the stream net during storms in New England, with growth downhill from saturated areas as well as upslope expansion. She describes extension as a process of integration and overflow of pools. However, there is not mention of piping being present in her area.

Pipe inlets may capture both Hortonian excess and saturation overland flow from primary or secondary source areas and speed up the drainage process 10-to 100-fold (cf. Table 15). Pipe outlets on hillslopes may augment saturation overland flow and extend the surface contributing area, typically in linear elements which may eventually cause rilling (Stagg, 1974; 1978). Indeed, Stagg (1978) provides interesting evidence which suggests that rill systems in the Grwyne Fechan basin, Black Mountains, originate from pipes actually draining groundwater through the soil from the outcrop of sandstone strata. This extends the 'contributing zone' substantially beyond established theoretical limits.

Up to the present, research has been confined to individual pipes or piped slopes. This has been partly because of the obvious demand on resources made by a complete pipe monitoring programme even for a relatively small catchment, and partly because of the theoretical argument that exploratory measurements should first be made at the lowest level of aggregation in order to separate source response from routing effects. The highest level of aggregation so far achieved has been by Gilman and Newson (1980) using the Institute of Hydrology (1978) model to predict outflow in the flush at the base of the Cerrig yr Wyn slope which forms the head of a first-order stream. But the authors claim 'only limited progress'. Unfortunately, extension to the catchment scale in this field area is bound to run up against the problem of disentangling the effects of artificial drainage works, as illustrated in the maps of Newson and Harrison (1978, figs. 2 and 3). In this respect, the current research programme on Maesnant, a 0.54 km² basin on the opposite face of Plynlimon, is fortunate to be located on unimproved mountain pasture. Intensive instrumentation here is aimed at providing data for a catchment model incorporating pipe response (Jones and Crane, 1979).

However, Rees (1979) has attempted to evaluate the effects of pipeflow activity on the response of a whole catchment without using pipeflow measurements. Instead, Rees uses the classic hydrological method of paired basins, one piped, the other nominally unpiped. He chose two subbasins of the upper Wye on Plynlimon where standard hydrometric data were available from the Institute of Hydrology, together with back-up information on soil parameters and the distribution and activity of soil pipes. The Gwy basin has dense pipe networks, including the Cerrig yr Wyn slope, and the adjacent Cyff basin rather fewer pipes, although it has quite a number of artificial drains (cf. Newson and Harrison, 1978, fig. 2-3). Rees used the split sample procedure on

Table 18. Measured pipeflow parameters.

Source	Pipe size (mm diam.)	Velocity $m\ s^{-1}$	Maximum discharge $l\ s^{-1}$	Storage per m length of pipe (litres)	% of rainfall	% of streamflow	Location
Knapp (1970a)	100		0.67-0.83				Hiraethog peaty podzols, Upper Wye, Plynlimon, side-slopes.
Weyman (1971; 1975)	25-50	0.1/0.3	> 1				50-150mm down in peaty horizon of podzols, East Twins, Mendip, headwater bowl.
Stagg (1974)	25-50	0.3	0.75(meas.) 0.62-1.39 (calc.)				
Gilman (1971)					4-72 ($\bar{x}$=34)		as Knapp
Jones (1975; 1978 a & c)	90	0.15	0.30		2-26	19-54 (incl. delayed) $\bar{x}$ = 39 10-40 (direct runoff) $\bar{x}$ = 30	Peaty gley podzol, Maesnant, Plynlimon, terrace.
Dovey (1976)	90	-	0.247 (s=70ml s^{-1})			19-54 (incl. delayed) $\bar{x}$ = 35	

Reference						Location
Wilson (1977)	60	0.2/0.8	1.5		1.8 - 2.8	Nant Cwmllwch, Brecon Beacons, side-slope.
Newson & Harrison (1978)	?c.50	0.1	1.0(artif.) 2.0(nat.)			Cerrig yr ŵyn, Plynlimon, side-slope.
Gilman and Newson (1980)						
McCaig (1979)	?	0.1				Peats to sandy rankers, Slitheroe Clough, Yorks., side-slope.
Howells (1980)	50-328			10-81 (total flow) $\bar{x} = 20$		Pendoylan gley soils, Nant Maescadlawr, S. Wales, stream-side seeps.

193

the Severn-Trent Water Authority's physically-based modelling
program, developed by Manley (1978). Model parameters were
optimised for the first half of the data period (July 1974-
December 1975) on the Cyff catchment and satisfactorily
tested over the second half (January 1976-December 1977)
for the same basin. The optimised parameters were then
transferred to the 'piped' basin. This worked well for the
earlier half of the Gẃy data, but not for the period covering
the great drought of 1976. Re-optimisation for the latter
period caused a significant shift in two important parameters
which could be related to piping, namely, the permeability
of the horizon boundary between upper and lower soil zones
was reduced from 156.94 mm h^{-1} to 107.52 mm h^{-1} and the interflow runoff from the upper horizon at saturation was
increased dramatically from 168.05 mm h^{-1} to 247.14 mm h^{-1}.
Bearing in mind that Rees set the boundary between upper and
lower zones at the mean depth of soil pipes found on Cerrig
yr Ẃyn, these results could be consistent with increased
pipe drainage of the upper zone, resulting also in less infiltration into the lower zone, and he notes that observations by Institute staff (*v.s.*) had indicated widespread
cracking of the peat in the Gẃy basin effectively feeding
existing pipes and increasing the density of the pipe network. Comparisons between the two basins during 3 known
pipeflow events which occurred during the spring of 1976
suggested that for 2 of these events streamflow peaks occurred a day ahead of the predicted peak in both basins until
the model was run with the re-optimised parameters.

Indicative though these results may be, Rees admits
that since he used daily not hourly flow data and basin response time is only a few hours, it is hardly surprising that
the results are not more definitive. Indeed, added to the
fact that the contrast between 'piped' and 'unpiped' catchments is far from perfect, it is remarkable that differences
appeared at all, and this may owe much to the unusual climatic events of 1976. Clearly, extension of the analysis on
an hourly basis or less and for more periods of known pipe
activity would be interesting.

The hydrological role of soil piping is still, therefore, a matter for active, current research. As pointed out
by Weyman (1973, 268), it is evident that 'water may move
rapidly through the profile along water-worn pipes or through
organically-created non-capillary routes without reference
to the hydraulic system of the main soil mass' and this must
therefore be considered in any measurements of the downslope
flow of water in the soil (Weyman, 1973) and indeed of the
soil properties relevant to flow (Jones, 1979). On the whole
current research is supporting the view that where piping is
reasonably extensive, pipeflow is a significant component
of the drainage process. The only truly negative conclusions come from Wilson (1977), in a case where piping does
not appear to be very extensive and where only ephemeral
pipes were considered. Finally, in the light of Wolman and
Gerson's (1978) recent contention that partial area activity
is normal in all climates, it is interesting to note the

the results of recent work on the role of pipeflow in arid and semi-arid regions.

7.2. Pipeflow in arid and semi-arid areas.

Few measurements have been made of pipe hydrology in semi-arid areas, although many speculations have been made. N.O. Jones (1968, 80) used the equation derived for the Walnut Gulch Experimental Catchment:-

$$Q = 1.95A^{-0.31}$$

to estimate mean annual water yield, Q (inches per square mile), from area of catchment, A, for pipes in the San Pedro Valley, Arizona. He concluded that pipe systems were most strongly developed below catchments with estimated annual discharges of 100-600 acre-feet year^{-1} (approx. 125 x 10^3 m^3 yr^{-1} - 740 x 10^3 m^3 yr^{-1}). The proportion of runoff entering a particular pipe system varied from storm to storm depending on the cutting and blocking of rills feeding the pipes from the surface. Looking at the rills suggested that not more than 33% of the mean runoff from any of the larger catchments flows to any given pipe system. This suggested a minimum of 20 acre-feet p.a. for development of the smallest systems since 1890, though extreme events may be more important than the mean. Jones also observed 5 storm events, noting in one case that all but 10% of the inflow into a pipe system came from rills, the remainder being sheetflow (*ibid.*, 87).

Drew (1972, 204) reports observations of a storm of 14mm in 2h in south Saskatchewan in which pyranine dye took 35 min. to pass through a system which appears from his map to be 40m long, i.e. a velocity of c. 1m min^{-1} or 0.02ms^{-1}. This is considerably slower than measured velocities in humid pipes. Drew noted the flow was rarely over 20cm deep (little more than 4% pipe-full according to his cross-sections), although 'rapid and turbulent'. This suggests that the pipe is adjusted to carry much higher discharges, perhaps more efficiently. However, as with pipes in humid areas, Barendregt observed that pipes in the Milk River Canyon, Alberta, were generally never filled to capacity except at the downstream end of small pipes during rare storms. (Barendregt, 1977; Barendregt and Ongley, 1977).

Drew suggests that most runoff in his area is initially overland, but a small proportion percolates through voids to create pipes which eventually capture most of the drainage.

The most intensive studies to date have been the field experiments conducted in Canada and Israel by Bryan and Yair and their colleagues (Bryan, Yair and Hodges, 1978; Yair, Lavee Bryan and Adar, 1980). Bryan *et al.* (1978) report sprinkler experiments and observations from four natural rainfall events on four plots in Dinosaur Park, Alberta, in which dye was used to identify the sources of pipeflow.

Fundamental differences were found in the patterns of overland and subsurface flow between sandstone and shale plots,
as well as within the shale units which were described as
stepped and rilled (plot 1), gentle concave and rilled (plot
2), steep and unrilled (plot 3). The sandstone had the
lowest threshold for runoff generation, which occurred as
overland flow over the whole slope. The piping occurred in
shale bedrock, albeit in the modified surface material in a
layer of loose shards below a dense crust 'caprock'. The
shards slake to form an impermeable basal layer. The monitored pipes had cross-sectional areas of up to $5 cm^2$, although
larger examples were present.

Bryan et al. (1978, 165) note that although the contribution of subsurface flow to storm hydrographs has been
widely discussed, all the work cited refers to humid areas.
They imply that although matrix interflow may not be important in dry areas, pipeflow does occur in these badlands and
claim that it differs from subsurface flow in humid areas
in so far as 1) it occurs in 'bedrock' or alluvial fill
rather than in soil (which is thin or absent), 2) it occurred
while the surrounding bedrock was dry, whereas in humid
areas subsurface flow occurs as a seepage front where the
soil is saturated, 3) pipeflow can occur anywhere on the
slope, irrespective of moisture content, 4) pipeflow comes
mostly from rill flow and may return to rill flow in a
sequence repeated many times on long slopes, 5) pipeflow
velocities of $0.15 ms^{-1}$ are similar to those for rill flow
and suggest little delay in contribution and the ability to
create distinct hydrograph peaks indistinguishable from rill
flow. Personal communication from Campbell (ibid) suggests
that even in extensive basins with very large pipes only
one peak is found; to the reviewer this implies that the
pipe/rill network is the dominant source of streamflow.
They conclude that partial contributing areas on the shale
units are controlled by material properties rather than
moisture regime. It is interesting that a similar conclusion has been reached by Jones (1978a;1979) for runoff in
humid areas. Although Jones does not deny the established
work on diffuse subsurface flow and overland flow, he attempts to redress the balance in terms of the role of discrete
sources of runoff from soil pipes and percolines. McCaig
(1979b) has independently expressed the same view. However,
the recent work in humid areas (q.v.) clearly requires some
modification of Bryan's second point; subsurface flow does
not occur simply as a 'seepage front'.

Yair, Lavee Bryan and Adar (1980) studied pipeflow in shale badlands in the Negev as a follow-up to the
previous research in Canada. They note that shale badlands
are less impermeable than is frequently assumed in view of
the evidence of piping, which was again found in an incoher-
ent, granular layer beneath a dense, sealed surface. The
pipes were typically 10 cm below the surface of the regolith
and up to 10cm in diameter. Again, contributing areas were
confined to pipes and rills, with the spatial pattern determined by the properties of the surface crust, particularly

by desiccation cracks. They noted that spatial non-uniformity can present problems for extrapolating from small sprinkler plots to large areas and repeated experiments on different sized areas, 1.5 m^2, 30 m^2 and 60 m^2. In all cases, most of the runoff was derived from rills and pipes. However, runoff coefficients fell dramatically on the larger plots, from 47% (for smallest plot on thick 20-30 cm impermeable crusted area) or 29% (for a similar plot in a thin, < 6mm porous crusted area) to 14% and 2% respectively for 30 m^2 plots in the same areas, and 2% for the combined thin and thick crusted areas. This they attributed to large cracks towards the interfluves. From the discussion pipes appear to be most common in the thick crusted area.

A unique feature of the work of Yair *et al.* (1980) is the observation of natural pipe development and destruction during the artificial rainfall experiments. High intensity rainfall (c. 50 mm h^{-1}) on a dry surface with a thick crust generated pipeflow after 15 min with new pipes developing in mid-slope, followed by general pipe collapse after 30 min. Recession was complete in one minute. Overland flow was negligible and most of the runoff was derived from rills and pipes.

The continuing research programmes of Yair and Bryan promise to yield very interesting information on pipeflow characteristics in badlands. Perhaps the most important discovery to date is that less overland flow may occur in badland areas than hitherto generally assumed, particularly because of the widespread presence of piping (cf. section 8.1). The not infrequent statements that the Hortonian infiltration model of hillslope hydrology represents an endmember of a continuum of models which applies to poorly-vegetated slopes with thin soil covers (e.g. Ward, 1975; Chorley, 1978a, 25) must therefore be taken with a pinch of salt. Evidence seems to be accruing to suggest that runoff processes in humid, semi-arid and arid areas are less fundamentally distinct than previously supposed.

8. PIPING AS AN AGENT OF EROSION

Piping erosion has mainly been studied in four environments, namely, 1) badlands, 2) loess and loessial soils, 3)
mid-latitude marine climates with either sodium-rich or
loess-derived subsoils and 4) tropical areas.   Needless to
say, these are the areas in which piping is most highly
erosive.   Piping does not normally occur alone and is commonly associated with other forms of erosion, not only rilling and gullying but mass movement and surface and subsurface sheetwash as well.   Downes's (1949) soil erosion maps
of Victoria indicated piping closely associated with ridge
areas suffering from severe sheet erosion and gullying,
extending into areas of moderate erosion, but absent from
areas where general soil erosion was minimal.

This chapter begins with a systematic approach in terms
of process associations and quantitative measurements,
followed by a brief review of observations of piping in
paleosols, and concludes with the problems of control and
prevention.

8.1. Piping, rilling and gullying.

There is a long-standing association between piping and
gully development in many parts of the world.   Morgan (1979,
10-11) has recently pointed out that the erosive effects of
subsurface flow through tunnel collapse and gully formation
are well known, quoting the reports of Berry and Ruxton
(1960), Berry (1970), Buckham and Cockfield (1950), Heede
(1971) and Zaborski (1972).   Like Heede (1976), he points
out the need to know whether surface or subsurface processes
are responsible for gully development in the design of control works.   Morgan also points out that of the three types
of gully network recognised by De Ploey (1974) as axial,
digitate and frontal, the former are associated respectively
with gravelly and clay loam soils, but the frontal variety
is associated with piping, especially in loamy sands with
columnar or prismatic structure.   According to De Ploey,
frontal gullies generally develop by progressive collapse of
pipes from the channel banks.   In the semi-arid badlands of
Tunisia piping is a dominant feature of 'pluvial erosion',
beginning along desiccation cracks which give the prismatic
structure to the weak horizontally layered consolidated and
cemented (calcareous) loamy sands and aided by dispersive
clays (De Ploey, 1974, 188).   He emphasises the importance
of topsoil structure to the type of hillslope erosion and
the fact that pluvial erosion, including piping and pseudokarst development, is the main category of hillslope process.   Piping may also be generated by gullying, and according to Drew (1972, 208) 'it is not clear which feature is
the response variable'.

Two basic types of gully can be recognised in terms of
process 1) the waterfall headcut gully and 2) gullies

developed by subsurface sapping, termed 'tunnel-gullies' in Australia and New Zealand. In general, the latter type has only been recognised more recently in the mainstream literature on gullying, although Ireland, Sharpe and Eargle (1939) did report 'seep caves' in a few gully-heads in their classic survey of south Carolina for the USDA. Back-trickle was by far the more dominant process there, but in some cases relatively minor extensions of the gully-head were due to water entering tension cracks behind the gully-head and emerging, sometimes in concentrated jets into a sapping cave in the gully-head. The cave walls tended to develop in weak C-horizon saprolite beneath a thick resistant B horizon (*ibid.*, 53-56, 132-133). Similar caving processes were reported by Bennett and Chapline (1928) and in the note by Walther Penck (1924) describing piping in tension cracks around streamheads (cf. Ollier and Mackenzie, 1975, 65).

However, the association between piping and gullying was noted particularly in Australia and New Zealand during the soil erosion surveys of the 1940s. Cumberland (1944) and the Soil Conservation and River Control Council (1944) pointed out that the cave-in of 'under-runners' frequently developed into serious gullying and ravine development in loess soils on steeper downland. Gibbs (1945) used the now familiar term 'tunnel-gully' erosion. In Victoria, Downes (1946, 287) reported gully formation when pipes in solonetzic soils reached 1.5-1.8 m wide and collapsed and in Canada Cockfield and Buckham (1946; Buckham and Cockfield, 1950) produced the first strong evidence that gullies conform by 'sinking of the ground' due to pipe collapse and suggested that the process might repeat itself after total collapse. More recently, Colclough (1965) has stated that much of Tasmania's severe gullying results from collapse of pipe systems. In South Africa, Downing (1968) divided donga heads into two categories: (i) long dongas containing a channel with lines of sinks at the head with tributary sinks and (ii) shorter funnel-shaped depressions with no permanent stream but tunnels.

Leopold, Wolman and Miller (1964, 446-447) reported tributary gullies extending by tunnel collapse for several hundreds of feet into ungullied alluvium in New Mexico. Between 1962 and 1968 Heede undertook an intensive study of piping and gullying on the upper Alkali Creek near Silt, Colorado (Heede, 1967; 1970; 1971; 1974; 1976), in which he studied the factors responsible for pipe development as well as measuring pipeflow and sediment yield. Heede (1976, 6) observes that piping may result in both headward extension of integrated gullies and the formation of discontinuous gullies by roof collapse. In the latter case, Hamilton (1970) has extended the ideas of Hadley and Rolfe (1956) on the role of effluent or positive seepage in the erosion of seepage steps to include piping as a mechanism of erosion of 'channel-scarps' in the beds of discontinuous gullies by successive roof collapse.

In fact, Cockfield and Buckham (1946) had earlier

suggested that gullies in the White Silts of Kamloops, B.C., were deepened by sinkhole formation in the beds. They noted that sinks were less common in broad, gently graded gullies than in steep, narrow ones. Aghassy (1973) documented a process of transition from rilling to gullying in badlands of the Negev, which passes through a piping phase. The complete cycle of pipe development usually lasts 2 to 3 rainy seasons. The process begins with cross-grading of rills, 4 to 8 rills creating every first order stream each typically 5-10 m apart. Deepening of beds is usually fast because roof collapse of pipes beneath the surface channels lowers the beds by 2-3m in each cycle. Typically several generations of piping exist at different levels in the gullies with 'channel scarps' tending to coincide with the walls of vertical inlets before collapse (*ibid.*, fig. 8). The process continued to be prevalent up to third order channels, but tailed-off in higher orders. In favourable cases, however, channel development by piping was observed in up to fifth order channels, and headward advances of 500-750 m had occurred into the surface of the upper plain. Colclough (1965) illustrates tunnel erosion capturing surface runoff from 'flow lines' (thalwegs or shallow dry valleys) under grassland in Tasmania and subsequent collapse of the tunnel line.

A similar association was reported in Britain by Jones (1975, 209; 1978a) with piping in lessivé brown earth a few centimetres below rilling extending downslope from a surface bog, although there is as yet no evidence of any erosional changes in the networks in this environment (Fig. 44). Stagg (1978) has also found rills associated with pipe outlets in a horizontal rather than vertical sense in South Wales, with the rills heading in pipe outlets and Morgan (1977; 1979) observed that rilling on agricultural land in England tended to begin at bursts from subsurface flow and not at Horton's (1945) supposed critical distance of overland flow. Blong (1965a) noted rilling feeding pipe inlets in ignimbrite deposits in Auckland and N.O. Jones (1968, 57) reported rapid pipe growth where inflow was from rilling and the subsoil was erodible, compared with only minor piping where inflow was from sheet flow and the subsoil less erodible.

Both gullying and rilling may develop from subsurface seepage without the presence of open pipes, as illustrated by Bunting (1964, 78) in Rossendale, England, where percolines concentrate pressure, and by Wigwe (1975) in Nigeria.

Heede (1974; 1976) postulates a cycle of piping similar to karst with:

1) Youth - development of pipes with diversion of surface drainage subsurface,

2) Maturity - with well developed pipes and only the major entrenched streams persisting at the surface as gullies,

and 3) Old age - collapse of pipes.

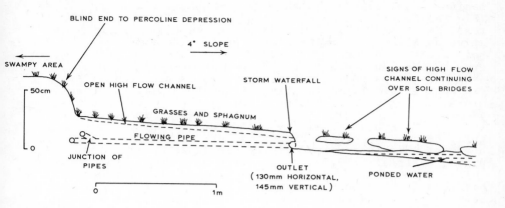

Fig. 44   Soil pseudokarst near head of piped percoline on
          Burbage Brook (Jones, 1975, fig. 6.2.7; 1978a,
          fig. 6).

He even suggests that discontinuous gullies are the youthful
stage of all gully development (Heede, 1974), and some
evidence for this is provided by N.O. Jones's observation
that discontinuous gullies on air photographs of 1935 were
continuous by 1954 (Jones, 1968, 13).   In general, gully
development depends on many trigger mechanisms similar to
those causing piping, such as devegetation, slope instability and rejuvenation (cp. Heede, 1976, 3).   Sometimes both
forms exist together with the one encouraging the other, as
at Alkali Creek. Sometimes they may be alternative responses
to similar stimuli.   This is suggested by Gibbs's (1945)
observation that no pipes, only open V-shaped gullies formed
in stony soils adjacent to his piped area, and by Downes's
(1949) observation of tunnelling and U-shaped gullying in
one group of soils in Victoria but only V-shaped gullying on
adjacent soils with more stable crumb structure and lacking
the dispersive A horizon of the other.   Given the stimulus,
the decision-tree of natural processes depends on the suitability of the soil material for pipe development.   In
Rhodesia, Stocking (1976a and b) noted that gullies occur
alone where ESP values are too low to cause the dispersion
necessary for pipe development, whereas 'extremely serious
piping and gully erosion' are one of the 'peculiar problems'
of sodium-rich soils (Stocking, 1978b).   Newman and Phillips
(1957) noted that in solodised solonetzic soils and brown
earths in New South Wales piping was encouraged by gullying
where B horizons were impermeable.   Deep piping associated
with gullying caused rapid drying of the soil, encouraging
cracking and more tunnelling.   References to gullies forming
the foci for subsidiary pipe development are really too
numerous to mention.

Stocking (1976a and b) reported very rapid development in pipe-gully systems observed between 1972 and 1975. Some monitored gullies passed through a complete cycle in 2 years (Fig. 45), whilst others remained static. In a wider context, Stocking (1978a) considered the relevance of quantitative assessment of rainfall energy to investigations of soil erosion in the light of his observations of gully growth. He concluded that 'the way in which the energy is used and the pathways that water takes is critical to an explanation of erosion'. Since pathways are rarely known, erosivity parameters should be seen as proxy variables, adopted on an empirical basis (*ibid.*, 149). In attempting to assess the best erosivity parameters for predicting gully head advance, Stocking studied 9 representative gully heads of which 6 involved piping, including 3 of mixed piping/waterfall origins (Fig. 46). He used correlation and multiple regression analysis (d.f. 20) to test parameters. His results have been reorganised to emphasise the difference between piped and waterfall headcuts in Table 19. For the piping headcuts inclusion of a 10-day antecedent rainfall index improved the correlation of storm rainfall parameters with the volume of soil removed from gully heads, whereas it made only a marginal improvement in waterfall cases. In fact, simple ranking of the coefficient (exponent) of antecedent rainfall in Stocking's regression equations in Table 20 shows that this parameter has the greatest weighting for the piped headcuts, whereas there is no distinction between waterfall and composite-process headcuts, although the fourth highest exponent does occur in the composite class. On the same basis, there is no difference in the effect of total storm rainfall between the three groups of gully head. Apart from the 10-day antecedent rainfall index, the best descriptors of both piped and waterfall headcut erosion were total storm precipitation, momentum or energy. In the composite heads, threshold intensity parameters are more important. Stocking does not speculate as to the reasons for these differences. However, the stronger relationship between erosion at piped fully heads and antecedent rainfall could be the result of a correlation between frequency of pipe flow activity and antecedent rainfall.

Crouch (1976) has tried to resolve the 'chicken-and-egg' relationship between piping and gullying in his division of tunnelling processes into three groups, which include one group which initiates gullies and another initiated by gullies (cf. chapter 2). The distinction is made on the assumption that the deep tunnels which initiate gullies are formed in cracks in the B_2 horizon, whereas tunnels caused by gullies are initiated by the boiling action of infiltrated water, as outlined by Parker (1963). Unfortunately, Crouch's formulation is too much of a special case to be generally applicable. On the one hand many references attribute desiccation-stress crack piping to improved drainage around gullies (e.g. Ireland, *et al.*, 1939; Parker and Jenne, 1967) and on the other, Parker (1963) and others have suggested that boiling can lead eventually to gully formation.

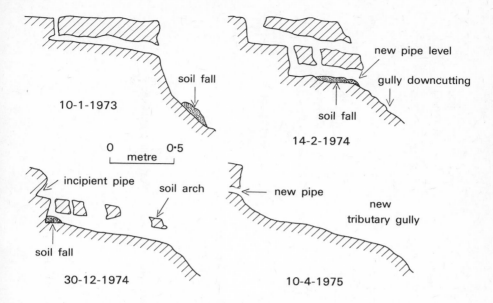

Fig. 45 Progressive development of a tunnel in Mhondoro,
Rhodesia, after Stocking (1976b, fig. 1).

Interesting observations of the collateral development
of small-scale piping and rilling have been made by Bryan,
Yair and Hodges (1978) in Canada and by Yair, Lavee Bryan
and Adar (1980) in Israeli badlands (cp. section 8.3).
Bryan et al. (1978) noted that pipes captured rill flow and
often returned flow to rills in repeated sequence on shale
badland slopes. Yair et al. (1980) observed both the
development and destruction of pipes during sprinkler experiments in Negev badlands. Fig.47 illustrates the regolith
profiles and location of piping on these slopes. The piping
developed after infiltration through desiccation cracks in
the surficial crust reached a thin non-coherent layer of
granular sand-size particles underlain by a compacted layer.
This underlayer was absent elsewhere and may be formed by
slaking of the granular layer. Rillflow began first,
within 90 sec of the start of the experiment on the main
piping plot. New pipes developed after 15 min and rill
bank and pipe collapse was general after 30 min.

On slopes where there was little evidence of pipe collapse and rill development, they report dense smooth crusts
downslope of pipe outlets, which presumably restrict rill
erosion and seem to have formed by the spreading out and
drying of broad, liquid mudflows coming from the pipes.

As with gullying, so rilling may commonly occur with or
without piping in adjacent environments depending on the
properties of the earth materials involved. This is apparent
in the contrast between opposite facing slopes in the study

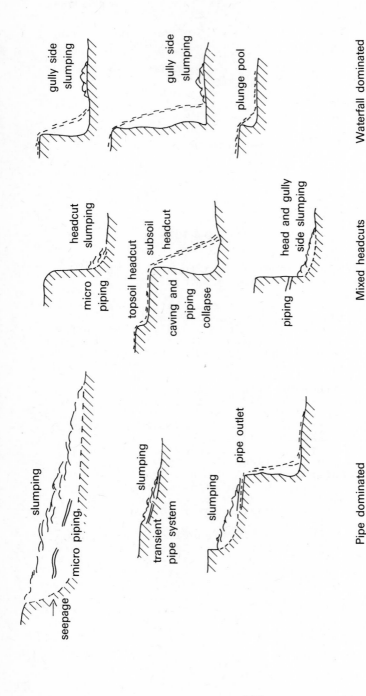
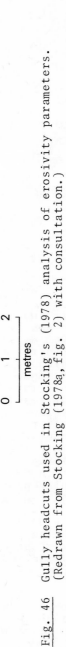

Fig. 46   Gully headcuts used in Stocking's (1978) analysis of erosivity parameters. (Redrawn from Stocking (1978a, fig. 2) with consultation.)

Table 19. Correlation between erosivity parameters and soil loss from piped, waterfall and composite gully headcuts in Rhodesia.

Erosivity Parameter	Piped head cuts						Composite waterfall and piped head cuts						Waterfall head cuts					
	1		2		3		4		5		6		7		8		9	
	(1)	(2)	(1)	(2)	(1)	(2)	(1)	(2)	(1)	(2)	(1)	(2)	(1)	(2)	(1)	(2)	(1)	(2)
P	0.281	0.679	0.343	0.781	0.615	0.933	0.696	0.729	0.657	0.688	0.632	0.636	0.793	0.804	0.544	0.581	0.764	0.891
P_{12}	0.038	0.626	0.268	0.791	0.523	0.938	0.858	0.862	0.814	0.814	0.746	0.765	0.785	0.788	0.462	0.495	0.603	0.761
P_{25}	0.008	0.623	0.230	0.780	0.454	0.901	0.820	0.824	0.751	0.751	0.815	0.838	0.621	0.627	0.477	0.503	0.599	0.733
E	0.153	0.649	0.330	0.803	0.576	0.947	0.788	0.806	0.735	0.739	0.745	0.756	0.765	0.774	0.505	0.543	0.674	0.824
$E_{>12}$	0.057	0.630	0.280	0.795	0.532	0.941	0.863	0.867	0.816	0.816	0.765	0.783	0.778	0.779	0.466	0.498	0.602	0.755
$E_{>25}$	0.006	0.626	0.208	0.781	0.421	0.896	0.852	0.856	0.742	0.742	0.812	0.843	0.644	0.650	0.461	0.493	0.599	0.732
M	0.243	0.676	0.356	0.806	0.594	0.944	0.749	0.776	0.716	0.725	0.712	0.720	0.779	0.790	0.539	0.578	0.729	0.872
$M_{>12}$	0.060	0.631	0.282	0.797	0.530	0.941	0.860	0.864	0.815	0.815	0.757	0.776	0.780	0.782	0.465	0.497	0.610	0.755
$M_{>25}$	-0.014	0.626	0.177	0.776	0.379	0.883	0.828	0.834	0.712	0.713	0.799	0.835	0.617	0.626	0.460	0.493	0.599	0.733
EI_5	0.103	0.634	0.316	0.796	0.558	0.934	0.843	0.851	0.793	0.793	0.800	0.812	0.746	0.751	0.517	0.545	0.641	0.761
EI_{15}	0.132	0.636	0.323	0.787	0.597	0.944	0.828	0.839	0.778	0.779	0.776	0.783	0.760	0.765	0.518	0.548	0.665	0.783
EI_{30}	0.156	0.648	0.333	0.801	0.583	0.945	0.800	0.816	0.746	0.750	0.772	0.781	0.756	0.763	0.505	0.541	0.670	0.805

Correlation between erosivity parameters and soil loss from piped, waterfall and composite gully headcuts in Rhodesia, based on Stocking (1978a, tab. 2) : (1) simple correlation, (2) multiple correlation of a regression including antecedent precipitation.

Significance levels : $\propto$ = 0.01 for $\geqslant 0.564$, $\propto$ = 0.001 for $\geqslant 0.672$, d.f. 20.

Parameters : P – Rainfall (mm), E – Energy of rainfall (Jm^{-2}) and M – momentum per millimetre of rainfall ($kg\ m^{-1}\ s^{-1}\ m^{-2}$) respectively divided into daily total (P, E and M), total for intensities $\geqslant 12\ mm\ h^{-1}$ (P_{12}, $E_{>12}$, $M_{>12}$), and for intensities $\geqslant 25\ mm\ h^{-1}$ (P_{25}, $E_{>25}$, $M_{>25}$).

EI – Energy–intensity index for 5, 15 and 30 min (EI_5, EI_{15}, and EI_{30}).

Table 20. Exponents of 10-day antecedent rainfall index and total storm rainfall in Stocking's (1978a, tab. 3) regression equations for volume of soil loss (m³) in piped and waterfall gully headcuts in Rhodesia.

Type of head cut	Antecedent rainfall	Storm rainfall
Piped	1.398	1.524
	1.097	2.105
	0.752	0.764
Composite piped/waterfall	0.508	3.454
	0.116	1.366
	0.067	1.382
Waterfall	0.397	1.222
	0.190	1.040
	0.183	2.405

by Yair *et al.* (1980) and in the contrast between sandstone and shale plots studied by Bryan *et al.* (1978). Barendregt and Ongley (1979) observed continuous rilling on sandstones compared with 'a form of interflow' and discontinuous rills in bentonite shales in the Onefour Badlands, Alberta; the interflow included piping in similar bentonite shales in the adjacent Milk River Canyon (Barendregt and Ongley, 1977).

Many descriptions exist of gullying formed by coalescence of circular or oval collapse depressions, similar to the evolution of true karst. The process was described in Vietnam by Saurin (1935; 1944) and Fontaine (1965). Fontaine mapped 'digitated depressions' ('grande depression digitée') with sand trails across the beds and natural bridges at the finger-tips leading to circular depressions, formed by the collapse of roofs of 'grottes souterraines' within the soil profile of calcareous 'black clays'. The lavakas of Malagasy (Segalen, 1949; Hénin, 1949) and owragi (ovraghi) of the Ukrainian steppe (Schomer, 1953) are similarly formed gullies, and Neboit (1971) has described the process in Southern Italy, again in calcareous clays, extending from the edge of 'fossi' because initial cracking was greatest there. Here small dolines, 10 m in diameter, form with one or more small tubes leading off to a level of 'conduits souterrains' with outlets in sandy soils near the base of the gully wall. Coalescence of these dolines leads to gullies with very irregular long profiles. Feininger (1969) described a dendritic pattern of several dozen depressions with 1 m holes in the bottoms and karstlike blind valleys in clayey saprolitic soils in Colombia. And in Poland, Maruszczak (1965, 95) described gullies connected to kettle-holes or pits in 'porous sediments' by mechanical suffosion and 'evorsion', and Zaborski (1972) has described the gradual coalescence of loess depressions, giving youthful gullies with irregular beds. Sautter (1951) noted similar forms in Congolese sands.

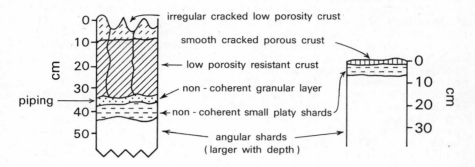

irregular cracked low porosity crust

smooth cracked porous crust

low porosity resistant crust

non - coherent granular layer

piping →

non - coherent small platy shards

angular shards
(larger with depth)

Fig. 47   Regolith profiles on pipes and non-piped semi-arid
          slopes used in the study by Yair *et al.* (1980).

Bogutski (1968) described two types of 'steppe dip' in
Russian loess:Type 1 is round and randomly located, mainly
on flat interfluves and river terraces and Type 2 occurs on
valley sideslopes in chains of 5-6 depressions, elliptical
in shape with long axes oriented downslope. The dips are
tens of metres (occasionally 200 m) in diameter and up to
2 m deep. Both types are formed by washing out of clay
particles and Type 2 act as water collecting basins for
intense gullying developed along fractures in the sandstones
or limestones underlying the loess. Grain-size analysis
indicate a winnowing out of fines in the depressions, with
greatest differences in the 0.25-0.05 mm range (21% in dips,
9% in surrounds) and finer than 0.005 mm (9% in dips, 16%
in surrounds).

8.2 Piping and mass movement.

The relationship between piping and major mass movement
is ambiguous. In some cases, pipes have been identified
as the cause of landslides or subsidence. In other cases,
pipes are thought to have developed as a result of the subsidence or landsliding, in the zones of tension and rupture
around the slipped mass, in the landslip scar due to increased hydraulic gradients, or else within the slipped mass,
perhaps at points of fracture or due to bursts under pressure. The exact relationships have never been closely re-
searched. Baillie (1975, 14) points out that the probability
of both landslips and piping is increased with increasing
throughflow, but that the relationship between them is complex. It is also clear that the role of piping is still
generally ignored in studies of mass movement and that by
and large such observations of the role of piping as have
been recorded in the literature have received relatively
little emphasis and have often been obscure.

A number of observations have been recorded from New
Zealand. Ward (1966a and b) noted that piping often occurs

on the intermittently moving bodies of 'incipient sheet-slides' and on displaced landslide masses in North Auckland. The 'incipient sheet-slides' resemble grassed over sheet-slide scars but there is no evidence of catostrophic sliding, and Ward suggests that subsurface eluviation of material, a process akin to piping (and, indeed, would probably be termed 'entrainment piping' by Parker and Jenne (1967)), allows the slide body to subside slowly, with crescentic scarps upslope and flanking it. Once the mass has settled to an angle less than 12°, pipes and shafts commonly develop in the surface horizons. Piping may also prepare slopes for sliding. Ward seems to suggest that 'embryonic pipes' (pseudopipes or 'slightly enlarged continuous voids') allow rapid throughflow let in by surface desiccation and tension cracks and that this can reduce soil cohesion and allow a build up in the weight of the slide body which may trigger a landslide. From observations also made in North Island, Jackson (1966) concluded that fissuring during dry periods increased landslip failures during rainy periods in yellow-grey podzolic soils developed on loess, and Bell (1968, 251) mentions interrelated 'piping and bank retreat by gravity slices', with each process promoting the other in the bad-lands of North Dakota. Here slippage is promoted by snowmelt with snowbanks limiting the penetration of frosts. He maintains that: 'Piping is an important factor in the process of landsliding which is in turn one of the prime agents in the development of Badland topography' (*ibid.*, 255). Landslips become honeycombed by pipes as they waste away. Bell suggests that each of Parker and Jenne's (1967) genetic types of piping are active during the landslide cycle:

i)   Desiccation-stress and ice wedge cracks form on the crests of slices and blocks.

ii) Entrainment occurs where release of high effective pressures on the down-gradient side of a block causes 'boiling'.

iii) Variable permeability subsidence occurs on the face of some blocks, with jetting along the toe of moving blocks where water is forced from com-pressed cavities.

Watson and Wright (1963) and Wright (1964) considered piping to be responsible for landslides and closed, lake-filled subsidence depressions in the Chuska Mountains, New Mexico. Piping is thought to have occurred in loose sands beneath a sandstone caprock and they state enigmatically that piping 'operates most effectively just beneath a relatively impervious cover' (Watson and Wright, 1963, 543), aiding the separation of ridge-blocks and gliding of slips by lubrication. They quote Terzaghi's (1931) observation of dramatic piping failure through a sand layer exposed in the east bank of the Mississippi near Memphis, Ward's (1948) description of landslides at Castle Hill, Newhaven, Sussex, after wave action exposed a fine sand layer and Haldeman's (1956) report of collapse and 'bottle-flow' in saturated ash beds, removing 25,000 ft^3 along a 2 ft. diameter pipe in the Rungwe volcanic area, Tanzania. Mears (1968) states that some disagree with Wright's hypothesis, although he admits the possibility of boiling assisting landsliding.

Suffosion has also been considered as being responsible for subsidence of lake-filled depressions in the Tbilisi area, Georgian SSR, by Alpaidze (1965), but in this case, and in a number of others in Eastern Europe, solution and leaching, i.e. chemical suffosion, in gypsum and potassium salt deposits appear to be responsible.

Polish writers in particular have noted how piping can aid landslips in loess. Kastory (1971) considered that the larger landslides at Godziszow, Poland, were due to piping. Walczowski (1971) has ascribed landslips on the edge of loess in the Vistula valley to piping in 1 m thick sands and gravels sandwiched between 15 m of loess and Tertiary clays. Continued landslips create boggy 'suffosive terraces' around the loess edge.

Piping may contribute to slipping by undermining or by allowing rapid access of water to slip-prone areas. However, Jackson points out that no definite cause and effect relationship can yet be inferred and that, since both tunnelling and slipping tend to develop under the same conditions, their association does not imply a causal relationship. Also in New Zealand, Crozier (1969) explained many earth flows during a storm of unknown but very high return period in Eastern Otago, by the presence of large pipes up to 30 cm diameter and 90 m long which run at the base of the regolith over incompetent sands and silty mudstones.

Johnson and Rahn (1970) give a good description of 'spring sapping' in Pennsylvania, which began with the fall of turf clods from the vertical bank above the spring outlet. As turf units cracked and shifted behind, the area of collapse increased until the water became turbid and eventually the water became incorporated in a sliding semi-fluid mass. Similar events may have been responsible for a number of slope failures reported in Britain. Following the lethal 1952 Lynmouth floods in Devon, Gifford (1953) noted a close association between land-slides and seepage lines, and reported water entering a landslide scar 'through a series of channels the size of rabbit holes' (*ibid.*, 16) in the stoney slope deposits. Bunting (1961) also described the close association between severe storms, seepage lines and landslides in Northern England, although without directly observing any features akin to piping.

Some association between slumps and piping has been found at the Institute of Hydrology experimental catchment on the Upper Wye, Plynlimon. Bell (1972) reported pipes in slump faces up to 40 cm in diameter, though more generally they were 5-10 cm across, but he could see no clear relationship between slumps and piping. Newson (1975, 24-25) describes sliding of soils and superficial deposits over a

lubricated stratum along intermittent pipes or seepage lines
during the severe floods of August, 1973. The pipes which
lubricated the slides are 'less regular' than the perennial
pipes of the flushes and tend to concentrate in stony areas
or where creep has caused tearing. In contrast, there is
no sliding associated with the ephemeral pipes which typify
the middle slopes in the peaty, gley podzols. Newson also
noticed 'bursts' along perennial pipes in the peat-filled
rushy flushes, with rupturing of the peat infill and sliding
of peat rafts during the floods. He concluded that the
subsurface movement of water and mass movement on slopes are
inseparable and that the peat infill of the flushes is probably moving 'glacier-like' down a bedrock gully particularly
during floods. There is a similarity here with the idea of
Walter Penck (1953, 113) of 'linear corrasion' valleys
formed by slow slippage of the soil mass with some evidence
of erosion by flowing water.

Jones (1975, 284; 1978a) noted that intermittent and
perennial pipes on the river terrace of Maesnant, Plynlimon,
tend to follow mass movement cracks caused by slumping along
seepage lines (Fig.48). Ephemeral pipes seem to have
developed to feed these in places. The indication is that
at least part of the network post-dates slumping.

Ward (1948) described a rather special case of mass
movement which appears to have involved what Parker and
Jenne (1967) call 'entrainment piping', in which the surface
peat layer of Borth Bog, mid-Wales, cracked and moved forward
on top of the clayey silt underlayer which was in a 'quick'
state.

Discussion of bursts invites comparison with bog bursts.
Colhoun, Common and Cruickshank (1965, 171) suggested that
bog bursts in Northern Ireland may be cyclical, involving
gradual growth of the peat mass to the point of instability.
Gully erosion appears to be an alternative and gullying was
thought to be less common there than in the Pennines, where
no bursts were reported. Without gully erosion, the lower
peat layers tend to become very soft as humification pro-
ceeds, so that either the bog becomes unstable as it grows
above this base or else water pressure increases in the
lower layers and ruptures the surface of the bog. Jones
(1975, 147) describes observations made by Miss J. Clarke on
the same bogs along the Glendun River, N. Ireland, in 1970,
which suggest that the more stable bogs contained piping;
piping was found along the Altaguire between Beaghs and
Owennaglush River, but not in the bogs with recent bursts
on Brockaghs Burn (1963), Beaghs Burn (1903) and Owennaglush
River (unknown date). Pearsall (1968, 274) noted a similar
relationship in that a bog 'will either flow or develop
its own internal drainage system'. Jones (*ibid*) suggests
that the presence of piping may stabilise the peat by preventing excessive build up of hydrostatic pressure, although
clearly it would always be possible for the pipe capacity
to be exceeded in the manner described by Newson (1975, 22);
and, in fact, Clarke did report minor piping on the edge of
the old Beaghs Burn burst.

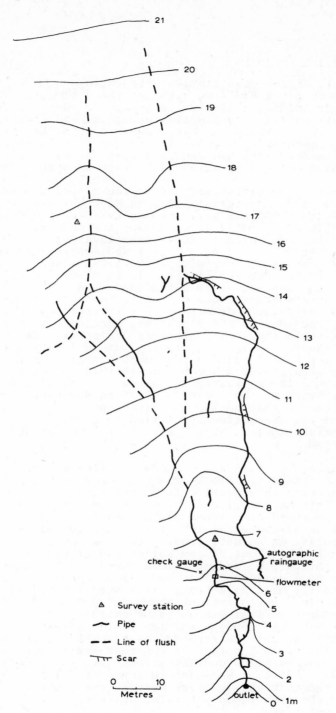

Fig. 48   Detail of the Maesnant pipe network showing relation to slump features (Jones, 1975, fig. 8.1.1;
1978a, fig. 5).

8.3. Pipeflow sediments

As yet, direct quantitative observations of the sediment yield of pipes are few, although there is quite a considerable amount of indirect evidence in the form of sediment streaks on hillsides, deltas and cones and fine linings in pipes. In all respects there seems to be marked differences in the quantity of material transported in British pipes as compared with badlands and with the highly erosive soils of parts of Australasia and Eastern Europe.

In Britain, Lewin *et al.* (1974) considered that sediment discharge from arterial pipes on Maesnant was very minor . Jones (1975, 252) reported only 40-80 ppm for combined dissolved and solid loads during the main discharge events in the same basin in the winter 1974-75, and Newson (1976a) reported a total of 500 g of sediment, mainly peat nodules, discharged by one pipe in Nant Gerig nearby during the whole of the same wet winter. It is therefore eminently reasonable to suppose, as do Gilman and Newson (1980), that these Welsh networks are quasi-stable. Theoretical justification for this also comes from the fact that the peak velocities recorded in the pipes (cf. section 7.1) are all basically within the limits of 'safe' velocities for clay loams and sandy clay loams quoted by Luthin (1966). This is not to say that local erosion cannot occur. Indeed, the author has seen gravel and cobbles (individual cobbles weighing up to 180 g) raised to the surface by pipe resurgences activated in the spring of 1979 on Maesnant. Newson (1975, 25) noted that during the Plynlimon floods of August, 1973, ephemeral pipes clearly carried large quantities of peaty material causing large black stains seen around bursts in the system.

Measurements of pipeflow and streamflow sediment concentrations on Burbage Brook 1968-69 suggested very low concentrations even in stormflow from stream bank outlets (Jones, 1975, tab. 7.2.1). Only 9ppm solids and 6ppm dissolved load were measured over an extended period of sampling in a dry weather discharge of 100 ml s^{-1} issuing from a major piped percoline in August, 1968. Low flow concentrations in the Brook were about 27 ppm solids and 61 ppm dissolved load. Stormflow pipe sediment load could not be estimated separately from dissolved load because of the small 190 ml sampling bottle used as a trap, but a very limited sample showed no joint concentrations above 400 ppm solid and dissolved in stormflow from midbank pipe outlets, compared with 3125 ppm solids and 160 ppm solutes in the stream. The measurements can only be regarded as a small pilot study, but Jones (*ibid,* 259) noted the apparent similarity between streamflow and pipeflow concentrations also reported by Lewin *et al.* (1974) on Maesnant in terms of solutes. However, he also noted that the low flow measurements taken after a 13-day drought on a piped percoline contained very little dissolved material.

McCaig (1979b, 7) noted the slow rate of change in a British pipe network in Yorkshire. Because of the very slow retreat of pipe outlets, he presumed that the plumes of mineral sediment below pipe outlets must be from bed erosion within the pipes. He listed the sources of sediment on his hillslopes as 1) pipe beds, 2) sides of pipe collapses, 3) active drainage lines especially below pipe outlets, which he estimated to be the main sources, and 4) slope surfaces, which deliver mainly organic matter. McCaig set up a number of sediment traps along piped seepage lines (c.12 modified Gerlach troughs). His efforts therefore potentially represent the most detailed investigation of pipe sediment load to date, beginning in October 1977, on a three year research programme. His results so far suggest that suspended sediment concentrations are very irregular in relation to discharge, because high concentrations are produced by discrete events such as roof collapse or erosion at micro-knickpoints. Unfortunately, McCaig has not yet published data from the pipe traps separately and we must eagerly await completion of his thesis.

One of the few quantitative studies in Europe, which seems to have been largely overlooked even in the German literature, comes from Hauser and Zötl (1955) in Austria, who studied the process of suffosion in the Sarmatian sandstone of the mountains of Graz. They recorded 0.5 kg of sand a week issuing from one resurgence with an average discharge of 0.16 1 s^{-1}, and estimated an annual erosion rate of 13 dm^3 in the region due to suffosion.

Fine sedimentary linings have been analysed from British pipes by Jones (1971, 607; 1975, 259-260; 1978a, 12) and Conacher and Dalrymple (1977, 21). Jones analysed three varieties of deposit on Burbage Brook: i) deltaic deposits in a streambank outlet, ii) fine veneers in similar pipes and iii) coarse deposits below the outlet of pipes on terrace scarps. Only 5% of streambank pipes (out of a total 138 surveyed) contained recognisable deltas inside the outlet and all were located above the mean height of outlets in the bank (44% bankfull height), whereas a similar proportion of outlets had erosion channels extending down the bank below the outlet and all these cases were located at below mean outlet level. This suggests that the deltaic deposits were formed in pipes which flowed for shorter periods, with insufficient pipeflow to remove the sediment after the stream stage fell below pipe level; conversely rilled outlets are evidence of longer continued drainage from greater soil volumes. The most significant differences between delta and bed soil were 1) no particles above 1mm diameter in deposits, against 5% over 1 mm in the surrounding soil, ii) this was balanced by 5% more fine sand to medium silt in deposits, and iii) deposits were much more dispersible (46% aggregates stability against 96%). Sampling of fine veneers on walls and beds of pipes suggested significantly less material over 100 μm diameter compared with the surrounding 'body' of lessivé brown earth and slightly better sorting. In a gleyed lessivé, a 10 cm diameter pipe was found completely blocked in a percoline 2.5 m from the stream bank. The blocking material was 68% organic and probably consisted basically of ferric iron bacteria. Excavation revealed a hard iron concretion c.1 mm thick lining roof, walls and bed. Conversely, coarse sand and gravel deltas were found on three 15-20 cm outlets on terrace scarps with evidence of renewal after most storms of 25 mm and over. Blong (1965, 84) also mentioned that veneers of fine sands and silts were usual in the New Zealand pipes he studied.

Conacher and Dalrymple (*loc. cit.*) reported silt-sized materials composed of rounded to subangular quartz grains, soil colloids composed essentially of iron oxide/clay complexes and redeposited iron oxides in linings on the walls and in unconformable bed deposits in pipes of 1-3 cm diameter in gleyed lessivés. In addition, bleached sand grains were washed into the larger cavities around. These pipes were clearly on the border of pseudopipes and 'true' pipes, rarely more than 25 cm long and clearly did not carry very great discharge. In some cases, discrete, seasonally active G horizons developed on landsurface units 2(2) above London clay had circular or ⌒-shaped cross-sections suggesting pipes blocked with translocated soil materials (*ibid.,26*).

Banerjee (1972) studied thin sections of iron oxide from soil-filled pipes in ferrallitic soils in Bengal and found large irregular quartz and silicate grains in a matrix of hematite and goethite, which he attributed mainly to desiccation and crystallisation of material from iron-rich seepage water. Clay skins have also been reported in paleo-piping in argillic horizons in desert soils by Gile and Grossman (1968, 12), whilst in K horizons laminar carbonate linings were observed (Gile, Peterson and Grossman, 1966).

Much higher sediment loads may be expected in badlands. N.O. Jones (1968, 81) suggested from qualitative observation that pipeflow concentrations were an order of magnitude higher than sediment concentrations in sheetflow. This is supported by his observation that bed deposits of 15 cm were cleared by 'even small flows' (*ibid.*, 90), and by a very rough estimate of 16000 ppm in one flow (*ibid.*, 85). However, blocks of crusted soils 1 m in diameter presented an immovable obstacle. Heede (1971, 7) reported measurements of pipe sediment loads in the semi-arid Alkali Creek watershed, Colorado (Table 21). The material deposited below one pipe after spring snowmelt flow (the pipes only flow after snowmelt) was 'relatively high in silt content'. Analyses of suspended sediment from two samples taken during the spring snowmelt flows of 1964 and 1965 indicated high ESP and pH with sodium concentrations of 140 ppm, calcium 10 ppm and magnesium 18 ppm, reflecting the characteristic content of the local 'youthful' solonetz soils.

Recent measurements of rate of sediment yield in another semi-arid area have been reported by Barendregt (1977) and Barendregt and Ongley (1977) in the Milk River Canyon, Alberta. They built a dam across the outlet from one system in 1975. Unfortunately, the sediment overtopped the dam in

Table 21. Chemical analyses of pipeflow sediments and source material based on Heede (1971, tab. 2): mean values.

Sample	ESP	SAR	Conductivity at 25°C (mmhos/cm)	Gypsum (me/ 100g)	pH	Sample Size
Sediment from pipeflow	7.2	6.2	-	-	8.6	2
Alluvial layers inside pipe	10.1	8.7	2.6	0.2	9.5	13
Deposited below pipes (after snowmelt flow)	10.3	8.7	0.9	0.1	9.4	10
Pipe inlet in soil	13.1	11.1	1.9	1.8	8.6	2
Shale parent material	17.9	15.7	0.8	0.9	9.7	4

a summer storm of 40 mm in 6h. Only about a quarter of the total sediment yield of c.112 tonnes was actually caught in the trap. Extrapolating on the basis that 5 similar storms occurred in the summer of 1975 they calculated that piping shifted 111×10^3 tonnes of sediment from the whole of the canyon walls and concluded that it is the main mechanism of valley wall recession.

In their study of processes in the Negev badlands, Yair et al. (1980) took regular samples of sediment loads during sprinkler experiments. Although pipe load was not measured separately, pipeflow and rillflow were the dominant sources of discharge on the slopes where pipes were present, and therefore probably of sediment. Sediment yields were greater from the piped slope (235 g m^{-2} of slope) than for the unpiped slope (85 g m^{-2}) or for the basin combining both slope faces (12.3 g m^{-2}) (Fig. 49). However, because of the greater discharge from the piped slope, sediment concentrations were greatest in the unpiped slope. It should be noted, however, that reliable sediment samples could only be taken on the piped slope up to the 24th minute, whereas general pipe collapse and mudflows did not occur on the piped slope until after 30 minutes, when sediment increased suddenly. The contrasts would have been greater if this had been measured.

Yair et al. (1980) also noted that sediment concentration was pulsatory and not directly related to runoff. This was probably due to collapsed unstable soil blocks in the rill and pipe system, similar to the situation described by McCaig (1979b) in Britain. The same pulsatory nature was observed in viscous rillflow on another plot, where subsequent investigations suggested that pulsatory sediment

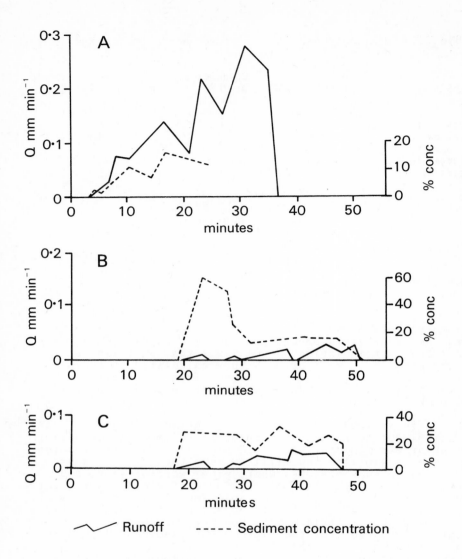

<u>Fig. 49</u> Runoff and sediment data for piped and non-piped
        semi-arid slopes obtained by Yair *et al*. (1980).

release might be explained by the time necessary for hydrodynamic pressure to build up to a threshold level necessary
to extrude the shards through the surface crust.   Finally,
they noted concentrations of salt after 50 min near the
wetting depth and around pipes, which suggest marked mobilisation of dissolvable salts as well.

        Earlier studies, although purely qualitative, indicate
the frequency of sediment discharge from pipes in many
different environments.   Delta deposits, downslope of outlets are a fairly frequently observed feature in Australia.

Downes (1946, plate 6) observed outlet deltas extending for many yards, giving white streaks on air photos in Victoria (also Marker, 1958), and Newman and Phillips (1957, fig. 5) observed deltas of fine sand and silt, most of the clay having been washed away. Bell (1968) also noted deltas in North Dakotan badlands.

Henkel, Bayer and Coutts (1938, 239) reported 'fan mounds' in Natal where outlets occur on hillslopes or in gullies which are colonised by grasses and sedges. They took the vigorous herbaceous growth on these mounds as indicative of faster subsurface than surface erosion. When the outlet occurs on flatter ground a cone is formed (Fig. 50). They remarked that large volumes of water poured out of the openings after heavy summer storms and continued to flow for some time after, but the quantity of material deposited at outlets was 'unexpectedly small', though spread over a wider area than first appeared. Downing (1968) observed the same features on bottomlands in Natal causing sufficient aggradation of the floodplain to shift river courses away from the cone-shaped mounds. Similar ejection cones have been reported by Polish and Russian writers in loess (Malicki, 1946; Subakov et al., 1967), on periglacial slopes in Spitzbergen (Czeppe, 1965) and by Kunsky (1957) downslope of 'fissure caves' on slopes in Czechoslovakia.

There is clearly considerable room for further field experiments in monitoring pipeflow sediments.

8.4. Pipeflow solutes.

Measurements of pipeflow solutes have formed part of a number of doctoral research projects in Britain during the 1970s, with studies in Wales (Lewin, Cryer and Harrison, 1974; Oxley, 1974; Cryer, 1978, 1980), the Mendips (Waylen, 1976; Finlayson, 1976, 1977) and the Pennines (McCaig, 1979a and b). However, the field is relatively new and no comparable data seem to be available from elsewhere, nor, indeed, have these British authors compared results amongst themselves.

Cryer took weekly samples from the outlets of three perennial pipes draining peaty gley podzols on a terrace of solifluction material in the Maesnant basin, mid-Wales, during 1971-73 (Lewin et al., 1974; Cryer, 1976; 1978; 1980). He found no significant difference between water samples taken from any of the three pipes or from the stream in terms of specific conductivity, pH, Ca, Mg, Na, K or Cl ion concentration based on sample sizes between 21 and 84 (Cryer, 1978, 269 ff; 1980). First difference correlation analysis showed that seasonal fluctuations at all four sampling points, which gave above average solute concentrations in winter, reflected the seasonal fluctuation in bulk precipitation quality (Cryer, 1976). This led Cryer to argue that pipeflow should be regarded as hydrologically identifiable with surface streams (Cryer, 1980), i.e. that the sources are the same.

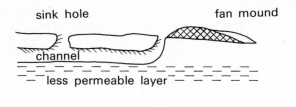

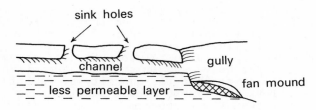

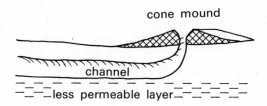

Fig. 50   Forms of pipeflow deposits observed by Henkel
et al. (1938, figs. 2-4).

Similar water quality analyses on samples from three observation wells in the peat and from two overland flow sites showed marked differences in chemistry between these sites and the pipeflow and streamflow sites (Table 22).   For every solute species except potassium mean and weekly concentrations were lower in the peat, whereas concentrations of all specific ions were higher in the surface flow water compared with pipeflow or streamflow.   The high concentrations in surface water were thought to be due to a combination of dry deposition of aerosols on the surface and ion production from decomposing plant litter (Cryer, 1978). The fact that the pipe and stream water was intermediate in quality between these two probable sources, the peat matrix water and overland flow, led Cryer (1980) to support Jones's (1975; 1978a) conclusion that pipeflow is derived from a combination of the two sources.

Unfortunately, Cryer did not measure pipeflow discharge, so that in order to compute solute rating curves he used Jones's (1978a) observation that pipeflow in the vicinity of Cryer's sampling sites varies very closely with stream discharge, which was measured.   By using a reciprocal transformation of discharge as a model for the mixing of two

Table 22 Relative chemical quality of pipeflow water as indicated by measurements on Maesnant taken by Cryer (1978; 1980).

	site	pH	Ca	Mg	Na	Cl	K	Spec. Cond.
Rainfall		4.90	1.65	0.37	2.27	4.18	0.14	27.5
OLF	1	4.09	1.19	1.45	4.19	10.0	2.14	78.9
	2	3.86	1.13	1.23	3.81	9.96	2.62	87.4
Pipeflow	1	4.62	0.75	0.62	3.48	5.87	0.14	34.8
	2	4.72	1.05	0.64	3.67	6.09	0.15	34.5
	3	4.51	1.13	0.69	3.77	6.15	0.15	40.2
Stream		5.12	0.92	0.67	3.41	5.55	0.13	32.6
Peat matrix	1	6.17	0.51	0.56	2.67	7.04	0.39	97.6
	2	6.44	0.54	0.51	3.02	5.51	0.36	77.0
	3	6.68	1.65	0.94	2.65	4.87	0.61	95.6

ionic concentrations in mg l^{-1}

specific conductivity in μmhos/cm at 25°C

chemically distinct water types, Cryer obtained statistically significant approximations to measured pipeflow quality, except for sodium. The solute ratings indicated increasing concentration with increasing discharge, which could be explained by increasing dominance of the surface water source in pipeflow at higher discharges. pH showed a reduction with increased discharge probably due to mixing of acid overland flow (pH 4), neutral peat water (pH 6) and rainfall (pH 5) in similarly differing proportions. This is also in agreement with the conclusions of Jones (1975; 1978a) based on hydrograph analyses.

Cryer's analyses suggested different residence times for different elements which indicate complex short term ionic exchange and storage, particularly affecting calcium and sodium. The short residence time of sodium might explain the dilution effect with increasing discharge, the reverse of the other elements. Deviations of pipeflow quality from the mixing model prediction could be largely explained by changes in the ionic concentrations of the rainwater, chemical buffering by the soil and a possible dilution effect from rainwater entering quickly via overland flow may also explain some deviations. However, the mixing model could not explain conductivity and potassium concentration which were lowest of all in pipeflow. Some unknown source of conductivity seems to be present in the peat and surface waters, perhaps due to dissolved organic compounds. Lewin *et al.* (1974) noted that the solute data obtained from these measurements did not indicate any active denudation. The main feature was that in passing through the soil the concentrations were smoothed and less variable than in the original rainwater.

Oxley (1974) also measured solute concentrations in a Welsh pipe in the Llyn Ebyr North Basin, Powys. He sampled a 2.5 cm diameter pipe at the base of a 10 cm A_o horizon in Denbigh Series acid brown earth overlying till, with a solid geology of Silurian greywackes, and he also suggested that rapid throughflow, especially pipeflow, was the major factor controlling the solute concentrations in the stream. Oxley reported higher solute concentrations in the pipe than in surface runoff, which may be due to high levels of humic and carbonic acids in the soil water derived from the A_o horizon, causing high rates of chemical weathering in the underlying mineral horizon at the base of the pipes. When pipeflow commences, it flushes out these solutes at higher concentration than in the stream baseflow. However, the concentrations fell as stormflow proceeded, presumably as stored water was replaced by fresh rainwater. A further fall in solute concentrations in the stream was noted after the cessation of contributions from pipeflow. By comparison, Lewin *et al.* (1974) and Cryer (1978, 269), despite the over-all similarity of pipe and stream water, reported invariably higher levels of Na and Cl in the pipes and lower Ca and Mg, possibly they suggest, because of stronger adsorption of divalent ions by the peat organic complex.

Finlayson (1976; 1977) took point samples of dissolved solids concentrations in the East Twins pipes mentioned in section 7.1. Neither the pipes nor his throughflow showed any relationship between discharge and dissolved concentrat-ions as shown by his graphs (1977, fig. 8), whereas the stream showed an inverse power law relationship between the two (*ibid.*, fig. 7). However, a total of only 9 measure-ments were plotted for the two shallow, ephemeral pipes running through the peaty horizon of the podzols in the upper basin, and some scatter could be caused generally by diff-erences in timing between the point sample and the storm runoff cycle within the pipes. Reading Finlayson's graphs, values of 50-100 mg 1^{-1} were obtained regardless of flow regime in the brown earth pipes, comparable to, though a little higher than, throughflow water and stream water. In the podzol pipes values were in the range 12-20 mg 1^{-1} and comparable to the rainfall, with a mean of 15.

More detailed chemical analyses were performed on pipe-flow in the same basin by Waylen (1976). Again his scheme of weekly sampling gave too few measurements for the ephem-eral pipes (cf. Table 23). Hence he based his main conclus-ions on sample sizes of 59 (4A) and 41 (4D) for the perennial pipes. He found higher concentrations of Mg, Ca and HCO_3 but lower K in water from the stream, seep and perennial pipes than in soil water samples, confirming the pattern observed at about the same time by Cryer in Wales (Table 22).

Table 23 Mean Chemical composition of pipeflow after Waylen (1976)

Soil	Flow regime	pH	Ca*	Mg	Na	K	HCO$_3$	Cl	SO$_4$	Si	Sample Size	Code Name
Brown Earth	perennial	6.54	2.09	3.55	5.31	0.86	9.09	8.70	12.16	2.77	59	4A
		7.14	5.70	6.17	5.75	0.81	38.93	8.31	9.70	3.85	41	4D
	ephemeral	6.81	2.43	3.20	5.30	0.76	8.69	8.33	12.0	2.67	15	4B
		5.84	1.69	2.88	5.25	0.72	3.97	8.48	12.30	2.43	9	4C
Podzol	ephemeral	5.30	1.83	2.91	5.16	1.48	4.84	10.06	11.92	1.99	4	P1
		3.60	0.93	0.96	4.03	1.20	0	7.1	5.83	1.26	10	P2
		3.76	1.05	1.35	4.81	1.26	0	9.56	6.44	1.57	2	P3

* all concentrations in mg l^{-1}

He regarded the stream and perennial pipe waters as having been in residence longer than the soil water and that the reactions had therefore progressed more fully. Overall solute concentrations were highest during the first storms following long dry spells, reaching a seasonal maximum in autumn and a minimum in winter. However, there were marked differences between the pipes (labelled 4A and 4D) only 10m apart. Ca concentrations in 4D were 2-3 times those in 4A ($8.9-3.5$ mg l^{-1} against $4.8-1.1$ mg l^{-1}), although both varied inversely with discharge. In comparison, at a nearby seep Ca concentration increased with discharge, in the range 1.5 to 4.0 mg l^{-1}. Concentrations of Mg also varied inversely with discharge in the pipes, and with higher levels in 4D than 4A ($7.59-5.22$ mg l^{-1} against $5.25-2.88$), whilst the seep showed a positive rating of between 1.16 and 3.9 mg l^{-1}. Bicarbonate concentrations showed an identical pattern with $53.13-24.78$ mg l^{-1} in 4D, $18.34-3.85$ in 4A and $10.22-4.15$ mg l^{-1} in the seep. In all three cases, potassium concentrations increased slightly with discharge but remained around the 1.0 mg l^{-1} level. Sodium also remained close to a constant level (5.5 mg l^{-1}) and showed less of a pattern, increasing with discharge in the seep, but showing no relation to discharge in 4D and an inverse relation in 4A. Silicon concentrations showed uniformly inverse ratings, but were rarely above the lower limit for solubility of quartz (3.27 mg l^{-1}) with maxima of 3.95 (4A), 4.9 (4D) and 3.35 (seep). Cl and SO_2 concentrations were very variable but determinations were regarded as imprecise (*ibid*, 284).

Waylen concluded that the close similarity of response between the two pipes suggested that their source areas were hydrologically similar, although the differences in concentrations suggested that they were chemically different. He linked this with the observation that certain beds of the Devonian Old Red Sandstone bedrock are more carbonaceous and suggested that pipe 4D might be fed from such a bed (*ibid*, 285). This seems to imply that Waylen considered a substantial part of the pipeflow to be derived from bedrock, a possibility raised by Stagg (1978) from work in Wales (cf. section 7.1). However, both Weyman (1970; 1971) and Waylen himself maintained that the East Twins stream is not fed by groundwater (*op. cit.*, 107), and in any case the short distance between pipes (10m) would imply very discriminating groundwater movements. Moreover, Waylen noted important contributions to pipeflow from saturation overland flow, especially from slump faces in piped percolines.

More recently, McCaig (1979a and b) has also measured higher solute concentrations in pipeflow than in streamflow in the Slitheroe Clough, Yorks, which he suggests 'may be related to the ephemeral nature of saturation in the source areas'. His streamflow total dissolved solids also failed to show the normal dilution effect with increased discharge and he attributed this to the increasing proportion of flow from the pipe network rather than from diffuse seepage at higher discharges, as illustrated in Fig.51. Although his results agree with Oxley's as regards the relative status of

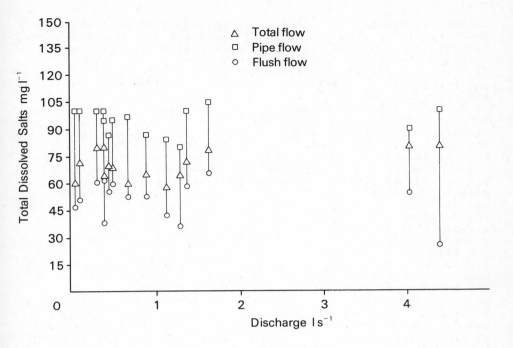

Fig. 51 Dissolved solids-discharge relationships for pipeflow and flush flow observed by McCaig (1979b,
fig. 7).

pipeflow TDS vis-a-vis streamflow, he did not observe the
'flushing' effect reported by Oxley. However, we must wait
until completion of his research programme for fuller
details.

    Few firm conclusions can yet be drawn as to the general
nature of pipeflow solute concentrations or, indeed, as to
the meaning of spatial and temporal variations in specific
ion concentrations. Current knowledge of the relationships
between residence times and solute uptake is still very im-
perfect (Trudgill, Laidlaw and Smart, 1980) and too few
measurements have yet been made on pipeflow solutes. To
summarise the results available within the limited context
of the British Isles, little difference has been found between stream and pipeflow TDS in the case of the perennial
pipes measured by Cryer, Finlayson and Waylen, although
there was a slight tendency for pipeflow to have higher TDS.
However, relationships differ in the case of ephemeral pipes.
Finlayson's ephemeral podzol pipes showed concentrations comparable with rainwater suggesting rapid collection and very
short residence times. Waylen found that all the ephemeral
pipes showed generally lower concentrations of Ca, Mg and
HCO$_3$ than the perennial pipes (Table 23). On the other hand,
McCaig and Oxley found higher TDS in ephemeral pipeflow
which could be attributed to tapping ephemerally saturated
areas and even trapping of solute-rich overland flow. Clearly

marked differences are likely between different environments.
Analyses of specific ion concentrations are only available
for perennial pipeflow and these show marked similarities,
although the reasons for this do not appear to be entirely
clear. Trudgill *et al.* (1980) suggested on the basis of
laboratory based dissolution experiments that high Mg levels
are typical of long residence time water and high Ca levels
of short residence time water. They observed a common pattern in responses from 5 footslope seeps on dolomitic limestone in N.W.Scotland, which comprised (i) a high Mg long
residence water peak immediately prior to a rainstorm, (ii)
general dilution of both Mg and Ca levels associated with
input of quick flow during the first two days of a stormy
period, due to overland flow and possible pipeflow inputs,
followed by (iii) a Ca-rich pulse of short residence water
and (iv) a return to Mg-rich water, possibly displaced long
residence water. To date, no equivalent data are available
for pipeflow itself since Cryer and Waylen only sampled on
a weekly basis, and Oxley and McCaig have not reported
specific ions. However, average Ca:Mg ratios greater than
1.0 in Cryer's Welsh pipes (Table 22) suggest a preponderance of short residence time water. In Waylen's Mendip pipes
the pattern was less marked, with if anything a tendency for
Mg to predominate, which might lend support to Waylen's
apparent suggestion that some water emanates from sandstone
bedrock there. It is possible that the short residence time
of pipe waters in an experimental basin near Oslo may in fact
have a positive environmental benefit in terms of directing
acid polluted rainwater more quickly through the soil and
reducing soil contact time (Nordseth, pers. comm., 1980).

There is no doubt that in the foreseeable future chemical
analyses of pipeflow will add significantly to our knowledge
of flow routes and residence times.

8.5 Rates of development

As might be expected, there are more observations of the
rates of gully extension along pipe systems than of the extension of pipe systems. For example, Thorbecke (1951, 38)
reported up to 9 m of gully extension in one season in the
Cameroons and Bishop (1962) reported 314.5 m extensions in
one storm in unconsolidated lacustrine deposits in Uganda.

N.O. Jones (1968, 42) considered that piping cycles
lasted about 20 years from initiation to destruction in the
San Pedro Valley, Arizona. He measured 2 ft. (0.6 m) lowering of pipe floors and headward advances of arroyos by
collapse of over 20 ft. (6 m) in 15 months (*ibid.*, 49).
Table 24 lists Jones's estimates of extension rates based
on air photo analysis of the larger networks. Maximum mean
headward extension rates for all pipe networks (based on
collapse features) were probably 10-25 ft. (3-7.6 m) p.a.,
whilst the more active headcut gullies advanced somewhat
faster, at rates of 40 to 60 ft. (12-18.3 m) p.a. He proposed
that rates of pipe extension are particularly limited by the
fact that excessive capture of drainage by one part of the

Table 24. Rates of extension of pipe systems in the San Pedro Valley, Arizona, based on N.O. Jones (1968, tab. 10) and derived largely from air photo interpretation.

Pipe system	period	extension (m)	extension rate (m y^{-1})
A1	1955-66	61.0	5.54
A1	1935-55	45.7	2.28
A2	1955-66	30.5	2.77
B2	1935-66	182.9	8.71
B4	1955-66	76.2	6.93
E1	1935-66	76.2	3.63
E1	1895-1935	213.4	5.33
F	1935-66	152.4	4.92
F	1895-1935	243.8	6.10
		mean rate	5.13

pipe system will lead to rapid self-destruction and extension of an open gully instead (*ibid.*, 141). Most of the small systems in the area were now developing only slowly, because a large proportion of the runoff was flowing in distinct channels (*ibid.*, 155). In contrast to the cyclical development in his area, Jones (*loc. cit.*, 157) suggested that piping may be a more permanent, steady-state feature in badland areas such as that described by Mears (1963) in the Painted Desert, where pipes develop along cracks in claystone below a horizontal sandstone cap.

Galarowski (1976) monitored pipe growth on a slope in the East Carpathians between 1971 and 1974. The slope was concave with a gradient of 13-16° and a double terrace scarp (29-34°) at the foot, covered in loam '0.002-0.05 mm in dimension in all layers' 0.8 m thick upslope increasing to 1.3 m downslope and overlying sandstone bedrock. Most of the development occurred during the early spring snowmelt and the wet late autumn period. The longest pipe system in the vicinity was initiated in the unusually wet autumn of 1973 (400 mm in 3 months) and by May 1974 had reached 150 m in length. The main system which he resurveyed expanded more slowly from being 25 m long in August 1971 to a length of 40 m by May 1973, although his evidence for extension is based only on the location of the furthest collapse funnel upslope. Over the next year, development was confined to coalescence and extension of the downslope collapse features.

He suggested that development slowed up in the thinner soils upslope and that pipes were destroyed where deep ploughing extended down to 50-80 cm in the area. Development was often ephemeral, the cycle being halted by choking up of the pipes. A private communication from Starkel who had studied an area nearby some years previously indicated that the pipe forms had been transformed in 10 years with the development of suffosion now finished, leaving 'dellen-like hollows'. Starkel (1960) had suggested that progressive suffosion during the Holocene had been responsible for valley development in the area.

The very interesting report by Floyd (1974) of pipe-development during conservation experiments in Australia is perhaps best reviewed under the heading of conservation and control in section 8.7.

8.6. Piping processes in paleosols and paleo sediments

Conacher and Dalrymple (1977, 35) have arbitrarily defined 'contemporary' landsurface process-responses as relating to processes which have operated within the past 50 years. However, most, if not all, landsurface units contain widely varying proportions of contemporary to 'paleo' properties, where 'paleo' presumably implies development under processes no longer operative. Naturally the inherent variability of climate, soils and geomorphic environment belies strict definition.

Within these limitations of definition, both paleo-piping and contemporary piping of paleo materials have been observed. Bell (1968, 250) observed infilled pipes in the Eocene Golden Valley Formation and inferred intermittent piping since Eocene times. Piping was slight 'until the semiarid climate with cold winters evolved and the modern piping process began' attacking the terraces probably originally formed by extensive flooding under a cold humid climate contemporaneous with glaciation further north. Smith (1968) envisaged piping as responsible for flushing periglacial valley deposits to leave boulder fields and stone rivers (cf. section 4.6(i)).

Gile, Peterson and Grossman (1966), Gile and Hawley (1966), Gile (1967), and Gile and Grossman (1968) have described paleopiping in argillic and petrocalcic (K) horizons in desert soils in New Mexico. In the clay horizons of buried soils developed during Pleistocene pluvials, Gile and Grossman (1968, 12) observed clay skins on pipes which had no connexion to the surface and were below the reach of present-day precipitation. They concluded that the pipes were formed before the soil was buried. Wet conditions at the time of formation would have prevented carbonate accumulation which is disruptive to clay skins.

The argillic horizon is described as having a moderate clay percentage and appreciable montmorillonite. In general, these pipes appear to be vertical both in argillic (*ibid.,* fig. 6) and K horizons (Gile *et al.*, 1966, fig. 7). Gile *et al.* (1966. 358-359) described a K horizon penetrated by funnel shaped vertical pipes 15-30 cm in diameter filled with soil and ending in the underlying clean sand and gravel. The smaller pipes had laminar carbonate linings. The A horizon and non-calcareous upper Bt horizon continue across the pipe inlet without change, but the K horizon in the pipes is peculiar to the pipes themselves consisting of a carbonate matrix with nodules of reddish-brown Bt horizon inclusions, the laminar carbonate horizon extending beneath this cellular zone. They suggest that carbonate infilling began after cessation of regular flushing by floodwater. Gile (1967, 270) described larger pipes 1.5-7.6 m in diameter at 'common intervals of several tens or scores of feet' into which the Bt horizon dips and thickens, generally without any surface expression, in a similar caliche horizon, and Gile and Hawley (1966) described 11 pipes of 22-60 cm diameter in a 152 mm transect, the pipe area accounting for 2% of the measured section in the same soils:

In the case of calcretes, erosional processes are likely to be dominantly chemical and therefore more akin to the karst than to the common definition of piping processes (chapter 2). However, Goudie (1973, 59) has noted that direct connections can be found between piping in soils overlying calcified, kafkalla in Cyprus and piping in the calcrete itself. This suggests that the water is eroding mechanically and chemically in sequence, and it highlights the common problem of drawing too fine a dividing line when studying nature. Extensive calcrete pseudokarst occurs in stranded beach ridges in Australia and in the Kalk Plateau, S.W. Africa (Goudie, 1973), and Kosack's (1952) world map of pseudokarst pinpoints many others.

There are numerous reports of piping and associated features beneath exposed lateritic duricrusts, although the term 'piping' has only recently been applied (Goudie, 1973). Gautier (1902, 60) recognised the features in the zone between laterite duricrusts and the clayey mottled zone and possibly another at the interface of the mottled zone and the sandier pallid zone (cf. Goudie, 1973, 46). Dixie (1920, 219) reported streams disappearing underground into caves beneath laterite in Sierra Leone and in Western Australia. Jutson (1934) described the progressive development of laterite pseudokarst from i) 'rock hollows' up to 5 m deep, hollowed out under the crust, to ii) 'blowholes', i.e. pipe inlets, and iii) collapse and 'natural bridge' formation. According to Goudie, the most detailed study of pseudokarst forms in laterite was produced by de Chélélat (1938) from work in French Guinea. Segalen (1949) also described undercutting of lateritic crusts in Madagascar by the entrainment of many cubic metres of particles by water infiltrating through cracks in the bare lateritic clays, and resulting in the formation of lavakas (circular depressions coalescing to form ravines). Hénin (1949, 43) repeated the essentials in a brief survey of Madagascan soils and McBeath and Barron (1954, 97) described undermining of laterite caprock in

Guyana, with sinkholes of 800 x 400 m. Aubert (1963) described caverns below lateritic capping in the Fonta Djallon
of West Africa and Patz (1965) mentions collapse depressions
in the laterite of the Buganda Plateau, Uganda. Thomas (1968)
mentions underground channels beneath duricrusts, hollows
and slips in profiles with lateritic ironstone overlying
stiff clay in the Sula Mountains of Sierra Leone. Goudie
(1973, 48) also describes ferricrete 'karst' on the Panch-
goni Tableland, south of Poona, in Western India. He observed slumping and undermining of the duricrust creating a
scalloped edge with small pipes radiating into the litho-
marge and clay substrate from 'rock shelters' beneath the
crust. Goudie (1973, 49) suggests that vermiform laterites
are more affected by pseudokarst than indurated pisolithic
laterites because they tend to be more permeable, and Baillie
(1975) also states that well developed piping is restricted
to the extensive permeable ferricretes of the older land-
surfaces of Africa and Australia in the humid and subhumid
tropics where the ferricrete provides a strong, though permeable, roofing.

Recently, Bowden (1979; 1980, and pers. comm.) has
been studying laterite pseudokarst features in the Kasewe
Hills, Sierra Leone, where piping and collapse is currently
active. The piping occurs in the mottled zone beneath
fersiallitic duricrusted slopes of $1-3^0$. Near the hills the
pipes run very near to the weathering front, but down 'pediment' they run well within the regolith. They run parallel
to the grain of the main hills and bedrock trend, whilst
the main surface streams radiate away from the Hills. Some
stream gorges return to the subsurface trend and may be due
to collapse of subsurface tunnels. Highly seasonal rainfall
results in water table shifts of 4 m, so that all surface
and most subsurface channels are dry for part of the year.
Bowden regards the piping as evidence of a changing water
table. Some pipes, caves and link pipes between different
levels are no longer active because of lowering of the
watertable and cross-sections of the pipes suggest a vadose
origin. (Six cross-sections provided by Bowden (pers. comm.,
1979) indicate mean widths of 1.45 m and heights of 0.83 m,
generally 3-6 m below the surface.)

It is interesting that McFarlane (1969; 1976, fig. 27)
sees water table lowering as a fundamental process in the
formation of laterite duricrusts. Perhaps the frequency of
recorded pseudokarst features beneath such duricrusts, and
in particular the evidence of vadose formation and 'fossil'
piping given by Bowden can be taken as further support for
McFarlane's thesis.

An interesting case is described by Banerjee (1972) of
'pipe structure' in ferrallitic soils in West Bengal. The
pipes are filled with soil and have a mean diameter of 10
cm, are near circular or oval/irregular and occur in the
beds and banks of tributary gullies generally in the mottled
loamy soil (B_1) beneath a yellow, nodular loam A horizon
and just above a semi-permeable plinthite (B_2). In places

the soils have a conglomerate duricrust surface. Some of
the pipes seem to have continued down to a thin hematite
pan. Small pipes of a few millimetres diameter occurred in
loose soils but the larger ones, up to 50 cm, had hard 1
cm thick, iron oxide rims (cp. section 8.3). Some pipes
had internal partitions of iron oxide, quartz and weathered
felspar, which he suggests is due to coalescing of pipes
followed by removal of divisions under reduced oxygen pressure in saturated conditions during the rainy season and
redeposition on the rims during other seasons. He suggests
that biotic holes, especially small insect holes, have been
expanded by flowing water to form the pipes.

Kunsky (1957) maintained that piping landforms soon disappear and that nothing older than the early Pleistocene
could survive. Piping in the loess of central Europe was
probably initiated in late Würm, the Holocene or Younger
Pleistocene. Halliday (1960, 110) quoted a C^{14} date of
2000 BP for tunnels incised into partially compacted soil
beneath a lava flow at Lake Cave, Washington, and on strati-
graphic evidence gives a very late Pleistocene to Recent
date for all similar tunnels in the western U.S.A. However
Bell's (1968) evidence suggests that some piping may be
older (v.s.).

8.7 Control and Prevention of piping erosion

Considerable attention has been paid to the problems of
containing and preventing pipe development by agricultural-
ists and dam engineers. Indeed, the practical need for such
measures has been the main stimulus for research into piping
erosion. This fact stands in stark contrast to the statement in a text on 'soil conservation' by Norman Hudson of
the National College of Agricultural Engineering in Britain
that piping is usually confined to badlands of little agricultural value, which he says is fortunate for there are
no effective control measures (Hudson, 1971, 41). Indeed,
many antipodean conservationists would also strongly chall-
enge Hudson's second point.

Soil conservation is very much a topical subject. Of
immediate concern to conservationists is the continued general decline in soil fertility and increasing soil loss
despite the conservation measures of the last 40 years (cp.
Stocking, 1978c). Concern in Australia is responsible for
the current nationwide soil conservation survey (Carder,
1978; Pauli, 1978) and tunnelling is clearly a major cause
of loss of agricultural land in some parts of Australia.
According to Colclough (1978, 66): 'Tunnel erosion is the
most frustrating and insidious form of erosion in Tasmania'.
With the aid of an Australian Extension Services Grant the
Department of Agriculture in Tasmania has established a
current programme of observational areas to monitor the
development of erosion patterns, including tunnel development, at 56 sites (*ibid*). Colclough ascribes the erosion
problem to Tasmania's historical role as granary to the Australian colonies. Monoculture impoverished the land leading

to critical sheet erosion, rilling of arable lands, degradation of grazing and massive gullying associated largely with tunnel collapse. Colclough describes the subsequent change to a more pastoral economy as a classic example of almost total land use change forced by factors including the absolute degradation of the basic soil resource and the associated impoverishment of the land-users.

Among specific agricultural practices that have encouraged pipe development apart from deforestation and overgrazing, Fletcher and Carroll (1948) noted that ploughing may open up coarser subsoils with downwash of fine material creating surface depressions, and that deep rooted crops like alfalfa may cause increased water loss and encourage cracking. ·

Changes in methods of husbandry appear to have had some remedial and perhaps some preventative effect. According to Mitchell (1978, 117) 'correct land use and adequate soil conservation techniques are now able to prevent tunnel erosion from occurring'. Hughes (1970; 1972) has suggested that the tunnel systems of the Banks Peninsula, New Zealand, have hardly expanded since the first aerial photography of 1941 owing to the improved techniques of pasture management. However, the elimination of existing tunnels requires more direct and often costly measures. As pointed out by Colclough (1978), problem areas are usually not brought to the attention of soil conservation officers until it is too late for economic reclamation. One factor leading to complacency, according to Stocking (1976; 224), is the fact that pipe-headed gullies may appear stable for many years before sudden and extensive collapse occurs. In such cases, only specific laws demanding soil conservation and institutional subsidies for reclamation projects, such as those available in mainland Australia, will encourage remedial work (Colclough, 1978).

Following his pioneer survey of soil erosion in New Zealand, Cumberland (1944) pointed out that: 'In all cases European occupance seems to have accentuated the already abrupt break between a topsoil with high percolation and a compact, silty subsoil' (ibid., 98-9) and that a general cure must be 'tying together again - as nature tied - previous topsoil and impervious subsoil'. Not only does the enhanced structural contrast concentrate water but the erosion hazard is increased by slipping of the topsoil and sod over the compact loess at the base of the podzol profiles (ibid., 184). With a similar idea in mind, Harris and Fletcher (1951) suggested that remedial measures should aim to decrease porosity in the upper layer and to increase it in the lower layer. Since they were concerned with a special case of piping developed by irrigation techniques, they also suggested lighter irrigation using sprinklers.

In early discussions of remedial treatments, Downes (1946) suggested control by suitable vegetation and contour furrowing to prevent quick runoff and Monteith (1954) suggested prevention of concentrations of water, establishing suitable vegetation, and soil conditioners including use of sawdust to improve soil structure with the aid of ammonium sulphate to prevent cracking.

A common solution has been mechanical break-up of the pipe systems by deep ploughing or 'contour ripping'. By destroying the pipes and ploughing up any hardpans or imped-ing layer beneath, ripping should result in more homogeneous permeability and less concentration of seepage, thus prevent-ing the built-up of erosive velocities. On the other hand, contour banking aimed at reducing sheet erosion by reducing surface velocities has been criticised as likely to encour-age piping erosion by creating surface ponds. Stocking (1976b) comments on misguided conservation work in the Mhon-doro Tribal Trust Land in Rhodesia, where contour-banks built in 1945 created a reservoir of water which eventually drained by piping failure through the dams. Similarly, many earth dams built as gully control structures have failed through pipe development (Forster and O'Meara, 1978).

Colclough (1973) sees the problem in rather different terms as basically one of water balance and has proposed that the land use changes which took place earlier this cen-tury and seem to have initiated severe tunnelling in Tasmania should be reversed by planting trees and shrubs, preferably deep rooted species, to increase evapotranspiration and re-duce the 'excess' water available. Colclough gives an example of a successful reclamation programme beginning with contour ripping followed by temporary protection of the sur-face with a rye-clover mixture and subsequent planting with *Pinus radiata*.

In his review of the experiments in pipe control set up after the report of Newman and Phillips (1957) in New South Wales, Floyd (1974) reported complete freedom from piping in the area subjected to rotation cropping with wheat and past-ure since 1960 (Table 25). However, tunnels began to re-establish within a period of 10 years in the areas sown to permanent pasture and in the natural pasture subjected to grazing control. It is interesting that although parts of the untreated pasture were severely tunnelled in 1957 (*ibid.*, fig. 1), after only 3 years of grazing control Floyd could report no tunnels in the whole area, *including* the natural pasture block (*ibid.*, 151). Re-inspection in 1970 showed that the most active tunnelling had apparently re-developed in the natural pasture area, but it was also present in the block sown to permanent pasture in 1962. No difference was found between ripped and chiselled areas or, he says, between gypsum-treated and untreated areas (in block B?). Ripping destroyed tunnels directly, but chiselling also tended to block them by weakening the roofs (cp. Hickey and Dortignac, 1964).

However, although research into conservation techniques is relatively well advanced in S.E.Australia, the techniques developed are not necessarily effectively exported to other areas. Stocking (1976a, 222) has emphasised the value of

Table 25. Plan of the Riverina tunnel erosion experiment, according to the description by Floyd (1974)

Site	Soil association and groundsurface*	State in 1957	Land Use Treatments (1960)**	Result
Urangeline	Red-Brown Earth (solodised) and Red Solodic Northcote class Dr. 2.33 parna and Dr. 2.32. K_2 over K_3.	Extensive minor and moderate tunnelling, small areas severe. pipes 20-30cm down 8-12cm diameter. on upper slopes, poor vegetation	Block A rotation cropping (wheat/pasture) — 50% of each block ripped to 30-35cm	1963 no piping 1970 piping started on B and C, C most active 1971-2 stopped in B.
	Yarabee groundsurface	minor rills and gullies	Block B permanent pasture** — 50% chisel ploughed to 15cm. Both areas treated with gypsum (at 670 kg ha^{-1}	
	Tywong soil association	Eucalyptus woolsiana alliance (Grasses: Stipa falcata and Danthonia spp.)	Block C untreated natural pasture, with controlled grazing	
Pleasant Hills	Solodised solonetz Northcote class Dy 3.43 K_3 Brucedale groundsurface Munyabla soil association.	severely tunnelled with large gully Eucalyptus albens aliance (Grasses: Stipa falcata and Danthonia spp.)	Gully and grazing control measures Sown pasture 1967, plus trees (Eucalyptus viridis)	No tunnels till 1972 on well vegetated area, but 2-5cm diameter pipes 6-10cm down in bare areas.

* Groundsurface and soil association classifications after Beattie, J.A., 1972: Groundsurface of the Wagga Wagga region, New South Wales. Soil pub. No. 28, CSIRO, Melbourne.
Soil profile classification after Northcote, K.H., 1971: A factual key for the recognition of Australian soils. Rellim Technical Pubns., Glenside, South Australia.

** Wheat crop in 1961, then permanent pasture.

ESP as an indicator of piping hazard in Rhodesia and as a guide to suitable reclamation treatments, but he points out the urgent need for comparative data from other areas to establish variations in threshold values in order to help decision-making or conservation procedures. Moreover, in Britain ESP is clearly an irrelevant factor, except perhaps on salt marshes (section 4.5(i)). In fact, the British case also provides an ironical postscript which illustrates that piping erosion is not always a bad thing. Gilman and Newson (1980) conclude their report by quoting the case of ploughing by the late Captain Bennett-Evans on Plynlimon, Wales, to re-seed and improve his upland pastures, where the resulting destruction of the natural pipe drainage has produced the only notable areas of Hortonian overland flow in the whole of the experimental catchment. They might have added that the captain threw his plough down a mine-shaft as a result, vowing he would never use it again (Newson, pers. comm.). They suggest that British farmers and foresters might begin to pay more heed to natural drainage and less to artificial networks and that the trend to surface treatments during re-seeding upland pastures clearly aids pipe preservation.

Some piping may also prove to be self-healing. Heede (1971, 13) describes a case of natural reclamation of piped solonetz soils in the semi-arid Alkali Creek basin in Colorado. Here field survey and laboratory analyses suggested that piped gully walls shear off during wet periods, depositing a soil mass on the opposite side forcing the ephemeral stream to undercut the piped wall more. The resulting lateral migration of the thalweg leaves a shallower slope on the vacated side which has higher soil moisture content and is probably more easily leached after disturbance, so that sodium is removed, the pH decreased and calcium becomes more soluble and available for plant growth. Vegetation begins to stabilise the deposit and the greater influence of calcium stimulates flocculation, which speeds up the leaching process. Moreover, the horizons with high ESP appeared to exist as lenses, so that once migration had worked through a lense, the banks were more stable.

Bell (1968) actually concluded that delta formation in the North Dakotan badlands marked the natural conclusion of a cycle for 'miniature retiform' piping with the outlet buried in sediment and the inlet obliterated by erosion.

There is no doubt that remedies are available for all but the most severe cases of piping erosion on agricultural land. At best these may consist of ploughing, reseeding and careful management of grazing, at worst perhaps levelling and planting with deep rooted tree species. Prevention, however, is always better than cure and it would appear that sufficient knowledge exists for effective prevention in the areas most intensively studied or in homologous regions. However, apart from possibly over-cautious land utilisation, it is difficult to formulate safe or effective procedures for other areas and a major gap still exists in our knowledge here.

9.   CONCLUSIONS

Piping clearly takes many forms. Probably the best classification scheme for piping processes and pipe genesis has been that produced by Parker and Jenne (1967), but this has been little used, perhaps partly because it appeared in a somewhat obscure publication. American engineers have been particularly prolific in introducing new terms for piping processes, in many cases, but by no means all, as a result of discovering a new process. Hence, the recently introduced term 'hydraulic fracturing' (Sherard *et al.*, 1972b) appears to refer to any form of dam failure due to piping. The emphasis in recent American engineering literature is on forms similar to Parker and Jenne's 'stress-desiccation crack piping', including failure by seepage water entering and extending 'rainfall tunneling' (itself principally developed in desiccation cracks) or extensions from bedrock fractures. On the other hand, settlement cracks and faster seepage in less consolidated or more permeable zones are also reported. The 'internal piping' defined by Reuss and Schattenberg (1972) is essentially a variation of the classic heaving/boiling process in which flow is in a 'confined aquifer' situation. Although currently receiving considerable attention, the role of dispersive clays in pipe initiation is clearly not universal and many other factors may be more critical in some situations.

Numerous suggestions have been made as to factors limiting the occurrence of piping to specific climatic regions or specific soils, and there is no doubt that many relationships can be substantiated in specific regions. The mechanics of piping commonly involve cracking, which may result from highly expansive clay minerals such as montmorillonite or from high organic content, but of course climate and local drainage also play a part; vertic, argillic and histic properties are most in evidence in the soils. In many instances, dispersive clays also encourage piping, these tending to be associated with aridisols particularly in Australia, Africa and North America, and marked by high exchangeable sodium percentages. However, high ESP may occur in mollisols and it has even been suggested that cyclic salt may affect dispersion without creating a recognisably saline soil (e.g. Charman, 1970a). But an important and almost universal factor is the presence of horizons or surfaces of restricted permeability, which cuts right across soil taxonomic boundaries. Given rates of infiltration greater than the permeability of a specific soil or bedrock interface, piping may develop if either the build up of hydrostatic pressure or the breakdown of soil resistance passes a critical stability threshold. The process is particularly likely if cracks or similar lines of weakness permit the development of significant hydrodynamic forces and erosive turbulent flow, and if rainfall distribution is irregular, particularly with sequences of desiccation followed by heavy downpours or snowmelt. These criteria are sufficiently widely met to make piping a worldwide process.

It does not seem profitable to distinguish between tunnelling and piping as some have done, nor to restrict the term piping to linear voids of purely hydraulic origin and exclude hydraulically-modified cracks and biotic holes. There is plainly a continuum of processes and genetic origins between these forms. Table 26 purports to distinguish recognisable types along this continuum. The continuum approximates to a gradient in terms of frequency of saturation and hence possibly also to a sequence in terms of depth in the solum or even of catenary landsurface units.

One question that this classification raises and which is also apparent from hydrological studies of piping is: where is the 'water table'? Above all, the one feature that distinguishes soil pipes from groundwater springs is their quick response to rainfall. Although 'infiltration-excess pipeflow' may occur (Jones, 1979), the overwhelming weight of evidence to date suggests that 'saturation pipeflow' is the norm. In most cases the saturation appears to be that of a 'perched soil - water table', generated by exceedence of the infiltration capacity of one or more layer in the solum, without necessarily being connected (however tenuously) with the classic zone of saturation of the bedrock matrix that is traditionally thought of as comprising the groundwater body. Indeed, Hewlett (1969) observed the rapid development of ephemeral, perched groundwater feeding storm runoff via intermittent channels, and Anderson and Burt (1978a) have measured a similar pattern in an instrumented hollow. However, many such phreatic bodies seem to last much longer than the few days observed by Hewlett (1969), sufficient, for example, to feed perennial pipes from the slow drainage of the hillslope. Furthermore, the fact that the concept of an integrate groundwater body has been challenged by many karst geomorphologists (cp. Jennings, 1971; Sweeting, 1972) has two important corrollaries worth considering in this context. First, not only may a number of perched saturation zones exist in a single soil profile under different storm intensity/antecedent soil moisture situations, as pointed out by Starkel (1976), but similarly it is conceivable that some shallower soil pipes may flow at times when deeper pipes do not. Secondly, the zone of matrix saturation surrounding a flowing pipe need not be very extensive nor indeed be integrated with the saturation zones surrounding adjacent flowing pipes. These are possibilities which may complicate studies of the hydrological behaviour of soil profiles.

For the time-being, however, evidence is beginning to accrue on the hydrological significance of soil pipes which should attract very serious attention from all those concerned to establish the true nature of runoff processes. Just as in many areas pipe-erosion and pipe-linked mass movement are highly significant and perhaps paramount components in soil erosion and landsurface evolution, wasting the land away from beneath the surface, so hydrologically many more areas probably respond more commonly by subsurface quickflow.

Table 26. Genetic forms of piping process

Dominant Initiating Factors	Hydrological environment	Type	Comments
Non-hydraulic	Vadose	(Desiccation crack piping (Biotic hole piping	
		Mass movement crack piping	due to subsidence or landsliding, perhaps triggered by deeper seepage - may extend to phreatic surface, and may aid phreatic piping where fractures extend below surface of saturation.
Hydraulic	Phreatic	Phreatic surface heaving	generated just below a surface of saturation, commonly above layer of lower hydraulic conductivity.
		Confined aquifer heaving	confined between zones of lower hydraulic conductivity. Might utilize infilled bedrock fissures for part of route.

Hewlett (1974) has emphasised that a modelling strategy
based primarily on the hydraulic conductivity of the soil,
such as that used by Freeze (1972), may be deficient in ignoring the effects of increased channel length and slope convergence concentrating subsurface stormflow. The significance of topographic hollows, suggested by Hack and Goodlett
(1960), was re-emphasised by Kirkby and Chorley (1967) and
has been further underlined, for example, by Beven (1978),
Anderson and Burt (1978b) and Beven and Kirkby (1979).
Whipkey and Kirkby (1978, 136) have described such areas as
'the nucleus of the dynamic contributing area', and Beven
(1976) has stated that hillslope convergence seems to be the
dominant topographic parameter affecting the hillslope
hydrograph. Beven, Gilman and Newson (1980) also note the
probable importance of the spatial distribution of such
source areas to the form of the channel hydrograph. At the
same time, the non-linear nature of hydrological response on
hillslopes has been noted, for example, by Beven (1976; 1978)

who concluded that initial conditions, especially in the
unsaturated zone, are paramount in governing the timing and
magnitude of peak discharge.

To each of these points must be added the role of soil
piping. Pipes do not necessarily follow 'topographically
sensible' routes and may both augment and/or compete with
hollows as contributing zones. Both the velocity and volumes
of pipeflow may far exceed those of return flow and saturation overland flow in hollows. And hollows may be fed by
pipes (cp. Anderson and Burt, 1978a). In such cases 'area
drained per unit contour length' as measured from a contour
map is no longer an appropriate basis for modelling erosion
potential. Pipes may extend the range of source areas well
beyond the riparian zone and adjacent hollows, over distances of 300 m and more. Nor is there any need to assume
either 'total infiltration except for precipitation falling
on the expanding channel', as did Hewlett and Nutter (1970),
or that pipe response depends on the speed of transmission
of water through the soil overburden, as did Whipkey and
Kirkby (1978, 124). The general concensus of current evidence is that many pipes collect water from overland flow or
surface bogs. Moreover, exceedence of thresholds within the
pipe systems, controlling the horizontal extent of the active pipe networks and even the activation of semi-discrete
vertical overflow systems, result in marked non-linearity of
response (cp. Jones, 1978a, c; 1979). This is analogous to
the extension of surface channels and zones of saturation
overland flow noted, for example, by Gregory and Walling
(1968), Calver, Kirkby and Weyman (1972) or Day (1978), but
complicated by the extra vertical dimension in pipe networks,
with networks at different levels and perhaps even overlain
within the same soil profile.

For the fluvial geomorphologist, the recognition of
'combining flow' systems extending well beyond the stream
channel domain must also radically alter the framework for
process-response models. The notion of distinct 'dominance
domains' within a drainage basin, defined by the dominance
of different processes, such as the hillslope domain and
the channel domain (Kirkby, 1978a), requires careful scrutiny.
The presence of piping on hillslopes effectively extends
channel processes outside their obvious\domain, in which
case, for example, indices of channel frequency might appro-
priately include pipe frequency. There is no doubt that in
some circumstances piping can become a dominant process of
hillslope erosion, and the evidence of relatively slow rates
of pipe erosion in Britain must be weighed against similar
evidence for many other forms of hillslope erosion (cp.
Carson and Kirkby, 1972). Although Kirkby (1978a) has stated
that 'one of the crucial thresholds in the landscape is that
associated with stream heads', in the 'pipeflow streamhead'
(McCaig, 1979b) or the 'tunnel-gully' (Gibbs, 1945) the only
threshold that is passed at the channel head is that of roof
stability. Hollows are not the only foci for channel extension (compare Jones (1971) with Smith and Bretherton (1972)),
and for the piped channel head sudden extension may occur

over considerable distances, with neither depth nor length of channel being proportional to the magnitude of storm event, in marked contrast to the relationships so far considered in modelling channel extension along hollows (e.g. Kirkby and Chorley, 1967; Calver, 1978).

The fluvial effects of piping can, of course, extend well beyond pipe networks directly connected to the channel net, both in terms of subsurface erosion and as a source for localised overland flow. Kirkby's (1978b) model of hillslope sediment transport incorporating the major advances of non-uniform generation of overland flow and of subsurface flow moves significantly nearer the field situation, perhaps not solely for humid, temperate regions (cp. *ibid.*, 360), and if this review has demonstrated nothing else, it is that piping can have an important role in both of these processes. Clearly, we may look forward to some change of emphasis in process studies in both hydrology and geomorphology in the coming years, which will supplement and broaden the older established views without entirely denying them.

APPENDIX I: LABORATORY TESTS OF PIPING POTENTIAL

It is appropriate only to provide a brief outline of
the available techniques at this stage and to refer the
reader to detailed descriptions and engineering drawings of
custom-made equipment given in the appropriate papers or in
the review of engineering experiments by Perry (1975). There
is also a very fine dividing line, if one at all, between
the general field of study of soil erodibility and the specific study of susceptibility to piping erosion. Only those
techniques so far used in piping studies will be related
here.

Yong and Sethi (1977) echo the current interest amongst
American engineers in the relationship between dispersivity
and piping and point out that tests are of two types, a)
physical/mechanical tests of performance such as pinhole
tests or rotating cylinder tests, and b) study of suspensions
in relation to particle interaction. The latter is concerned
to establish the mechanisms responsible for dispersivity,
the former with the relative performance of different soils
which may be empirically related to piping susceptibility.
Work at the McGill Soil Mechanics Research Laboratory by
Yong and Sethi represents one of the very few instances of
category (b).

Middleton (1930) defined a Dispersion Ratio:-

DR = percent in suspension after 40 seconds of shaking.

Values <10% indicate non-erosive soils. Brown (1961;
1962) found all his piping soils fell within the highly erodible category with Dispersion Ratios >20%. Downes (1946)
defined a Mean Dispersion Percentage:

$$\text{MDP} = 100 \times \frac{\text{\% silt-clay indicated by stirring in distilled water for 1 min}}{\text{\% silt-clay obtained by stirring in a dispersing agent for 20 min}}$$

Downes quotes MDP's of > 80% for his piped soils. This
approach has been refined by the Soil Conservation Service
of New South Wales. Charman has summarised the Australian
SCS criteria (Table 27) and, more recently, Crouch (1976),
has reviewed American and Australian literature and listed
some pragmatic criteria based on the range of values found
in the published literature. Similar summarises of test
criteria have been published covering the American SCS practice by Decker and Dunnigan (1977) and covering the experience
of the Australian CSIRO by Ingles (1968). The tests are
outlined at the end of this appendix. Nevertheless,
Decker (1972) pointed out that some dispersive clays in
Mississippi and Tennessee (c. 30%) cannot be identified by
the US SCS Lab Dispersion Test or the Crumb Test (Perry,
1975).

239

Table 27. Recent published criteria for susceptibility
of soils to piping.

Charman (1969,   Ritchie Dispersal Index <3 = susceptible,
  1970a and b)   DI <3 and Volume Expansion >10% = highly suscept-
                         ible.

Crouch (1976)   Dispersal Index <3 (after Ritchie, 1963)
                       Permeability <1 cm h^{-1} (<0.0028 mm s^{-1}) (from
                         evidence of Fletcher *et al*. (1954), Brown
                         (1962), Heede (1971)).

                       ESP > 0.1 (lowest noted by N.O. Jones (1968))
                         or more commonly 12 (after Fletcher and
                         Carroll (1958), Heede (1971)).

Decker and   Percent Dispersion by Volk Test:
  Dunnigan (1977)
                       >25% for inorganic low plasticity (ML) and
                         silty clay (SC) soils.

                       >35% for inorganic soils of low to medium
                         plasticity (CL)

                       >40% for inorganic clays of high plasticity (CH).

Ingles (1968)   Permeability $\geqslant 10^{-2}$ cm s^{-1} for silt or silt-size
                         aggregates.

                       Permeability $\geqslant 10^{-5}$ cm s^{-1} for dispersive clay
                         soils.

                       ESP > 12% indicating dispersive clay soils, but
                         dispersibility depends on complex function of
                         exchange cations, salt concentration of water
                         (both total salts and salt composition) and
                         clay minerals.

Stocking (1976)   ESP > 15% indicating dispersive clay soils.

    Jones (1971; 1975; 1978a) used a modified technique of
water-stable aggregates/aggregate stability analysis, in
which aggregates were wetted under vacuum in a modified
desiccator in order to avoid breakup by entrapped air and
to better simulate subsurface conditions (Kemper and Chepil,
1965, 499-510; Kemper, 1965, 511-519).

    The US Bureau of Reclamation (Gibbs, 1962) sought to
relate Atterberg limits and erodibility, and observed a
major division at Plasticity Index = 10 and Liquid Limit =
30, below which soils were erosive. Sherard (1952) also
used Atterberg Limits specifically in relation to piping
potential (cf. section 5.2(ii)), but the current concensus
is that these measurements are irrelevant to piping (Sherard
and Decker, 1977).

    Considerable research has been directed by engineers to

laboratory experiments into the design of protective filters against piping. An early experiment by Davidenkoff (1955) employed soil mounted behind a plate with a note which simulated the filter material. Similar experiments were made by Czyzewski *et al.* (1961), Ranganathan and Zacharias (1968), Wolski *et al.* (1970), Kulandaiswamy *et al.* (1971) and Landau (1976). Zaslavsky and Kassiff (1965) developed an apparatus similar to Davidenkoff's but with a multiple-holed base, and Sherard (1973) has conducted experiments with a 'pinhole' apparatus. Use of the pinhole test appears to be increasing and US Army Waterways Experiment Station at Vicksburg has been engaged in an equipment evaluation programme since 1973. Haliburton, Petry and Hayden (1975) have developed a 'physical erosion test' (PET) for measuring the erodibility of compacted dispersive clay soils in a similar instrument (Perry, 1975).

Summaries of tests

AMERICAN SCS DISPERSION TEST.

Volk (1957) developed the original test of dispersiveness used by the American Soil Conservation Service. With minor modifications this is still used (details in Sherard *et al.* 1976; Decker and Dunnigan, 1977). Volk's test calculated the percentage of the total fraction of ≤0.005 mm diameter that slakes free from an air-dry clod immersed in quiet, distilled water under a vacuum. This was found to be more indicative of piping potential than Middleton's (1930) ratio which was similarly defined but in relation to the ≤0.05 mm fraction. Volk dispersion ratios less than 10% indicate nondispersive soils, highly dispersive soils have over 30%. A variant of the test has also been developed for field use (Ryker, 1977; Decker and Dunnigan, 1977). Recent work suggests that air-drying affects the results (except for naturally arid soils) and natural moisture contents are recommended (Decker and Dunnigan, 1977).

AMERICAN SCS PINHOLE TEST.

This test resulted from investigation of rainwater piping on the surface of earth dams by Sherard and associates in and after 1970 (Sherard, Dunnigan and Decker, 1976). A good correlation was found between this and the SCS Dispersion Test (Decker and Dunnigan, 1977). The test models the erosional performance of soils by submitting a small cylindrical soil sample held in a container with circular perforations in the base to increasing hydraulic gradients. Piping potential is indicated by development of small vertical pipes above the perforations. It is regarded by Decker and Dunnigan (1977) as the best test yet available for identifying dispersive soils. Marshall and Workman (1977) evaluated a number of tests and found this and the crumb test the most reproducible.

NEW SOUTH WALES SCS TESTS:

RITCHIE'S DISPERSAL INDEX.

Ritchie (1963) defined this index as the ratio of soil particles <0.004 mm after dispersion with sodium polymetaphosphate solution to the weight of particles <0.004 mm using no dispersive agent and shaking for just 10 min in water.

If more than 33% of <0.004 mm disperses after 10 min shaking in water, then soil is susceptible. Charman (1969) found that not all 'susceptible' soils are actually tunnelled, presumably because other factors are not conducive.

EMERSON'S CRUMB TEST.

Emerson (1967) developed this test to define aggregate stability and it compares the degree of dispersion of natural soil peds when placed in distilled water to the amount of dispersion occurring after treatment with a dispersing agent (cf. Ingles and Metcalf, 1973).

FLOYD'S STICKY POINT TEST.

Developed by SCS New South Wales, tested by Floyd (1974) and still under evaluation (Crouch, 1977). Soil brought to the sticky limit is remoulded into 3 5mm diameter aggregates immersed in 3 cm of distilled water for 10 min, 1 h and 24 h and time of commencement of dispersion is noted. Commencement after 10 min indicates highly dispersible, 1 h moderately, 24 h slightly, > 24 h nondispersible.

Keys to symbols on Figures 5 (p.36) and 30A (p.156)

Key to Climatic regions on Figure 5

I. Climates dominated by equatorial and tropical air masses.

1. Rainy tropics
2. Monsoon tropics
3. Wet-and-dry tropics
4. Tropical arid climate
5. Tropical semi-arid climate

II. Climates dominated by tropical and polar air masses.

6. Dry summer subtropics
7. Humid subtropics
8. Marine climate
9. Mid-latitude arid climate
10. Mid-latitude semi-arid climate
11. Humid continental warm summer climate
12. Humid continental cool summer climate

III. Climates dominated by polar and arctic-type air masses.

13. Taiga
14. Tundra
15. Polar climate

IV. Climates having altitude as the dominant control.

16. Highland climates

Key to symbols on Figure 30A

Precipitation (gross rainfall)	P	Horton overland flow	q_h
Channel precipitation	P_c	Saturated overland flow	q_s
Precipitation intensity	i	Return flow	q_r
Evapotranspiration	e_t	Pipe flow	t
Canopy interception loss	e_c	Pipe storage	T
Interception and canopy storage	l	Unsaturated throughflow	m_u
Stemflow and drip	s	Saturated throughflow	m_s
Litter flow	l	Soil-moisture storage	M
Litter interception loss	e_l	Seepage into bedrock	s_b
Litter storage	L	Interflow in bedrock	a
Evaporation	e	Aeration zone storage	A
Depression storage	R_p	Deep seepage	d
Detention storage	R_T	Baseflow	b
Infiltration	f	Groundwater storage	B

243

BIBLIOGRAPHY

Adams, W.A., and Raza, M.A., 1978: The significance of truncation in the evolution of slope soils in mid-Wales. *Journal Soil Science,* 29: 243-297.

Aghassy, J., 1973: Man-induced badlands topography. Chap. 6 in *Environmental geomorphology and landscape conservation Vol. III: non-urban,* edited by D.R. Coates, "Benchmark Papers in Geology", Dowden, Hutchinson and Ross Inc., Stroudsberg, Pennsylvania, 483 pp., p. 124-136.

Aisenstein, B., Diamant, E., and Saidoff, I., 1961: Fat clay as a blanketing material for leaky reservoirs. *Proceedings of 5th International Conference Soil Mechanics and Foundation Engineering,* Paris, 2: 523.

Aitchison, G.D., 1960: Discussion to Cole and Lewis (1960). *Proceedings of 3rd Australia - New Zealand Conference on Soil Mechanics and Foundation Engineering.,* 230.

Aitchison, G.D., Ingles. O.G., and Wood, C.C., 1963: Post-construction deflocculation as a contributory factor in the failure of earth dams. *Soil Mechanics Section, C.S.I.R.O., Research Paper* No. 52, and *Proceedings of 4th Australia - New Zealand Conference on Soil Mechanics and Foundation Engineering* (Adelaide), 275-279.

Aitchison, G.D. and C.C. Wood, 1965: Some interactions of compaction, permeability, and post-construction deflocculation affecting the probability of piping failure in small earth dams. *Proceedings of the Sixth International Conference on Soil Mechanics and Foundation Engineering,* Canada, 2: 442-446. (Also: C.S.I.R.O., Div. of Soil Mechanics Res.Pap. No.61.)

Aitchison, G.D., Ingles, O.G., Wood, C.C., and Lang, J.G., 1968: Piping failure in earth dans. In 'Soils and Rock Engineering', Research Report of Division of Soil Mechanics, C.S.I.R.O., Project 2.3, p. 16.

Alizadeh, A., 1974: *Amount and type of clay and pore fluid influences on the critical shear stress and swelling of cohesive soils.* Unpub. Ph D thesis, University of California, Davis.

Alpaidze, V.S., 1965: Role of suffosion in the formation of lakes in the Tbilisi area. *Tbilisi University Szromebi Trudy,* Vol. III: 311-325. (In Georgian, Russian summary.)

Anderson, M.G., 1979: On the potential of radiography to aid studies of hillslope hydrology. *Earth Surface Processes,* 4(1): 77-83.

Anderson, M.G., and Burt, T.P., 1978a: Time-synchronised stage recorders for the monitoring of incremental discharge inputs in small streams. *Journal of Hydrology,* 37: 101-109.

Anderson, M.G., and Burt, T.P., 1978b: The role of topography in controlling throughflow generation. *Earth Surface Processes,* 3: 331-344.

Arnett, R.R., and Conacher, A.J., 1973: Drainage basin expansion and the nine unit landsurface model. *Australian Geographer,* 12: 237-249.

Atkinson, T.C., 1978: Techniques for measuring subsurface flow on hillslopes. Chap. 3 in *"Hillslope Hydrology"*, edited by M.J. Kirkby, Wiley, 389 pp., p. 73-120.

Aubert, G., 1963: Soil with ferruginous or ferrallitic crusts of tropical regions. *Soil Science,* 95(4): 235-242.

Baillie, I.C., 1975: Piping as an erosion process in the uplands of Sarawak. *Journal Tropical Geography,* 41: 9-15.

Banerjee, A.K., 1972: Morphology and genesis of 'pipe structure' in ferrallitic soils of Midnapore District, West Bengal, *Journal Indian Society Soil Science,* 20(4): 399-402.

Barendregt, R.W., 1977: *A detailed geomorphological survey of the Pakowski - Pinhorn area of southeastern Alberta.* Unpub. Ph.D. thesis, Queens University, Kington, Ontario.

Barendregt, R.W,, and Ongley, E.D., 1977: Piping in the Milk River Canyon, southeastern Alberta - a contemporary dryland geomorphic process. In *"Erosion and solid material transport in inland waters",* Proceedings of the Paris Symposium, I.A.H.S. - A.I.S.H. Pub. 122, p. 233-243.

Barendregt, R.W., and Ongley, E.D., 1979: Slope recession in the Onefour Badlands, Alberta, Canada - initial appraisal of contrasted moisture regimes. *Canadian Journal of Earth Science,* 16(2): 224-229.

Beare, J.A., 1965: Erosion by water in Australia. In *"Soil erosion by water: some measures for its control on cultivated lands",* FAO Agricultural Development Paper No. 81, Rome, pp.195-211.

Beckedahl, H.R., 1977: Subsurface erosion near the Olivier-shoek Pass, Drakensberg. *The South African Geographical Journal,* 59(2): 130-138.

Bell, G.L., 1968: Piping in the Badlands of North Dakota. *Proceedings 6th Annual Engineering Geology and Soils Engineering Symposium,* Boise, Idaho, pp.242-257.

Bell, J.P., 1972a: A preliminary appraisal of the soil hydrology of the Wye catchment on Plynlimon. *Welsh Soils Discussion Group Report* No. 13: 107-125.

Bell, J.P., 1972b: The soil hydrology of the Plynlimon experimental catchments. *Natural Environment Research Council Institute of Hydrology,* Wallingford, Berks., Report No. 8.

Bennett, H.H., and Chapline, W.R., 1928: Soil erosion a national menace. *U.S.D.A. Circular* 33.

Bennett, H.H., 1939: *Soil Conservation.* McGraw Hill Series in Geography, New York and London, 993 pp.

Berg, L.S., 1902: The morphology of the shores of the Aral Sea. (In Russian.) *Russian Yearbook of Geology and Mineralogy,* 5: 6-7.

Berry, L., 1964: Some erosional features of semi-arid Sudan. *20th International Geographical Congress, London, Abstracts of Papers, Supplement,* p. 3.

Berry, L., 1970: Some erosional features due to piping and sub-surface wash with special reference to the Sudan. *Geografiska Annaler,* 52A(2): 113-119.

Berry, L., and Ruxton, B.P., 1959: Notes on weathering zones and soils on granitic rocks in two tropical regions. *Journal of Soil Science,* 10(1): 54-63.

Berry, L., and Ruxton, B.P., 1960: The evolution of Hong Kong Harbour basin. *Zeit. für Geom.,* 4(2): 97-115.

Bertram, G.E., 1940: *An experimental investigation of protective filters.* Harvard University, Graduate School of Engineering, Soil Mechanics Series, No. 7.

Bertram, G.E., 1967: Experience with seepage control measures in earth and rockfill dams. *Transactions 9th Congress on Large Dams,* Istanbul, 3: 91.

Beven, K., 1976: Hillslope hydrograph by the finite element method. University of Leeds, Department of Geography, Working Paper No. 143, 20 pp.

Beven, K., 1978: The hydrological response of headwater and sideslope areas. *Hyd. Sciences Bulletin,* 23(4): 419-437.

Beven, K., Gilman, K., and Newson, M., 1979: Flow and flow routing in upland channel networks. *Hyd. Sciences Bulletin,* 24(3): 303.

Beven, K., Iredale, R., and Kew, M., 1977: The hydrological
    response of headwater and sideslope areas in the
    Crimple Beck catchment. University of Leeds, Depart-
    ment of Geography, Working Paper No. 184, 15 pp.

Beven, K., and Kirkby, M.J., 1979: A physically based,
    variable contributing area model of basin hydrology.
    Hyd. Sciences Bulletin, 24(1): 43-70.

Bishop, W.W., 1962: Gully erosion in the Queen Elizabeth
    National Park. *Uganda Journal,* 62(2): 161.

Black, C.A., *et al.* (ed.), 1965: *Methods of Soil Analysis.*
    American Society Agronomy and American Society for
    Testing and Materials, Agronomy No. 9, Madison,
    Wisconsin, 2 vols.

Blench, T., 1962: Discussion of "Resistance to flow in
    alluvial channels", by Simons and Richardson.
    Trans. American Society of Civil Engineers, 127(i):
    927-1006.

Blong, R.J., 1965a: Subsurface water as a geomorphic agent
    with special reference to the Mangakowhiriwhiri catch-
    ment. *Auckland Student Geographer (N.Z.),* 1(2):
    82-95.

Blong, R.J., 1965b: *Landsurface morphology and process in
    Mangakowhiriwhiri catchment.* Unpub. M.A. thesis, Univ-
    ersity of Auckland, New Zealand.

Bogutski, A.B., 1968: Steppe dips of the Volyno-Podol'ski
    elevation and their origin. *Izvest. Wses. Geogr.,*
    100(2): 125-128. (In Russian.)

Bond, R.M., 1941: Rodentless rodent erosion. *Soil
    Conservation,* 10: 269.

Bosazza, V.L., 1950: Cut a donga through your vlei. *Farm.
    South Africa,* August, 4: 7-9.

Bowden, D.J., 1980: Sub-laterite cave systems and other
    pseudokarst phenomena in the humid tropics: the example
    of Kasewe Hills, Sierra Leone. *Zeit. für Geomorph.*
    24(1): 77-80. (Also unpub. paper delivered to Institute
    of British Geographers Annual Conference, 1979.)

Bower, M.M., 1961: The distribution of erosion in blanket
    peat bogs in the Pennines. *Trans. Institute of British
    Geographers,* 29: 17-30.

Bremer, H., 1972: Flussarbeit, Flächen - und Stufenbildung
    in den feuchten Tropen. *Zeit. für Geomorph,* Suppl.
    Bd., 14: 21-38.

Bremer, H., 1973: Der Formungsmechanismus im tropischen
    Regenwald Amazoniens. *Zeit. für Geomorph,* Suppl.
    Bd., 17: 195-222.

Brewer, R., 1964: *Fabric and mineral analysis of soils.* Wiley, New York, 470 pp.

Brewer, R., and Sleeman, J.R., 1963: Pedotubules: their definition, classification, and interpretation. *Journal of Soil Science,* 14: 156-166.

Brown, G.W., 1961: *Some physical and chemical soil properties associated with piping erosion in Colorado.* Unpub. MSc thesis, Colorado State University.

Brown, G.W., 1962: Piping erosion in Colorado. *Journal of Soil and Water Conservation,* Nov. - Oct.: 220-222.

Bryan, K., 1919: Classification of springs. *Journal of Geology,* 27: 522-561.

Bryan, K., 1925: The Papago Country, Arizona. *U.S. Geological Survey, Water Supply Paper,* 499: 121-123.

Bryan, R.B., Yair, A., Hodges, W.K., 1978: Factors controlling the initiation of runoff and piping in Dinosaur Provincial Park badlands, Alberta, Canada. *Zeit. für Geomorph.,* Suppl. Bd. 29: 151-168.

Buckham, A.F., and Cockfield, W.E., 1950: Gullies formed by sinking of the ground. *American Journal of Science,* 248: 137-141.

Büdel, J., 1948: Die Klima-morphologischen zonen der Polarländer. *Erdkunde,* 2: 22-53.

Bunting, B.T., 1961: The role of seepage moisture in soil formation, slope development and stream initiation. *American Journal of Science,* 259: 503-518.

Bunting, B.T., 1964: Slope development and soil formation on some British sandstones. *Geographical Journal,* 130: 73-79.

Buraczynski, J., and Wojtanowicz, J., 1971: Development of the ravines in loess around Dzierzkowice, Lublin plateau, as a result of the violent rains of June 1969. *Annales Universitatis Mariae Curie - Skłodowska,* 26: 135-168. (In Polish.)

Burton, C.K., 1964: The Older alluvium of Johore and Singapore. *Journal of Tropical Geography,* 18: 30-42.

Burton, J.R., 1964: Modes of failure in small earth dams - review of selected case histories. *C.S.I.R.O. Soil Mechanics Division Paper 13, Colloquium on Failure of Small Earth Dams.*

Butzer, K., 1978: *Geomorphology from the Earth.* Harper and Row, London, 463 pp.

Calver, A., 1978: Modelling drainage headwater development. *Earth Surface Processes*, 3: 233-241.

Calver, A., Kirkby, M.J., and Weyman, D.R., 1972: Modelling
hillslope and channel flows. In R.J. Chorley (ed.)
"Spatial Analysis in Geomorphology", Methuen, London,
393 pp.

Campbell, D.A., 1950: Types of soil erosion in New Zealand.
Trans. 4th International Congress Soil Science, 2:
196-199.

Campbell, I.A., 1974: Measurement of erosion on badland
surfaces. *Zeit. für Geom.,* Supplement Band 21: 123-
137.

Capper, P.L., and Cassie, W.F., 1961: *The mechanics of
engineering soils.* Spon, London.

Carder, D.J., 1978: Managing Western Australia's land
resources. *Journal of Soil Conservation Service,
New South Wales,* 34(2): 73-80.

Carmichael, K.M., 1970: Applying the lessons of the 1960's
to farm dams in the 1970's. In "Water Problems in
Inland Southern Queensland", Report No. 34, pp.
2.1-2.8, Water Research Foundation of Australia.

Carroll, P.H., 1949: Soil piping in south-eastern Arizona.
*U.S. Department of Agriculture Regional Bulletin 10,
Soil Series 13,* (Soil Conservation Service, Region 6,
Albuquerque, New Mexico), 21 pp.

Carson, M.A., and Kirkby, M.J., 1972: *Hillslope form and
process.* Cambridge University Press, 476 pp.

Casagrande, A., 1937: Seepage through dams. *Journal of
New England Water Works Association,* 51: 131-172.

Casagrande, A., 1950: Notes on the design of earth dams.
Boston Society Civil Engineers, 405.

Cedergren, H., 1967: *Seepage, drainage and flow nets.*
Wiley, N.Y., 489 pp New Edition, 1977.

Chapman, T.G., and Dunn, F.X. (ed.), 1975: *Predicition in
catchment hydrology.* Australian Academy of Science,
498 pp.

Charman, P.E.V., 1967: An outline of soil conservation work
in the Bombala District. *Journal of Soil Conservation
Service, New South Wales,* 23(3): 174-186.

Charman, P.E.V., 1969: The influence of sodium salts on
soils with reference to tunnel erosion in coastal areas
Part I: Kemsey Area. *Journal of Soil Conservation
Service, New South Wales,* 25(4): 327-342.

Charman, P.E.V., 1970a: The influence of sodium salts on soils with reference to tunnel erosion in coastal areas. Part II - Grafton Area. *Journal of Soil Conservation Service, New South Wales,* 26(1): 71-86.

Charman, P.E.V., 1970b: The influence of sodium salts on soils with reference to tunnel erosion in coastal areas. Part III - Taree Area. *Journal Soil Conservation Service, New South Wales,* 26(4): 256-275.

Chevalier, A., 1949: Points de vue nouveaux sur les sols d'Afrique tropicale sur leur degradation et leur conservation. *Bulletin Agricole du Congo belge,* 40: 1057-1092.

Childs, E.C., 1969: *An introduction to the physical basis of soil water phenomena.* Wiley, London.

Chilingar, G.V., 1970: Discussion on paper by Ingles and Aitchison (1970). *International Association of Science Hydrology,* Special Pub. 89: 353.

Chorley, R.J., 1978a: The hillslope hydrological cycle. Chapter 1 in *"Hillslope Hydrology"* edited by M.J. Kirkby, Wiley, 389 pp., p. 1-42.

Chorley, R.J., 1978b: Glossary of terms. In *"Hillslope Hydrology"* edited by M.J. Kirkby, Wiley, 389 pp., p. 365-376.

Clayton, R.W., 1956: Linear depressions (Bergfussniederungen) in Savannah landscapes. *Geographical Studies,* 3(2): 102-126.

Coates, D.R. (ed.), 1971: *Environmental Geomorphology.* Dowden, Hutchinson and Ross, Stroudsburg, Pa.

Coates, D.R. (ed.), 1973: *Environmental Geomorphology and Landscape Conservation,* Dowden, Hutchinson and Ross, Stroudsburg, Pa.

Coates, D.R., 1977: Landslide perspectives. In *"Landslides"*, edited by D.R. Coates, Geological Society American Reviews in Engineering Geology, No. 3: 3-28.

Coates, D.R., 1979: Subsurface influences. Chapter 10 in *"Man and Environmental Processes"*, edited by K.J. Gregory and D.E. Walling, Dawson/Westview Press, Folkstone/Boulder, Colorado, pp. 163-190.

Cockfield, W.E., and Buckham, A.F., 1946: Sink-hole erosion in the White Silts at Kamloops. *Royal Society of Canada Trans.,* 3rd Series, 40 (Section 4): 1-10.

Colclough, J.D., 1965: Tunnel erosion. *Tasmanian Journal of Agriculture,* 36(1): 7-12.

Colclough, J.D. , 1967: Tunnel erosion control. *Tasmanian Journal of Agriculture,* 38(1): 25-31.

Colclough, J.D. , 1971: Contour ripping. *Tasmanian Journal of Agriculture,* 42(2): 114-125.

Colclough, J.D. , 1973: Soil conservation and soil erosion control in Tasmania: vegetation to control tunnel erosion. *Tasmanian Journal of Agriculture,* 44(1): 65-70.

Colclough, J.D. , 1978: Soil conservation in Tasmania. *Journal of Soil Conservation Service, New South Wales,* 34(2): 63-72.

Cole, B.A. , Chalaw Ratanasen, Pramote Maiklad, Liggins, T.B. , Suphon Chiraponto, 1977: Dispersive clay in irrigation dams in Thailand. In *"Dispersive clays, related piping, and erosion in geotechnical projects",* edited by J.L Sherard and R.S. Decker, ASTM STP 623, p. 25-41.

Cole, D.C.H. , and Lewis J.G. , 1960: Piping failure of earthen dams built of plastic materials in *arid* climates. *Proceedings 3rd Australia - New Zealand Conference on Soil Mechanics and Foundation Engineering.*

Cole, R.C. , *et al.* 1943: The Tracy Area, California. *U.S. Department of Agriculture Soil Survey Series,* 1938, No. 5, 83 pp.

Colhoun, E.A. , Common, R. , Cruickshank, M.M. , 1965: Recent bog flows and debris slides in the north of Ireland. *Scientific Proceedings of Royal Dublin Society,* Series A, 2(10): 163-174.

Collis - George, N. , and Smiles, D.E. , 1963: An examination of cation balance and moisture characteristic methods of determining the stability of soil aggretates. *Journal of Soil Science,* 14: 21.

Commonwealth Bureau of Soils, 1961: Bibliography on tunnel and piping erosion 1959 - 1945, No. 407, 3 pp., 14 refs.

Conacher, A.J. , 1975: Throughflow as a mechanism respon- sible for excessive soil salinisation in non-irrigated, previously arable lands in the Western Australian wheatbelt: a field study. *Catena,* 2: 31-68.

Conacher, A.J. , and Dalrymple, J.B. , 1977: The nine unit land surface model: an approach to pedogeomorphic research. *Geoderma,* 18(1/2): 1-154.

Condon, R.W. , and Stannard, M.E. , 1957: Erosion in western New South Wales. Part IV: Classification and survey of erosion. *Journal of Soil Conservation Service, New South Wales,* 13: 76-82.

Cooke, R.U., and Warren, A., 1973: *Geomorphology in Deserts*. Batsford, London.

Corbel, J., 1957: Les Karsts du Nord-Ouest de l'Europe et de quelques Regions de Comparaison: Etude sur le Rôle du Climat dans l'Erosion des Calcaires. *Institut des Etudes Rhodaniennes de l'Université de Lyon, Memoires et Documents* No. 12.

Coumoulos, D.G., 1977: Experience with studies of clay erodibility in Greece. In *"Dispersive clays, related piping, and erosion in geotechnical projects"* edited by J.L. Sherard and R.S. Decker, ASTM STP 623, 486 pp, p. 42-57.

Craze, B., 1974: The soils of Reedy Creek catchment area. *Journal of Soil Conservation Service, New South Wales,* 30(2): 103-111.

Crouch, R.J., 1976: Field tunnel erosion - a review. *Journal of Soil Conservation Service, New South Wales,* 32(2): 98-111.

Crouch, R.J., 1977: Tunnel-gully erosion and urban development: a case study. In *"Dispersive clays, related piping, and erosion in geotechnical projects"* edited by J.L. Sherard and R.S. Decker, A.S.T.M. S.T.P. 623, p. 58-68.

Crozier, M.J.,1969: Earthflow occurrences during high intensity rainfall in Eastern Otago (New Zealand). *Engineering Geology,* 3(4): 325-334.

Cruickshank, M.M., 1980: The cause of peat erosion. In *"Geographical Approaches to fluvial processes"*, edited by A.F. Pitty.

Cryer, R., 1976: The significance and variation of atmospheric nutrient inputs in a small catchment system. *Journal of Hydrology,* 28: 121-137.

Cryer, R., 1978: *A study of the sources and variations of major solutes in selected mid-Wales catchments.* Unpub. Ph.D. thesis, University of Wales.

Cryer, R., 1980: The chemical quality of some pipeflow waters in upland Mid-Wales and its implications. *Cambria,* 6(2): 1-19.

Culling, W.E.H., 1960: Analytical theory of erosion. *Journal of Geology,* 68: 336-344.

Cumberland, K.B., 1944: *Soil erosion in New Zealand.* Soil Conservation and River Control Council, Wellington, 288 pp.

Cussen, L., 1888: Notes on the Waikato River Basin. *New Zealand Institute Transactions and Proceedings,* 21: 406-416.

Czeppe, Z., 1960: Suffosional phenomena in slope loams of the Upper San drainage basin. *Instytut Geologiczny Biul. (Warsaw),* 9: 297-332.

Czeppe, Z., 1965: Activity of running water in south-western Spitsbergen. *Geographia Polonica,* 6: 141-150.

Czyzewski, K., Skrzynski, J., and Wolski, W., 1961: The sealing of earth and earth-rock dams made of Carpathian Flysch. *Trans. 7th International Congress on Large Dams,* 4: 441-465.

Dalrymple, J.B., Blong, R.J., Conacher, A.J., 1968: An hypothetical nine unit land surface model. *Zeit. für Geomorph.,* 12(1): 60-76.

Davidenkoff, R., 1955: The composition of filters in earth dams. *Trans. 5th Congress on Large Dams,* 1: 385-401.

Davis, R.A., 1972: Surface hydrological features in the Nant Gerig. Institute of Hydrology, Subsurface Section, internal report no. 48, 6 pp., Unpub. typescript.

Davis, S.N., and De Wiest, R.J.M., 1966: *Hydrology.* Wiley, New York, 463 pp.

Davis, W.M., 1927: Channels, valleys and intermontane detrital plains. *Science,* 66: 272.

Davis, W.M., 1930: Origin of limestone caverns. *Geological Society America Bulletin,* 41: 475-628.

Day, D.G., 1978: Drainage density changes during rainfall. *Earth Surface Processes,* 3: 319-326.

De Chéletat, E., 1938: Le Modelé laterique de l'ouest de la Guinée francaise. *Rev. de geog. phys. et de geol. dyn.,* 11(1): 5-120.

Decker, R.S., and Dunnigan, L.P., 1977: Development and use of the Soil Conservation Service Dispersion Test. In "*Dispersive clays, related piping, and erosion in geotechnical projects*", edited by J.L. Sherard and R.S. Decker, A.S.T.M. S.T.P. 623, p. 94 - 109.

Del Castillo, R., 1973: Estudio de Fallas por Tubificación de Materiales Arcillosos. Informe presentado a la Secretaria de Recursos Hidraulicos, Parte 1, Instituto de Ingenieria, Mexico.

Denisov, N.I., Bally, R.I., and Antonescu, I.P., 1960: Fenomene de Prabusire a Unor Canale de Irigatie in Lunca Dunarii de Jos. *Cer. Dom. Construct. Hidrotechnice,* 2: 3.

De Ploey, J., 1974: Mechanical properties of hillslopes and their relation to gullying in Central semi-arid Tunisia. *Zeit. für Geomorph.*, 21: 177-190.

Dickinson, H.R., 1950: Soil erosion in Tasmania: Part 1. *Tasmanian Journal of Agriculture,* 21(1): 106-114.

Dixey, F., 1920: Notes on lateritization in Sierra Leone. *Geological Magazine,* 57: 211-220.

Dobrovolny, E., 1962: Geologia del Valle de La Paz. *Dept. Nacion. de Geol. (La Paz) Boletin,* 3: 1-153.

Douglas, I., 1977: *Humid Landforms.* M.I.T. Press, Cambridge, Mass., 288 pp.

Dovey, W., 1976: *The significance of natural soil pipes.* Unpub. M.Sc. dissertation, University of Birmingham, 63 pp.

Downes, R.G., 1946: Tunnelling erosion in North-Eastern Victoria. *Journal of Comm. Science and Ind. Res.,* 19(3): 283-292.

Downes, R.G., 1949: A soil, land-use, and erosion survey of parts of the Counties of Moira and Delatite, Victoria. *Bulletin No. 243, C.S.I.R.O.,* Australia, 89 pp.

Downes, R.G., 1954: Cyclic salt as a dominant factor in the genesis of soils in south eastern Australia. *Australian Journal of Agricultural Research,* 5: 448.

Downes, R.G., 1956: Conservation problems on solodic soils in the State of Victoria (Australia). *Journal of Soil and Water Conservation,* 11: 228-232.

Downes, R.G., 1959: The ecology and prevention of soil erosion. In: *"Biogeography and Ecology in Aistralia",* edited by A. Kearst, R.L. Crocker and C.S.Christian, Junk, The Hague, p. 472-486.

Downing, B.H., 1968: Subsurface erosion as a geomorphological agent in Natal. *Trans. Geological Society of South Africa,* 81(2): 131-134.

Dregne, H.E., 1967: *Inventory of research on surface materials of desert environments.* Office of Arid Lands Studies, University of Arizona, 91 pp.

Dregne, H.E., 1978: *Soils of Arid Regions.* Elsevier.

Drew, D.P., 1972: Geomorphology of the Big Muddy Valley area, Southern Saskatchewan, with reference to the occurrence of piping. In *"Southern Prairies Field Excursion Background Papers"* edited for I.G.C. by A.H. Paul, E.H. Dale and H. Schichtmann, Department of Geography, Regina Campus, University of Saskatchewan, p. 197-212.

Dunne, T., and Black, R.D., 1970a: An experimental investigation of runoff production in permeable soils. *Water Resources Research*, 6(2): 478-490.

Dunne, T., and Black, R.D., 1970b: Partial area contributions to storm runoff in a small New England watershed. *Water Resources Research,* 6(5): 1296-1311.

Elliott, G.L., 1977: Amelioration of tunnelling susceptible soils in the Hunter Valley, New South Wales, by modification of clay double-layer interactions. In *"Dispersive clays, related piping, and erosion in geotechnical projects"* edited by J.L. Sherard and R.S. Decker, A.S.T.M. S.T.P. 623, p. 110-120.

Embleton, C., Brunsden, D., and Jones, D.K.C., 1978: *Geomorphology: Present problems and future prospect.* Oxford University Press, 281 pp.

Emerson, W.W., 1959: The sealing of earth dams. *C.S.I.R.O. Division of Soils Technical Memo.* 4/59.

Emerson, W.W., 1960: The sealing of earth dams. *Proceedings Arid Zone Conference, Warburton, Victoria,* vol. 1, Paper 38.

Emerson, W.W., 1967: A classification of soil aggregates based on their cohesion in water. *Australian Journal of Soil Research,* 5: 47-57.

Engelen, G.B., 1973: Runoff processes and slope development in Badlands National Monument, South Dakota. *Journal of Hydrology,* 18: 55-79.

Eyles, R.J., 1968: Stream net ratios in West Malaysia. *Geological Society of America Bulletin,* 79(6): 701-712.

Evans, O.F., 1927: Origin of certain stream valleys of the Interior Plains region of the United States. *American Journal of Science,* 13: 259.

Fairbridge, R.W., editor, 1968: *Encyclopedia of Geomorphology,* Encyclopedia of Earth Science Series volume 3, Reinhold, N.Y., 1295 pp.

Feininger, T., 1969: Pseudokarst on quartz diorite, Colombia. *Zeit. für Geomorph.,* 13(3): 287-296.

Fenneman, N.M., 1922: Physiographic provinces and sections in Western Oklahoma and adjacent parts of Texas. *U.S. Geological Survey Bulletin,* 730-D: 115-134.

Ferrar, H.T., 1934: The Geology of the Dargaville-Rodney Subdivision, Hokianga and Kaipara Divisions. *NZ DSIR Geological Survey Bulletin,* No. 34.

Finlayson, B.L., 1976: *Measurement of geomorphic processes in a small drainage basin.* Unpub. Ph.D. thesis, University of Bristol.

Finlayson, B.L., 1977: Runoff contributing areas and erosion. University of Oxford School of Geography Research Paper No. 18, 40 pp.

Fletcher, J.E., and P.H. Carroll, 1948: Some properties of soils that are subject to piping in southern Arizona. *Soil Science Society of America Proceedings,* 13: 545-547.

Fletcher, J.E., and Harris, K., 1952: Soil piping. *Progressive Agriculture in Arizona, 4:* 7.

Fletcher, J.E., Harris, E.., Peterson, H.B., and Chandler, V.N., 1954: Piping. *Trans. American Geophysical Union.* 35(2): 258-262.

Floyd, E.J., 1974: Tunnel erosion - a field study in the Riverina. *Journal Soil Conservation Service, New South Wales,* 30(3): 145-156.

Fontaine, H., 1965: Mode particulier d'erosion dans les formations meubles. *Archives Geologiques de Viet-Nam,* 7: 34-37.

Food and Agriculture Organisation, 1965: Soil erosion by water: Some measures for its control on cultivated lands. *FAO Agricultural Development Paper* No. 81, Rome, 284 pp.

Food and Agriculture Organisation, 1968: Definitions of soil units for the soil map of the world. *World Soil Resources Report UNESCO,* 33, Rome.

Foreman, N.G., 1963: Soil conservation in the Grafton sub-district. *Journal of Soil Conservation Service, New South Wales,* 19(3): 130-137.

Forster, B.A., and O'Meara, G.A., 1978: Muddy waters to grassed moonscapes - the changing scene of soil conservation in the Australian capital territory. *Journal of Soil Conservation Service, New South Wales,* 34(2): 82-87.

Forsythe, P., 1977: Experiences in identification and treatment of dispersive clays in Mississippi dams. In *"Dispersive clays, related piping, and erosion in geotechnical projects",* edited by J.L. Sherard and R.S. Decker, ASTM STP 623, p. 135-155.

Francq, J., and Post, G., 1977: Laboratory testing on high-sodium nondispersive clays as related to the repair of a clay dam in Algeria. In *"Dispersive clays, related piping, and erosion in geotechnical projects",* edited by J.L. Sherard and R.S.Decker, ASTM STP 623, p.156-171.

Franzle, O., 1976: Ein morphodynamisches Grundmodell der Savannen- und Regenwaldgebiete (A morphodynamic model of the savanna and rain forest regions). *Zeit für Geomorph.*, Suppl. Bd., 24: 177-184.

Freeze, R.A., 1972a: Role of subsurface flow in generating surface runoff-1. Base flow contributions to channel flow. *Water Resources Research,* 8(3): 609-623.

Freeze, R.A., 1972b: Role of subsurface flow in generating surface runoff - 2. Upland source areas. *Water Resources Research,* 8(5): 1271-1283.

Freeze, R.A., 1974: Stream flow generation. *Reviews of Geophysics and Space Physics,* 12(4): 627-647.

Fuller, M.L., 1922: Some unusual erosion features in the loess of China. *Geographical Review,* 12(4): 570-584.

Funkhowser, J.W., 1951: "Soil caves" in tropical Ecuador. *National Speleological Society (America) News,* 9(5): 4 (letter to editor).

Gaiser, R.N., 1952: Root channels and roots in forest soils. *Proceedings Soil Science Society, America,* 16: 62-65.

Galarowski, T., 1976: New observations of the present-day suffosion (piping) processes in the Bereznica catchment basin in the Bieszczady Mountains (The East Carpathians). *Studia Geomorphologica Carpatho-Balcanica (Krakow),* 10: 115-122.

Gautier, E.F., 1902: *Madagascar - Essai de Geographie physique.* Paris.

Gerson, R., 1971: *Geomorphic processes in Mt. Sdom.* Unpub. Ph.D. thesis, The Hebrew University, Jerusalem. (Hebrew with English summary.)

Ghuman, O.S., Allen, R.L., and McNeill, R.L., 1977: Erosion, corrective maintenance and dispersive clays. In *"Dispersive clays, related piping, and erosion in geotechnical projects"*, edited by J.L. Sherard and R.S. Decker, ASTM STP 623, p. 172-190.

Gibbs, H.S., 1945: Tunnel-gully erosion on the Wither Hills, Marlborough, New Zealand. *New Zealand Journal of Science and Technology,* 27 section A(2): 135-146.

Gifford, Joyce, 1953: Landslides on Exmoor caused by the storm of 15th August, 1952. *Geography,* 37: 9-17.

Gilbert, M.J., 1921: Geology of the Waikato Heads District and the Kawa unconformity. *New Zealand Institute Trans. and Proceedings,* 53.

Gile, L.H., 1967: Soils of an ancient basin floor near Las Cruces, New Mexico. *Soil Science,* 103(4): 265-276.

Gile, L.H., and Grossman, R.B., 1968: Morphology of the argillic horizon in desert soils of southern New Mexico. *Soil Science,* 106(1): 6-15.

Gile, L.H., and Hawley, J.W., 1966: Periodic sedimentation and soil formation on an alluvial fan piedmont in southern New Mexico. *Proceedings Soil Science Society of America,* 30: 261-268.

Gile, L.H., Peterson, F.F., and Grossman, R.B., 1966: Morphological and generic sequences of carbonate accumulation in desert soils. *Soil Science,* 101(5): 347-360.

Gilman, K., 1971a: Qualitative investigation into the nature and distribution of flow processes in Nant Gerig. Appendix I: An instrument for measurement of flow on hillslopes. NERC Institute of Hydrology, Subsurface Section, internal report no. 28, 5 pp, unpub. typescript.

Gilman, K., 1971b: A semi-quantitative study of the flow of natural pipes in the Nant Gerig sub-catchment. NERC Institute of Hydrology, Subsurface Section, internal report no. 36, 16 pp, unpub. typescript.

Gilman, K., 1972: Pipe flow studies in the Nant Gerig. NERC Institute of Hydrology, Subsurface Section, internal report no. 50, 7 pp, unpub. typescript.

Gilman, K., 1977: Instrument system for natural pipe flow. In "*Selected measurement techniques in use at Plynlimon experimental catchments*" Institute of Hydrology Report No. 43.

Gilman, K., and Newson, M.D., 1980: *Soil pipes and pipe-flow - a hydrological study in upland Wales.* British Geomorphological Research Group Monograph No. 1, 114 pp.

Glossop, R., 1945: Soil mechanics in foundations and excavations. In "*The principles and application of soil mechanics*", London lecture series, Institution of Civil Engineers, p. 63-90.

Glukhov, I.G., 1956: Seepage of water from canals in loess formations and subsidence phenomena in irrigated areas. *Gidroteckhnika i Melioratsiya,* 8(10): 9-18, issued in translation by the Israel Program for Scientific Translations, Jerusalem, OTS 60 - 21152.

Goudie, A., 1973: *Duricrusts in tropical and subtropical landscapes.* Clarendon Press, Oxford, 174 pp.

Gregory, K.J., 1976: Drainage networks and climate. Chap. 10 in *"Geomorphology and climate"* edited by E. Derbyshire, Wiley, 512 pp., p. 289-315.

Gregory, K.J., 1979a: Hydrogeomorphology: how applied should we become. *Progress in Geography,* 3(1): 84-101.

Gregory, K.J., 1979b: Fluvial geomorphology: Progress Report. *Progress in Geography,* 3(2): 274-282.

Gregory, K.J., and Walling, D.E., 1968: The variation of drainage density within a catchment. *Bulletin International Association Scientific Hydrology,* 13(2): 61-68.

Gregory, K.J., and Walling, D.E., 1973: *Drainage basin form and process.* Arnold, London, 456 pp.

Grim, R.E., 1962: *Applied Clay Mineralogy.* McGraw-Hill, New York.

Guthrie-Smith, H., 1921: *Tutira: the story of a New Zealand sheep station.* Blackwood, London and Edinburgh, first edition, 399 pp.

Gvozdeski, N.V., 1954: *Kärst.* Moscow.

Gyamarthy, A., 1962: The use of chemicals to improve highly dispersible dam materials. *Soil Conservation Authority, Victoria,* Technical Note 1/1962.

Hack, J.T., and Goodlett, J.C., 1960: Geomorphology and forest ecology of a mountain region in the Central Appalachians. *U.S. Geological Survey Prof. Paper,* 347, 66 pp.

Hadley, R.F., and Rolfe, B.N., 1955: Development and significance of seepage steps in slope erosion. *Trans. Amer. Geophys. Union,* 35(5): 792-804.

Haldeman, E.G., 1956: Recent landslide phenomena in the Rungwe volcanic area, Tanganyika. *Tanganyika Notes and Record,* no. 45, 14 pp.

Haliburton, T.A., Petry, T.M., and Hayden, M.L., 1975: Identification and treatment of dispersive clay soils. Report of U.S. Department of Interior, Bureau of Reclamation, Contract No. 14-06-D-7535, School of Civil Engineering, Oklahoma State University, Stillwater, Oklahoma. (Cp. Perry, 1975).

Halliday, W.R., 1960: Pseudokarst in the United States. *National Speleological Society (America) Bulletin,* 22(2): 109-113.

Hamilton, D.H., and Meehan, R.L., 1971: Ground rupture in the Baldwin Hills. *Science,* 172(3981): 333-344.

Hamilton, T.M., 1970: Channel-scarp formation in western North Dakota. *USGS Professional Paper,* 700-C, p. C 229-232.

Hardy, R.M., 1950: Construction problems in silty soils. *The Engineering Journal* (Montreal), 33: 775-779 and 782.

Harr, M.E., 1962: *Groundwater and seepage.* McGraw-Hill, London, 315 pp.

Harris, K., and Fletcher, J.E., 1951: *Report of the so called soil piping, Yolo Ranch, Yoma Country, Arizona.* Soil Conservation Service Research, unpub. paper, Arizona State Office, Phoenix, Arizona.

Hauser, A., and Zötl, J., 1955: Die morphologische Bedeutung der unterirdischen Erosion durch Gesteinausspülung. *Petermans Mitt.,* 99: 18-21.

Haworth, E., 1897: Physiography of Western Kansas. *University Geological Survey of Kansas,* 2: 11-49.

Heede, B.H., 1970: Morphology of gullies in the Colorado Rocky Mountains, *Bulletin of International Association Scientific Hydrology,* 15(2): 79-89.

Heede, B.H., 1971: Characteristics and processes of soil piping in gullies. *U.S. Department of Agriculture, Forest Service Research Paper,* RM-68, 15 pp.

Heede, B.H., 1974: Stages of development of gullies in western United States of America. *Zeit.für Geomorph.,* 18(3): 260-271.

Heede, B.H., 1976: Gully development and control: the status of our knowledge. *USDA Forest Service, Rocky Mountain Forest and Range Experimental Station, Fort Collins, Research Paper,* RM-169.

Heindl, L.A., and Feth, J.H., 1955: Discussion of Symposium on Land Erosion, "Piping" by J.E. Fletcher et al. (1954). *Transactions American Geophysical Union,* 36(2): 342-345.

Heinzen, R.T., and Arulanandan, K., 1977: Factors influencing dispersive clays and methods of identification. In *"Dispersive clays, related piping, and erosion in geotechnical projects",* edited by J.L. Sherard and R.S. Decker, ASTM STP 623, p. 202.

Henderson, J., and Grange, L.I., 1926: The geology of the Hunty-Kawhia subdivision. *New Zealand Department of Mines, Geological Survey Branch Bulletin,* No. 28, 112pp.

Hénin, S., 1949: Madagascan Soils. *Technical Communication of Commonwealth Bureau of Soils* (Harpenden), 46: 40-43.

Henkel, J.S., Bayer, A.W., and Coutts, J.R.H., 1938: Subsurface erosion on a Natal Midlands farm. *South African Journal of Science,* 35: 236-243.

Henkin, E.N., 1967: *Piping and consolidation as factors influencing small loess dam failures.* Unpub. MSc thesis, Technion University, Haifa. (In Hebrew.)

Hewlett, J.D., 1961: Soil moisture as a source of base flow from steep mountain watersheds. USDA Forest Service, Southeastern Forest Experiment Station, Asheville, North Carolina, Station Paper No. 132, 11 pp.

Hewlett, J.D., 1969: Tracing storm and base flow to variable source areas on forested watersheds. Technical Report No. 2, School of Forest Resources, University of Georgia, Athens, Georgia, 21 pp.

Hewlett, J.D., 1974: Comments on letters relating to "Role of subsurface flow in generating surface runoff. 2, Upland source areas" by R.A. Freeze. *Water Resources Research,* 10(3): 605-607.

Hewlett, J.D., and Nutter, W.L., 1970: The varying source area of streamflow from upland basins. *Proceedings of Symposium on Interdisciplinary Aspects of Watershed Management,* Montana State University, Bozeman, American Society, Civil Engineers, New York, pp. 65-83.

Hickey, W.C., Jr., and Dortignac, E.J., 1964: An evaluation of soil ripping and soil pitting on runoff and erosion in the semi-arid south west. *International Association Science Hydrology, Pub.* 65: 22-33.

Hillel, D., 1971: *Soil and water: physical principles and processes.* Academic Press, New York/London, 288 pp.

Horton, R.E., 1933: The role of infiltration in the hydrological cycle. *Trans. American Geophysical Union,* 14: 446-460.

Horton, R.E., 1936: Maximum ground-water levels. *Trans. American Geophysical Union,* 17: 344-357.

Horton, R.E., 1945: Erosional development of streams and their drainage basins: hydrophysical approach to quantitative morphology. *Geological Society of America Bulletin,* 56: 275-370.

Hosking, P.L., 1962: *Loess, and its erosion on the Port Hills, Banks Peninsula.* Unpub. MA thesis, University of Canterbury.

Hosking, P.L., 1967: Tunneling erosion in New Zealand. *Journal of Soil and Water Conservation,* 22(4): 149-151.

Howells, K.A., 1980: *Pipeflow and its contribution to streamflow in a small upland catchment.* Unpub. BSc dissertation, University College Swansea, 113 pp.

Hudson, N., 1971: *Soil Conservation.* Batsford, London, 320 pp.

Hughes, P.J., 1970: *Tunnel erosion in the loess of Banks Peninsula.* Unpub. MSc thesis, University of Canterbury.

Hughes, P.J., 1972: Slope aspect and tunnel erosion in the loess of Banks Peninsula, New Zealand. *Journal of Hydrology (New Zealand),* 11(2): 94-98.

Humphreys, B., 1978: *A study of some of the geomorphological and hydrological properties of natural soil piping.* Unpub. BSc. dissertation, University of East Anglia, 35 pp.

Hursh, C.R., 1944: Report of the subcommittee on subsurface flow. *Trans. American Geophysical Union,* Part V: 743-746.

Imeson, A.C., 1978: Slope deposits and sediment supply in a New England drainage basin (Australia). *Catena,* 5(2): 109-130.

Ingles, O.G., 1964 a: The effect of lime treatment on permeability-density-moisture relationships for three montmorillonitic soils. Paper No. 30, *Water Research Foundation of Australia Ltd. and Soil Mechanics Section, C.S.I.R.O., Colloquium on failure of small earth dams.* 16-19 Nov., 3 pp.

Ingles, O.G., 1964b: The water-soil regime in three dams sampled during failure. Paper No. 31, *Water Research Foundation of Australia Ltd. and Soil Mechanics Section, C.S.I.R.O., Colloquium on failure of small earth dams,* 4 pp.

Ingles O.G., 1968: Soil chemistry relevant to the engineering behaviour of soils. In *"Soil Mechanics - selected topics",* edited by Lee, Butterworth, London.

Ingles, O.G., and Aitchison, G.D., 1970: Soil-water disequilibrium as a cause of subsidence in natural soils and earth embankments. IASH-UNESCO, International Symposium on Land Subsidence, 1969, Tokyo, IASH Pub. 89, 342-353. Also: CSIRO, Division of Soil Mechanics Research Paper No. 131.

Ingles, O.G., Lang, J.G., and Richards, B.G., 1968: Pre-equilibrium observations on the reconstructed Flagstaff Gully Dam. *Symposium Earth and Rockfill Dams,* Talwarra (India), 1: 162.

Ingles, O.G., Lang, J.G., Richards, B.G., and Wood, C.C., 1969: Soil and Water observations in Flagstaff Gully Dam, Tasmania. *CSIRO, Division of Soil Mechanics, Technical Report* No. 11, 34 pp.

Ingles, O.G., and Metcalf, J.B., 1973: *Soil stabilization.* Wiley.

Ingles, O.G., and Wood, C.C., 1964a: The contribution of soil and water cations to deflocculation phenomena in earth dams. *Proceedings of 37th Congress, Australia-New Zealand Association for Advancement of Science,* Section H, Canberra.

Ingles, O.G., and Wood, C.C., 1964b: The recognition of failure in earth dams by aerial survey. *The Australian Journal of Science,* 26(11): 355 (and CISRO Soil Mechanics Section, Research Paper No. 53).

Ingles, O.G., and Wood, C.C., 1965: The passage of a wetting front through the earth retaining wall of the Flagstaff Gully Dam, Hobart. Paper presented to 38th Congress of Australia - New Zealand Association for Advancement of Science, Hobart, 16-20 August, 10 pp.

Inglis, C.C., 1949: *The behaviour and control of rivers and canals.* Research Pub. Poona, India, No. 13, 2 vols.

Ingram, H.A.P., 1967: Problems of hydrology and plant distribution in mires. *Journal of Ecology,* 55: 711-724.

Institute of Hydrology, 1972: *Research 1970-71.* Natural Environment Research Council, Institute of Hydrology, Wallingford, Berks, 54 pp.

Institute of Hydrology, 1974: *Research 1972-73.* Natural Environment Research Council, Institute of Hydrology, Wallingford, Berks., 66 pp.

Institute of Hydrology, 1977: *Research Report 1974-76.* Natural Environment Research Council, Institute of Hydrology, Wallingford, Oxfordshire, 107 pp.

Institute of Hydrology, 1977: *Selected measurement techniques in use at Plynlimon experimental catchments.* Natural Environment Research Council, Institute of Hydrology, Wallingford, Berks., Report No. 43.

Institute of Hydrology, 1979: *Research Report 1976-78.* Natural Environment Research Council, Institute of Hydrology, Wallingford, Oxfordshire, 124 pp.

Ireland, H.A., Sharpe, C.F.S., and Eargle, D.H., 1939: Principles of gully erosion in South Carolina. *USDA Technical Bulletin,* 633.

Jacks, G.V., and Whyte, R.O., 1936: Erosion and Soil Conservation. *Imperial Bureau of Soil Science, Tech. Comm.,* 36.

Jackson, R.J., 1966: Slips in relation to rainfall and soil characteristics. *Journal of Hydrology (New Zealand),* 5(2): 45-53.

James, W.R., and Krumbein, W.C., 1969: Frequency distribution of stream link lengths. *Journal of Geology,* 77(5): 544-565.

Jefferson, J.H.K., 1953: A note based on field experience in planning Hafir excavation programmes. *Ministry of Agriculture, Sudan Government Mem.* No. 4.

Jennings, J.E., and Knight, K., 1956: Recent experiences with consolidation tests as a means of identifying heaving or collapse of foundations on partially saturated soils. *Transactions Southern African Institute of Civil Engineers,* 6(8): 255-256.

Jennings, J.E., and Knight, K., 1957: The additional settlement of foundations due to a collapse of structure of sandy subsoils on wetting. *Proceedings of 4th International Conference on Soil Mechanics and Foundation Engineering* (London), 1: 316-319.

Jennings, J.N., 1971: *Karst,* MIT Press (Australia), 252 pp.

Jessup, W.E., 1964: Baldwin Hills Dam failure. *Civil Engineering,* 34(2): 62-64.

Johnson, A.M., and Rahn, P.H., 1970: Mobilisation of debris flow. *Zeit. für Geomorph.,* Supplement Band 9: 168-186.

Johnson, R.H., 1957/8: Observations on stream patterns on some peat moorlands in the southern Pennines. *Mem. and Proceedings Manchester Phil. and Lit. Society,* 99(7): 110-127.

Johnson, W.D., 1901: The High Plains and their utilization. *U.S. Geological Survey, 21st Annual Report,* 609-741.

Jones, J.A.A., 1971: Soil piping and stream channel initiation. *Water Resources Research,* 7(3): 602-610.

Jones, J.A.A., 1975: *Soil piping and the subsurface initiation of stream channel networks.* Unpub. Ph.D thesis, University of Cambridge, 467 pp.

Jones, J.A.A., 1978a: Soil pipe networks: distribution and discharge. *Cambria,* 5(1): 1-21.

Jones, J.A.A., 1978b: The spacing of streams in a random-walk model. *Area,* 10(3): 190-197.

Jones, J.A.A., 1978c: The hydrological significance of soil piping. Paper delivered to Welsh Geographers' Colloquium, Aberystwyth, May, 1978, unpub. mimeo.

Jones, J.A.A., 1979: Extending the Hewlett model of stream runoff generation. *Area,* 11(2): 110-114.

Jones, J.A.A., and Crane, F.G., 1979: The contribution of pipeflow to stream runoff in an upland catchment. First Report, NERC Research Grant GR3/3683, July, 12 pp.

Jones, N.O., 1968: *The development of piping erosion.* Unpub. Ph.D. thesis. University of Arizona, 163 pp.

Jongerius, A., 1964: *Soil Micromorphology,* Elsevier, Amsterdam, 540 pp.

Jumikis, A.R., 1962: *Soil Mechanics,* Van Nostrand Pub. Co., Princeton, N.Y., 791 pp.

Jutson, R.T., 1934: The physiography (geomorphology) of western Australia. *Bulletin Geological Survey, Western Australia, (Perth),* v.95.

Kassiff, G., 1956: *The engineering properties of Negev loess as applied to embankments.* Unpub. MSc. thesis, Technion University, Haifa. (In Hebrew.)

Kassiff, G., and E.N. Henkin, 1967: Engineering and physico-chemical properties affecting piping failure of low loess dams in the Negev. *Proceedings of the Third Regional Conference on Soil Mechanics and Foundation Engineering,* Israel, 1: 13-16.

Kassiff, G., Zaslavsky, D., and Zeitlen, J.G., 1965: Analysis of filter requirements for compacted clays. *Proceedings of 6th International Conference Soil Mechanics and Foundation Engineering,* Canada, 2: 495-499.

Kastory, L., 1971: The Godziszow landslides. *Czasopismo Geograficzne,* 42(3): 253-259. (In Polish.)

Kaszowski, L.A., Kotarba, M., Niemirowski, and Starkel, L., 1966: Maps of contemporaneous morphogenetic processes in Southern Poland. *Bulletin of Academy of Sciences (Poland)* ser. geol/geog., 14: 113-118.

Kemper, W.D., and Chepil, W.S., 1965: Size distribution of aggregates. In C.A. Black *et al.* (ed.), "*Methods of Soil analysis*", part 1, pp 499-510.

Kesel, R.H., and Smith, J.S., 1978: Tidal creek and pan formation in intertidal salt marshes, Nigg Bay, Scotland. *Scottish Geographical Magazine,* 159-168.

Khobzi, J., 1972: Erosion chimique et mecanique dans le genese de depressions 'pseudo-karstiques' souvant endoreiques. *Revue de Geomorphologie Dynamique,* 21(2): 57-70.

King, F.J., 1899: Principles and conditions of the movement of ground-water. *U.S. Geological Survey,* 19th Annual Report, Part 2.

King, L.C., 1942: *South African Scenery: a textbook of geomorphology.* Oliver and Boyd, London, 379 pp. Re-published 1951.

Kingsbury, J.W., 1952: Pot hole erosion on the western part of Molokai Island - Territory of Hawaii. *Journal of Soil and Water Conservation,* 7(4): 197-198.

Kirkby, M.J., 1978a: The streamhead as a significant geomorphic threshold. University of Leeds, Department of Geography, Working Paper No. 216, 18 pp.

Kirkby, M.J., 1978b: Implications for sediment transport. In M.J. Kirkby (ed.), *"Hillslope Hydrology",* Wiley.

Kirkby, M.J., and Chorley, R.J., 1967: Throughflow, overland flow and erosion. *Bulletin International Association of Scientific Hydrology,* 12(3): 5-21.

Kirkby, M.J., and Weyman, D.R., 1972: Measurements of contributing area in very small drainage basins. University of Bristol, Department of Geography, Seminar Paper Series B, No. 3, 12 pp.

Kirkham, D., 1951: Seepage into drain tubes in stratified soil. *Trans. American Geophysical Union,* 32: 422-441.

Kirkham, D., and Powers, W.L., 1972: *Advanced soil physics.* Wiley, London, 534 pp.

Klimaszewski, M., 1961: *Geomorfologia ogolua.* Warsaw.

Klimontov. A.M., 1972: Karstowo-suffozjounyje jawlenija w priedgornoj zonie Siewiero-Wostocznogo Altaja. *Ucz. zap. Barnaulsk. gos. ped. un-t.,* 25: 75-78.

Knapp, B.J., 1970a: A note on throughflow and overland flow in steep mountain watersheds. *Reading Geographer* (University of Reading, England), 1: 40-43.

Knapp, B.J., 1970b: *Patterns of water movement on a steep upland hillside, Plynlimon, Central Wales.* Unpub. Ph.D. thesis, University of Reading, 213 pp.

Knapp, B.J., 1974: Hillslope throughflow observations and the problem of modelling. In *"Fluvial Processes in instrumented watersheds",* edited by K.J. Gregory and D.E. Walling, Institute of British Geographers, Special Publication 6: 23-33.

Knapp, B.J., 1979: *Hydrology in Physical Geography,* George Allen and Unwin, 85 pp.

Knight, K., 1959: The microscopic study of the structure of collapsing sands. *Proceedings of 2nd Southern African Regional Conference on Soil Mechanics and Foundation Engineering,* Lourenco Marques, pp. 69-72.

Kosack, H.P., 1952: Die Verbeitung der Karst und Pseudo-karsterscheinungen uber die Erde. *Petermans Geog.Mitt.,* 96 Jahr, 1st Quart: 16-21.

Kral, V., 1975: Sufoze a jeji podil na soucasnych geomor-pologickych procesech v Cechach. *Acta Universitatis Carolinae 1975/1-2 Geographica,* pp. 23-30.

Kratkaja Geographical Encyclopedia 1960 vol. 1 563 pp vol. 4 1964 448 pp., Moscow.

Kriechbaum, E., 1922: Studia nad morfologia lessu w poludni owej części powiatu chełmskiego. *Prz. geogr.,* No. 2., Warsaw.

Kühn, A., 1963: Geologiczno-Inzynierska charakterystyka obszaru katastrofalnych szkod budowlanych w Klodzku. *Inst. Geol. Biulet.* (Poland), 182(2): 27-43.

Kulandaiswamy, V.C., Sakthivadivel, R., and Thanikachalam, V., 1971: Mechanics of piping in cohesive soils. Report No. 6, Hydraulics and Water Resources Department, College of Engineering, Madras, India.

Kunsky, J., 1957: Types of pseudokarst phenomena in Czechoslovakia. *Czechoslovensky Kras,* 10(3): 111-125.

Landau, H.G., 1974: *Internal erosion of compacted cohesive soil.* Unpub. Ph.D. thesis, Purdue University.

Landau, H.G., and Altschaeffl, A.G., 1977: Conditions causing piping in compacted clay. In *"Dispersive clays, related piping, and erosion in geotechnical projects",* edited by J.L. Sherard and R.S.Decker, ASTM STP 623, p. 240-259.

Lane, E.W., 1935: Security from under-seepage: masonry dams on earth foundations. *Trans. American Society of Civil Engineers,* 100: 1257.

Lazinski, W., 1909: Prsyklad tworzenia sie doliny wskutek podziemnuch zapadiec w W. Ks. Krakowskim. *Spraw. Kom. Fizjogr.* Ak. Um. T. 43 Krakow.

Lear, D.L., 1976: *Some aspects of soil piping on the Dovey Marshes.* Unpub. B.Sc. dissertation, University College of Wales, Aberystwyth, 36 pp.

Legget, R.F., 1939: Soil and soil mechanics. In *"Geology and Engineering"*, McGraw-Hill, N.Y., p.538-583.

Lehr, H., and Stănescu, E., 1967: Investigation of the efficiency of safeguards against piping in earth dams. *Proceedings of 3rd Asian Regional Conference on Soil Mechanics and Foundation Engineering*, Haifa, 1: 152-154.

Leopold, L.B., and Langbein, W.B., 1962: The concept of entropy in landscape evolution. *U.S. Geological Survey Professional Paper*, 500-A, 20 pp.

Leopold, L.B. and Miller, J.P., 1956: Ephemeral streams: hydraulic factors and their relation to the drainage net. *U.S. Geological Survey, Professional Paper*, 282-A, 37 pp.

Leopold, L.B., and Wolman, M.G., 1960: River Meanders. *Geological Society America Bulletin*, 71: 769-794.

Leopold, L.B., Wolman, M.G., and Miller, J.P., 1964: *Fluvial processes in geomorphology*. Freeman and Co., San Francisco, 522 pp.

Lewin, J., Cryer, R., and Harrison, D.I., 1974: Sources for sediments and solutes in mid-Wales. In K.J. Gregory and D.E. Walling (ed.), *"Fluvial-processes in Instrumental Watersheds"*, British Geomorphological Research Group, Special Publication, 6, p. 73-85.

Lewis, D.A., and Schmidt, N.O., 1977: Erosion of unsaturated clay in a pinhole test. In *"Dispersive clays, related piping, and erosion in geotechnical projects"*, edited by J.L. Sherard and R.S. Decker, ASTM STP 623, p.260-273.

Liao, K.H., and Scheidegger, A.E., 1969: Branching-type models of flow through porous media. *Bull. Inst. Ass. Scientific Hydrology*, 14(4): 137-145.

Lilienberg, D.A., 1955: Forms of underground erosion of the relief on the S.E. Caucasus. *USSR Academy of Sciences Georgian Institute*, Pub. 65.

Linsley, R.K., and Franzini, J.B., 1964: *Water Resources Engineering*, McGraw-Hill, N.Y., 654 pp.

Linsley, R.K., Kohler, M.A., and Paulhus, J.L.H., 1949: *Applied Hydrology*, McGraw-Hill, New York.

Löffler, E., 1974: Piping and pseudokarst features in the tropical lowlands of New Guinea. *Erdkunde*, 28(1): 13-18.

Logan, J.M., 1959: Erosion problems on salt affected soils. *Journal of Soil Conservation Service, New South Wales*, 14(3): 229.

Lowdermilk, W.C., 1934: The role of vegetation in erosion control and water conservation. *Journal of Forestry,* 32: 529-536.

Lustig, L.K., 1970: Appraisal of research on the geomorphology and surface hydrology of desert environments. In *"Deserts of the World"*, edited by W.G.E. McGinnies, University of Arizona Press, 787 pp.

Luthin, J.N., 1966: *Drainage Engineering,* Wiley, New York, 250 pp.

Luthin, J.N., and Reeve, R.C., 1957: The design of a gravel envelope for tile drains. In *"Drainage of Agricultural Lands"*, edited by J.N. Luthin, American Society Agronomy, p. 339-344.

Maack, R., 1956: Karst-like forms resulting from climatic and structural conditions in the State of Parana. Resumés des Communications, 18th IGC Brazil, p.36-37.

MacFadyen, W.A., 1950: Soil and vegetation in British Somaliland. *Nature,* 165: 121.

Malicki, A., 1935: Przyczynek do znajomózci zjawisk krasowych w obszarce lessowym. *Czas. Geogr.,* 13.

Malicki, A., 1938: Kras gipsowy Podola Pokuckiego. *Prz. Geogr. (Lwówek),* 18.

Malicki, A., 1946: Kras loessowy. *Annales Universitatis Mariae Curie-Skłodowska, Lublin, Poland, Section B,* 1(4): 131-155.

Malinowski, J., 1963: Uwagi o wspol czynniku makroporowatosci lessọw w Polsce. *Inst.Geol.Biul.(Poland),* 182(2): 5-24.

Mandel, S., 1967: A conceptual model of karstic erosion by groundwaters. Proceedings Dubrovnik Symposium on *"Hydrology of Fractured Rocks"*, IASH - AIHS, 2: 662-664.

Manley, R.E., 1978: Simulation of flows in ungauged basins. *Hydrological Sciences Bulletin,* 23(1): 85-101.

Marker, M.E., 1958: Soil erosion in relation to the development of landforms in the Dundas area of western Victoria, Australia. *Proceedings Royal Society Victoria,* 71(2): 125-136.

Marker, M.E., 1976: Soil erosion 1955 to 1974: A review of the incidence of soil erosion in the Dundas Tableland area of Western Victoria, Australia. *Proceedings Royal Society Victoria,* 88(1 and 2): 15-22.

Marshall, A.F., and Workman, J.P., 1977: Identification of dispersive clays in the Texas Gulf Coast Area. In "*Dispersive clays, related piping, and erosion in geotechnical projects,*" edited by J.L. Sherard and R.S.Decker, ASTM STP 623, p. 274-286.

Martin, C.S., 1970: Effect of a porous sand bed on incipient sediment motion. *Water Resources Research,* 6(4): 1162-1174.

Maruszczak, H., 1965: Development conditions of the relief of loess areas in East-Middle Europe. *Geographia Polonica,* 6: 93-104.

McBeath, D.M., and Barron, C.N., 1954: Report on the ore deposit at Iron and Wamara Mountains, Berbice. *Report of Geological Survey Department, British Guiana.* for 1953, p. 87-100.

McCaig, M., 1979a: The pipeflow streamhead - a type description. Unpub. paper delivered to Institute of British Geographers Annual Conference, Manchester, January.

McCaig, M., 1979b: The pipeflow streamhead - a type description. *University of Leeds, School of Geography, Working Paper,* 242, 15 pp.

McCaskill, L.W., 1973: *Hold this Land.* A.H. and A.W. Reed, 274 pp.

McCullagh, P., 1978: *Modern concepts in Geomorphology,* Science in Geography. Series, No. 6, Oxford University Press.

McFarlane, M.J., 1969: *Lateritisation and landscape development in parts of Uganda.* Unpub. Ph.D. thesis, University of London.

McFarlane, M.J., 1976: *Laterite and Landscape.* Academic Press, New York, 151 pp.

McGee, W.J., 1911: Soil Erosion. *U.S. Department of Agriculture, Bureau of Soils, Bulletin,* No. 71, 60 pp.

McGinnies, W.G.E., (ed)., 1970: *Deserts of the World,* University of Arizona Press, 787 pp.

Mears, B., Jr., 1963: Karst-like features in badlands of the Arizona Painted Desert. *Wyoming University Contributions to Geology,* 2(1): 101-104.

Mears, B., Jr., 1968: Piping. In "*The Encyclopedia of Geomorphology*", edited by R.W. Fairbridge, Reinhold, New York, 1295 pp., p. 849-850.

Middleton, H.E., 1930: Properties of soil which influence soil erosion. *U.S. Department of Agriculture Technical Bulletin,* 178.

Ministry of Water Development (Rhodesia), 1967: Dam failure by piping. *Soil Testing Designs Branch Report,* September.

Mitchell, A., 1978: Development of soil conservation in Victoria. *Journal of Soil Conservation Service,New South Wales,* 34(2): 117-123.

Monteith, N.H., 1954: Problems of some Hunter Valley soils. *Journal of Soil Conservation Service, New South Wales, Australia,* 10: 127-134.

Morawetz, S., 1969: Kleinmorphologische Beobachtungen auf einer Haldenboschung. *Mitteil. der Österreich.Geog. Gesell.,* 111(2-3): 257-259.

Morgan, A.L., 1976: *An investigation of the location, geometry and hydraulics of ephemeral soil pipes on Plynlimon, Mid-Wales.* Unpub. BSc. dissertation, University of Manchester.

Morgan, R.P.C., 1972: Observations on factors affecting the behaviour of a first-order stream. *Trans. Institute British Geographers,* 56: 171-185.

Morgan, R.P.C., 1977: Soil erosion in the United Kingdom: field studies in the Silsoe area, 1973-75. *National College Agricultural Engineering, Silsoe, Bedfordshire, Occasional Paper* No. 5.

Morgan, R.P.C., 1979: *Soil Erosion.* "Topics in Applied Geography" Series, Longmans, London, 113 pp.

Morisawa, M., 1968: *Streams: their dynamics and morphology.* McGraw-Hill, New York, 175 pp.

Murdoch, G., 1968: Soils and land capability in Swaziland, Part 1. *Swaziland Ministry of Agriculture (Mbahane) Bullein* 23.

Musierowicz, A., 1951: *Gleboznawstwo ogolne.* Warsaw.

Neboit, R., 1971: Morphogenèse récente des formations tendres en Lucanie (Italie du Sud). *Méditerranée,* 2(7): 701-719.

Newman, J.C., and Phillips, J.R.H., 1957: Tunnel erosion in the Riverina. *Journal of Soil Conservation Service, New South Wales,* 13: 159-169.

Newman, P., 1976: *Sources, routes and velocities of runoff in a peaty catchment, Plynlimon, mid-Wales.* Unpub. BSc. dissertation, University College of Wales, Aberystwyth, 39 pp.

Newson, M.D., 1975: The Plynlimon Floods of August 5th/6th, 1973. *Natural Environment Research Council Institute of Hydrology, Wallingford, Berkshire, Report* No. 26, 58 pp.

Newson, M.D., 1976a: The physiography, deposits and vegetation of the Plynlimon catchments. *Natural Environment Research Council, Institute of Hydrology, Wallingford, Berkshire, Report* No. 30.

Newson, M.D., 1976b: Soil piping in upland Wales: a call for more information. *Cambria*, 1, 33-39.

Newson, M.D., 1980: The geomorphological effectiveness of floods - a contribution stimulated by two recent events in mid-Wales. *Earth Surface Processes*, 5: 1-16.

Newson M.D., and Harrison, J.G., 1978: Channel studies in the Plynlimon experimental catchments. *Natural Environment Research Council Institute of Hydrology, Wallingford, Berkshire, Report* No. 47, 61 pp.

Nickel, S.H., 1977: A rheological approach to dispersive clays. In *"Dispersive clays, related piping, and erosion in geotechnical projects,"* edited by J.L. Sherard and R.S. Decker, ASTM STP 623, p. 303-312.

Nogushi, T., Takahashi, R., and Tokumitsu, Y., 1970: Small sinking holes in limestone area with special reference to drainage of coal mines. IASH-UNESCO, *International Symposium on Land Subsidence*, 1969, Tokyo, IASH Pub. 89, 1: 467-474.

Northcote, K.H., 1971: A factual key for the recognition of Australian soils. Rellim Technical Pub., Glenside, South Australia.

Ollier, C.D., and Mackenzie, D.E., 1975: Subaerial erosion of volcanic cones in the Tropics. *Journal of Tropical Geography*, 39: 63-71.

Oxley, N.C., 1974: Suspended sediment delivery rates and the solute concentration of stream discharge in two Welsh catchments. In *"Fluvial Processes in Insturmented Watersheds"* edited by K.J. Gregory and D.E. Walling, British Geomorphological Research Group, Special Publication 6, 141-154.

Paaswell, R.E., 1973: Causes and mechanisms of cohesive soil erosion: The State of the art. *U.S. Highway Research Board, Special Report*, 135: 52-74.

Parizek, R.R., 1970: Land use problems in Pennsylvania's ground-water discharge areas in soils. *Pennsylvania Geological Survey Report*.

Parizek, R.R., 1971: Impact of highways on the hydrogeologic environment. In *"Environmental Geomorphology"*, edited by D.R. Coates, Stroudsborg, Pa., 151-199.

Parizek, R.R., 1973: Impact of highways on the hydrogeologic environment. In *"Environmental Geomorphology and Landscape Conservation"*, edited by D.R.Coates, Stroudsberg, Pa., 137-159.

Parker, G.G., 1963: Piping, a geomorphic agent in landform development of the drylands. *International Association Scientific Hydrology Publication* 65: 103-113.

Parker, G.G., and Jenne, E.A., 1967: Structural Failure of Western U.S. highways caused by piping. *U.S. Geological Survey Water Resources Division,* 27 pp.

Parker, G.G., Shown, L.M., and Ratzlaff, K.W., 1964: Officer's Cave, a pseudokarst feature in altered tuff and volcanic ash of John Day Formation in eastern Oregon. *Geological Society America Bulletin,* 75(5): 393-402.

Passarge, S., 1929: *Morphologie der Erdoberfläche,* Breslau.

Patz, M.J., 1965: Hill-top hollows - further investigations. *Uganda Journal,* 29(2): 225-228.

Pauli, H.W., 1978: Soil conservation in Queensland - Problems and progress. *Journal of Soil Conservation Service, New South Wales,* 34(2): 106-116.

Pavlov, A.P., 1898: Concerning the contour of plains and its change under the influence of the action of subterranean and surface waters. *Zemlevedeniye,* 5: 91-147. (In Russian.)

Payne, D., 1954: Some factors affecting the breakdown of soil crumbs on rapid wetting. *Transactions of 5th International Congress Soil Science,* 2: 53-58.

Pazzi, J.J.O., 1963: The building of farm dams. *South African Department of Agriculture, Technical Services Bulletin,* No. 365, Pretoria.

Pearsall, W.H., 1968: *Mountains and Moorlands.* Collins, London, 415 pp. (Original edition, 1950.)

Penck, W., 1924: *Morphological analysis of land forms.* English translation, by H. Czech and K.C. Boswell, Macmillan, London, 1953.

Perry, E., 1975: Piping in earth dams constructed of dispersive clay: literature review and design of laboratory testing. *U.S.Corps of Engineers, Technical Report. 5 - 75 -15,* U.S. Army Engineer Waterways Experiment Station, Vicksburg, Mississippi.

Peterson, D.E., 1962: *Earth fissuring in the Picacho area, Pinal Co., Arizona*. Unpub. MS thesis, University of Arizona.

Peterson, H.V., 1954: Discussion of 'Piping" by Fletcher et al., *Transactions American Geophysical Union*, 35(2): 263.

Peterson, R., and Iverson, N.L., 1953: Study of several low earth dam failures. *Proceedings of 3rd International Conference Soil Mechanics and Foundation Engineering*, 2: 273.

Pethick, J.S., 1971: *Salt Marsh Morphology*. Unpub. Ph.D. thesis, University of Cambridge, 314 pp.

Philips, J.T., 1977: Case histories of repairs and designs for dams built with dispersive clays. In *"Dispersive clays, related piping, and erosion in geotechnical projects"*, edited by J.L. Sherard and R.S.Decker, ASTM STP 623, p.330-340.

Pinczes, Z., 1968: Vonales erozio a Tokaji-hegy loszen. *Foldrajzi Kozlemenyek*, 16(2): 159-171.

Pitty, A.F., 1971: *Introduction to Geomorphology*. Methuen, London, 526 pp.

Pond, S.F., 1971: The occurrence and distribution of subsurface pipes in upland catchment areas. NERC Institute of Hydrology Subsurface Section, internal report No. 47, unpub. typescript.

Purves, W.D., and Blyth, W.B., 1969: A study of associated hydromorphic and sodic soils on redistributed Karroo sediments. *Rhodesian Journal of Agricultural Research*, 7: 99-109.

Quinlan, J.F., 1966: Classification of karst and pseudo-karst types. *Geological Society America Special Papers*, 101: 448-450.

Quirk, J.P., and Schofield, R.K., 1955: The effect of electrolyte concentration on soil permeability. *Journal of Soil Science*, 6(2): 163-178.

Radley, J., 1962: Peat erosion on the high moors of Derbyshire and West Yorkshire. *East Midland Geographer*, 3(17): 40-50.

Rallings, R.A., 1964: Failure modes - review of selected case histories. *Proceedings Colloquium on Failure of small earth dams, CSIRO, Soil Mechanics Section, Melbourne*.

Rallings, R.A., 1966: An investigation into the causes of failure of farm dams in the Brigalow Belt of Central Queensland. *Water Research Foundation of Australia Bulletin,* No. 10.

Ranganathan, B.V., and Zacharias G., 1968: Interaction of density, soil type and time on piping resistance of cohesive soils. *Proceedings of 3rd Budapest Conference on Soil Mechanics and Foundation Engineering,* pp. 237-246.

Rao, K.L., 1960: Failures of earth dams. *Proceedings of 1st Asian Regional Conference, International Society of Soil Mechanics and Foundation Engineering,* New Delhi.

Rathjens, C., 1973: Subterrane Abtragung (Piping). *Zeit. für Geomorph.,* Suppl. Bd. 17: 168-176.

Redlich, K.A., Terzaghi, K., and Kampe, R., 1929: *Ingenieursgeologie.* Springer, Vienna.

Rees, D.H., 1979: *An investigation into the possible significance of soil piping to modelling studies.* Part I Postgraduate Course in Water Resources Technology Project Reports, Department of Civil Engineering, University of Birmingham, 151 pp.

Reniger, A., 1950: Proba oceny nasilenia i zasiegow potencjalnej erozji gleb w Polsce. *Poczniki Nauk Rolniczych (Polish Agricultural Annual),* 54: 1-59.

Reséndiz, D., 1977: Relevance of Atterberg limits in evaluating piping and breaching potential. In *"Dispersive clay, related piping, and erosion in geotechnical projects",* edited by J.L. Sherard and R.S. Decker, ASTM STP 623, p. 341-353.

Reuss, R.F., and Schattenberg, J.W., 1972: Internal piping and shear deformation. Victor Braunig Dam, San Antonio, Texas. *Proceedings ASCE Special Conference on Performance of Earth and Earth Supported structures,* 1: 627-651.

Richards, L.A. (ed.), 1954: *Diagnosis and improvement of saline and alkali soils.* U.S. Department of Agriculture, Handbook No. 60, Washington, D.C., 160 pp.

Riley, P.B., 1977: Dispersion in some New Zealand clays. In *"Dispersive clay, related piping, and erosion in geotechnical projects",* edited J.L. Sherard and R.S. Decker, ASTM STP 623, p. 354-361.

Ritchie, J.A., 1963: Earthwork tunnelling and the application of soil-testing procedure. *Journal of Soil Conservation Service, New South Wales,* 19(3): 111-129.

Ritchie, J.A., 1965: Investigations into earthwork tunnelling and mechanical control measures using small scale model dams. *Journal of Soil Conservation Service, New South Wales,* 21(2): 80-89.

Road Research Laboratory (U.K.), 1957: *Soil Mechanics for Road Engineers,* HMSO, (Second edition 1964, 541 pp).

Rodda, J.C., Downing, R.A., and Law, F.M., 1976: *Systematic Hydrology,* Newnes-Butterworth, London.

Rosewell, C.J., 1967: Soil investigations into the control of failure in earthworks used in soil erosion control. *Proceedings of Soil Conservation Colloquium,* University of New England, Armidale.

Rosewell,C.J., 1970: Investigations into the control of earthwork tunnelling. *Journal of Soil Conservation Service, New South Wales,* 26(3): 188-203.

Rosewell, C.J., 1977: Identification of susceptible soils and control of tunnelling failure in small earth dams. In *"Dispersive clays, related piping, and erosion in geotechnical projects",* edited by J.L. Sherard and R.S. Decker, ASTM STP 623, p.362-369.

Rowell, D.L., 1963: Effect of electrolyte concentration on the swelling of orientated aggregates of montmorillonite. *Soil Science,* 96: 368.

Ruxton, B.P., 1958: Weathering and sub-surface erosion in granite at the piedmont angle, Balos, Sudan. *Geological Magazine,* 95(5): 353-377.

Ryker, N.L., 1977: Encountering dispersive clays on Soil Conservation Service projects in Oklahoma. In *"Dispersive clays, related piping, and erosion in geotechnical projects",* edited by J.L. Sherard and R.S. Decker, ASTM STP 623, p. 370-389.

Rubey, W.W., 1928: Gullies in the Great Plains formed by sinking of the ground. *American Journal of Science,* 15: 417-422.

Saurin, E., 1935: Etudes géologiques sur l'Indochine du Sud-Est. *Bulletin Service Geological Indochine,* 22(1), 419 pp.

Saurin, E., 1944: Etudes géologiques sur le Centre Annam meridional. *Bulletin Service Geologique Indochine,* 27, 210 pp.

Sautter, G., 1951: Note sur l'erosion en cirque des sables au Nord de Brazzaville. *Bulletin Inst. Et. Centraf.,* N.S. no. 2: 49-61.

Savarensky, F.P., 1940: Significance of the works of A.P. Pavlov for engineering geology. *Bulletin Moscow Society Scientists, Geological Division,* 18(3-4): 69-74. (In Russian.)

Scheidegger, A.E., 1970: *Theoretical Geomorphology.* Prentice Hall, second edition. (First edition, 1961.)

Schomer, R., 1953: Trockenschluchten (Owragi) in der Ukraine. *Annales Universitatis Saraviensis (Saarbrucken), Sciences,* 2: 118-123.

Schumm, S.A., 1960: The shape of alluvial channels in relation to sediment type. *USGS Professional Paper,* 325-B.

Segalen, M.P., 1949: L'érosion des sols à Madagascar. *Bulletin Agric. Congo Belge,* 40(2): 1127-1137.

Sencu, V., 1970: Karstic phenomena in tectonical breccia. *Rev.Rouman. Geol. Geophy. et Geogr.,* 14(2): 277-280.

Sharpe, C.F.S., 1938: What is soil erosion? *U.S. Department of Agriculture Miscellaneous Pub.,* 286: 61.

Sherard, J.L., 1953: Influence of soil properties and construction methods on performance of homogeneous earth dams. *U.S.Department of Interior Bureau of Reclamation, Design and Construction Division, Technical Memo.* No. 645.

Sherard, J.L., 1971: Study of piping failures and erosion damage from rain in clay dams in Oklahoma and Mississippi. Unpub. report prepared for USDA Soil Conservation Service, Washington, D.C.

Sherard, J.L., and Decker, R.S., (Editors), 1977: *Dispersive clays, related piping, and erosion in geotechnical projects.* American Society for Testing Materials, Special Technical Publication No. 623, 486 pp. Including Introduction (p.1-2) and Summary (p.467-479) by Sherard and Decker.

Sherard, J.L., Decker, R.S., and Ryker, N.L., 1972a: Piping in earth dams of dispersive clay. *Proceedings, ASCE Spec. Conference on Performance of Earth and Earth-supported structures,* 1: 589-626.

Sherard, J.L., Decker, R.S., and Ryker, N.L., 1972b: Hydraulic fracturing in low dams of dispersive clay. *Proceedings ASCE Spec. Conference on Performance of Earth and Earth-supported structures,* 1: 653-689.

Sherard, J.L., Dunnigan, L.P., and Decker, R.S., 1976: Identification and nature of dispersive soils. *Journal of Geotechnical Engineering Division, Proceedings ASCE,* 102(GT4): 287-301.

Sherard, J.L., Dunnigan, L.P., and Decker, R.S., 1977: Some engineering problems with dispersive clays. In *"Dispersive clays, related piping, and erosion in geotechnical projects"*, edited by J.L. Sherard and R. S. Decker, ASTM STP 623, p. 3-12.

Sherard, J.L., Woodward, R.J., Gizienski, S.F., and Clevenger, W.A., 1963: *Earth and earth-rock dams*. Wiley, New York.

Sherman, L.K., 1932: Streamflow from rainfall by unit-graph method. *Engineering News Record,* 108: 501-505.

Shiobara, T., 1970: Consideration about the compaction mechanism of stratum lying at the deeper horizon in Tokyo lowland. IASH-UNESCO Int. Symp. on Land Subsidence, 1969, Tokyo IASH Pub. No. 89, 1: 315-321.

Shreve, R.L., 1972: Movement of water in glaciers. *Journal of Glaciology,* 11(62): 205-214.

Skvaljeckij, E.N., Halmatov, E., and Hasanov, I.R., 1971: Sinking karst-suffosion phenomena on the shores of the Selburski Reservoir. In "Symposium on the lithology and genesis of loess", *Fan(Tashkent)*: 135-144.

Sleeman, J.R., 1965: Cracks, peds and their surfaces in some soils of the Riverina Plain, New South Wales. *Australian Journal of Soil Research,* 1(1): 91-100.

Smith, D.I., and Stopp, P., 1979: *The river basin.* Cambridge University Press.

Smith, H.T.U., 1949: Periglacial features in the Driftless Area of South Wisconsin. *Journal of Geology,* 57: 196-215.

Smith, H.T.U., 1953: The Hickory Run boulder field, Carbon County, Pennsylvania. *American Journal of Science,* 251: 625-642.

Smith, H.T.U., 1968: "Piping" in relation to periglacial boulder concentrations. *Biul. Periglacialny,* 17: 195-204.

Smith, H.T.U., 1969: *Photo-interpretation studies of desert basins in Northern Africa.* Geology Department, University of Massachusetts, for Air Force Cambridge Research Laboratories, USAF, Bedford, Massachusetts, AFCRL-68-0590, 77 pp.

Smith, T.R., and Bretherton, F.P., 1972: Stability and conservation of mass in drainage basin evolution. *Water Resources Research,* 8: 1506-1524.

Soil Conservation and Rivers Control Council, New Zealand, 1944: The menace of soil erosion in New Zealand. *Bulletin No. 1*, 16 pp.

Sokolov. D.S., 1960: On the content and meaning of the concept of 'karst'. *Zemlevedeniye*, 5.

Stagg, M.J., 1974: *Storm runoff in a small catchment in the Mendip Hills*. Unpub. M.Sc. thesis, University of Bristol.

Stagg, M.J., 1978: Rill patterns derived from air photographs of the Grwyne Fechan catchment, Black Mountains. *Cambria*, 5(1): 22-36.

Stapledon, D.H., and Casinader, R.J., 1977: Dispersive soils at Sugarloaf Dam site, near Melbourne, Australia. In *"Dispersive clays, related piping, and erosion in geotechnical projects"*, edited by J.L. Sherard and R.S. Decker, ASTM STP, p. 432-466.

Starkel, L., 1960: Rozwój rzezby Karpat fliszowych w holocenie. *Prace geogr.*, IGPAN 22, Warsaw, ss. 293.

Starkel, L., 1972a: The role of catastrophic rainfall in the shaping of the relief of the Lower Himalayas (Darjeeling Hills). *Geogr. Polonica*, 21: 103-147.

Starkel, L., 1972b: The modelling of monsoon areas of India as related to catastrophic rainfall. *Geogr. Polonica*, 23: 151-173.

Starkel, L., 1976: The role of extreme (catastrophic) meteorological events in contemporary evolution of slopes. In *"Geomorphology and Climate"*, edited by E. Derbyshire, p. 203-246.

Statham, I., 1977: *Earth surface sediment transport*. Contemporary Problems in Geography Series, Oxford University Press.

Statton, C.T., and Mitchell, J.K., 1977: Influence of eroding solution composition on dispersive behaviour of a compacted clay shale. In *"Dispersive clays, related piping, and erosion in geotechnical projects"*, edited by J.L. Sherard and R.S. Decker, ASTM STP 623, p. 398-407.

Steers, J.A., 1960: *Scolt Head Island*. Heffers, Cambridge, 269 pp.

Stephens, C.G., 1962: *A manual of Australian soils*. CSIRO, Melbourne, Australia, 61 pp. Third edition.

Steward, V.I., Adams, W.A., and Abdulla, H.H., 1970: Quantitative pedological studies on soils derived from Silurian mudstones. The parent material and the significance of the weathering process. *Journal of Soil Science,* 21: 242-247.

Stocking, M.A., 1972: Relief analysis and soil erosion in Rhodesia using multivariate techniques. *Zeit. für Geomorph.,* 16: 432-443.

Stocking, M.A., 1976a: *The erosion of soils on Karroo Sands in central Rhodesia with particular reference to gully form and process.* Unpub. Ph.D. thesis, University of London.

Stocking, M.A., 1976b: Tunnel erosion. *Rhodesia Agricultural Journal,* 73(2): 35-39.

Stocking, M.A., 1978a: The measurement, use and relevance of rainfall energy in investigations into erosion. *Zeit. für Geomorph.,* Supplement Band, 29: 141-150.

Stocking, M.A. 1978b: The prediction and estimation of erosion in subtropical Africa: problems and prospects. *Géo - Eco - Trop.* 2: 161-174.

Stocking, M.A., 1978c: A dilemma for soil conservation. *Area,* 10(4): 306-308.

Stocking, M.A., 1979: A catena of sodium-rich soil. *Journal of Soil Science,* 30(1): 139-146.

Subakov, V.M., Nemchinova, E.E., and Sumochkina, T.E., 1967: Karst-suffosion phenomena in the Karshinski Steppe. *Izvest. Uzbekistan Geogr.,* 10: 79-84. (In Russian.)

Sweeting, M.M., 1972: *Karst Landforms.* Macmillan, London, 362 pp.

Taylor, N.H., 1938: Land deterioration in the heavier rain districts of New Zealand. *New Zealand Journal of Science and Technology,* 19(11): 657-681.

Temple, P.H., and Rapp, A., 1972: Landslides in the Mgeta area, Western Uluguru Mountains, Tanzania. *Geografiska Annaler,* 54A(3-4): 157-194.

Terzaghi, K., 1922: Der Grunbruch an Stauwerken und seine Verhütung. *Forch-heimer - Nummer der Wasserkraft,* p. 445.

Terzaghi, K., 1931: Earth slips and subsidences from underground erosion. *Engineering News Record,* 107: 90-92.

Terzaghi, K., 1943: *Theoretical Soil Mechanics.* Chapman and Hall, London/J. Wiley, New York, 510 pp.

Terzaghi, K., and Peck, R.B., 1966: *Soil Mechanics in Engineering Practice,* Wiley, New York (first ed. 1948).

Thomas, A.S., 1960: The tramping animal. *Journal of British Grassland Society,* 15: 89-93.

Thomas, M.F., 1968: Some outstanding problems in the interpretation of the geomorphology of tropical shields. *BGRG Occasional Paper,* 5: 41-49.

Thomas, T.M., 1956: Gully erosion in the Brecon Beacons, South Wales. *Geography,* 41: 99-107.

Thorbecke, F., 1951: *Im Hochland von Mittel-Kamerun.* Teil 4, halfe 2: Physische Geographie des Ost-Mbamlandes, University of Hamburg, Abhandlungen aus dem Gebiet des Auslandskunde.

Thorp, J., 1936: *Geography of the soils of China.* Nat. Geological Survey of China.

Tomlinson, R.W., 1980: Water levels in peatlands and some implications for runoff and erosional processes. In *"Geographical Approaches to Fluvial Processes",* edited by A.F. Pitty, Geo Books, pp. 149-162.

Tricart, J., 1965: *Principles et Methodes de la Géomorphologie,* Masson and Co., Paris, 496 pp.

Trudgill, S.T., Laidlaw, I.M.S., and Smart, P.L., 1980: Soil water residence times and solute uptake on a dolomite bedrock - preliminary results. *Earth Surface Processes,* 5: 91-100.

Truesdale, V.W., and Howe, J.M., 1977: Water level recorder and weir tank. In "Selected measurement techniques in use at Plynlimon experimental catchments". *Institute of Hydrology Report* No. 43, pp. 33-37.

U.S. Bureau of Reclamation, 1947: Laboratory tests on protective filters for hydraulic and static structures. *Earth Material Laboratory Report,* EM - 132, Denver, Col.

U.S. Department of Agriculture, 1960: Soil classification a comprehensive system; 7th Approximation. Govt. Printing Office, Washington D.C.

Van Olphen, H., 1963: *An introduction to clay colloid chemistry.* Interscience Publishers, New York.

Viessman, W., Jr., Knapp, J.W., Lewis, G.L., Harbaugh, T.E., 1977: *Introduction to Hydrology.* Second Edition, IEP-Dun-Donnelley, New York, 704 pp.

Villegas, F., 1977: Experiences with tests of dispersibility of fine-grained soils for dams in Colombia. In *"Dispersive clays, related piping, and erosion in geotechnical projects",* edited by J.L. Sherard and R.S. Decker, ASTM STP 623, p. 408-418.

Volk, G.M., 1937: The method of determination of degree of dispersion of the clay fraction of soils as used in the investigation of abnormal characteristics of soils in region 8 of the Soil Conservation Service. *Soil Science Society of American Proceedings,* 2: 561-565.

von Richthofen, F., 1886: *Fuhrer für Forschungsreisende.* Hanover.

Walczowski, A., 1962: Utwory czartorzedowe w okolicach Rakowa i Łagowa. *Kwart. Geol. (Warsaw),* 6(3).

Walczowski, A., 1971: Procesy suffozji w okolicach Pacanowa. *Biul. Inst. Geol. (Poland),* 9(242): 105-135.

Ward, A.J., 1966a: Pipe/shaft phenomena in Northland. *Journal of Hydrology (New Zealand),* 5(2): 64-72.

Ward, A.J., 1966b: *Mass movement and interrelated phenomena in the Paroti Area of Northland.* Unpub. MSc. thesis, University of Auckland, 111 pp.

Ward, R.C., 1975: *Principles of Hydrology.* McGraw Hill, London, Second edition.

Ward, R.C., 1978: *Floods.* Macmillan, London.

Ward, W.H., 1948: A slip in a flood defence bank constructed on a peat bog. *Proceedings 2nd International Conference of Soil Mechanics and Foundation Engineering,* Rotterdam, 2: 19-23.

Watson, R.A., and Wright, H.E., Jr., 1963: Landslides on the east flank of the Chuska Mountains, northwestern New Mexico. *American Journal of Science,* 261: 525-548.

Waylen, M.J., 1976: *Some aspects of the hydrochemistry of a small drainage basin.* Unpub. Ph.D. thesis, University of Bristol, 386 pp.

Wendelaar, F.E., 1976: Field identification of sodic soils. *Rhodesia Agricultural Journal,* 73(3): 77-82.

Weyman, D.R., 1970: Throughflow on hillslopes and its relation to the stream hydrograph. *Bulletin International Association Scientific Hydrology,* 15(2): 25-33.

Weyman, D.R., 1971: *Surface and subsurface runoff in a small basin.* Unpub. Ph.D. thesis, University of Bristol, 243 pp.

Weyman, D.R., 1973: Measurements of the downslope flow of water in the soil. *Journal of Hydrology,* 20: 267-288.

Weyman, D.R., 1974: Runoff processes, contributing area and streamflow in a small upland catchment. In K.J. Gregory and D.E. Walling (ed.): *"Fluvial processes in instrumented watersheds"*, British Geomorphological Research Group Special Publication, 6, p. 1433-1443.

Weyman, D.R., 1975: *Runoff processes and streamflow modelling,* Oxford University Press.

Weyman, D.R., and Weyman, V., 1977: *Landscape processes: an introduction to geomorphology.* George Allen and Unwin, London.

Whipkey, R.Z., 1965: Subsurface stormflow from forested slopes. *Bulletin Internation Association Scientific Hydrology,* 10(3): 74-85.

Whipkey, R.Z., 1969: In *"Floods and their computation Vol. II"*, International Association Science Hydrology Publication 85, 2: 773-779.

Whipkey, R.Z., and Kirkby, M.J., 1978: Flow within the soil. In *"Hillslope Hydrology"*, edited by M.J. Kirkby, Wiley, 389 pp.

White, E.M., and Agnew, A.F., 1968: Contemporary formation of patterned ground by soils in South Dakota. *Geological Society America Bulletin,* 79(7): 941-944.

White, L.P. and Law, R., 1969: Channelling of alluvial depression soils in Iran and Sudan. *Journal of Soil Science,* 20(1): 84-90.

Wieckowska, H., 1953: Zjawiska suffozyjne w okolicy Nidzicy. *Prz. geol.* (Warsaw), 25.

Wigwe, G.A., 1975: The laterite landscape of the Share area of Ilorin, Nigeria. *Journal of Tropical Geography,* 38: 61-78.

Wilson, C.M., 1977: *The generation of storm runoff in an upland catchment.* Unpub. Ph.D. thesis, University of Bristol.

Wolman, M.G., and Gerson, R., 1978: Relative scales of time and effectiveness of climate in watershed geomorphology. *Earth Surface Processes,* 3: 189-208.

Wolski, W., et al., 1970: Protection against piping of dams, cores made of flysch origin cohesive soil. *Trans 10th International Congress on Large Dams,* 1: 575-585.

Wood, C.C., Aitchison, G.D., and Ingles, O.G., 1964:
    Physicochemical and engineering aspects of piping
    failures in small earth dams. *Paper No. 29, Water
    Research Foundation of Australia Ltd., and Soil
    Mechanics Section, C.S.I.R.O., Colloquium on Failure
    of Small Earth Dams, 16-19 November, 1964,* 13 pp.

Woodward, H.P., 1961: A stream piracy theory of cave
    formation. *National Speleological Society (America)
    Bulletin,* 23(2): 39-58.

Worrall, G.A., 1959: The Butana Grass patterns. *Journal
    of Soil Science,* 10: 34-53.

Worrall, G.A., 1960: Patchiness in vegetation in the
    northern Sudan. *Journal of Ecology,* 48: 107-115.

Wright, H.E. Jr., 1964: Origin of the lakes in the Chuska
    Mountains, Northwestern New Mexico. *Geological Society
    America Bulletin,* 75: 589-598.

Yair, A., 1973: Theoretical considerations on the evolut-
    ion of convex hillslopes. *Zeit. für Geomorph.,*
    Supplement Band, 18: 1-9.

Yair, A., and Lavee, H., 1976: Runoff generative process
    and runoff yield from arid talus mantled slopes.
    Earth Surface Processes, 1: 235-247.

Yair, A., Bryan, R.B., Lavee, H., and Adar, E., 1980:
    Runoff and erosion processes and rates in the Zin
    Valley badlands, Northern Negev, Israel. *Earth Surface
    Processes.*

Yapp, R.H., Johns, D., and Jones, O.T., 1917: The salt
    marshes of the Dovey estuary: Part II The salt marshes.
    Journal of Ecology, 5(2), 65-103.

Yong, R.N., and Sethi, A.J., 1977: Turbidity and zeta
    potential measurements of clay dispersibility. In
    *"Dispersive clays, related piping, and erosion in geo-
    technical projects",* ASTM STP 623, p. 419-431.

Young, A., 1972: *Slopes,* Longmans, London, 296 pp.

Zaborski, B., 1926: O zjawiskach podobnych do krasowych w
    lessach. Ksiega Pamiatkowa XXI Zjazdu Lekarzy i
    Przyrodnikow Polskich w roku 1925 r T.I. Warszawa.

Zaborski, B., 1972: On the origins of gullies in loess.
    Acta Geographica Debrecina, (1971), 10: 109-111.

Zaslavsky, D., and Kassiff, G., 1965: Theoretical formulat-
    ion of piping mechanism in cohesive soils. *Geotechnique,*
    15(3): 305-316. (Also appeared as: Fac. of Civ.
    Engg., Israel Institute of Technology, Haifa, Pub.
    No. 40, 1964.)

Zeitlinger, J., 1959: Observations of sub-surface erosion in loam. *Mitt. der Österreich Geog. Gesell.*, 101(1): 94-95.

Zolnierczyk, A., 1956: Zjawiska suffozji w gruniach. *Gosp. Wodna*, 16: 12.

Abdulla, H.H.   87
Adams, W.A.   87
Adar, E.   5, 27, 52, 73, 87,
  88, 154, 195-197, 203, 206,
  207, 215, 216
Aghassy, J.   47, 51, 200
Agnew, A.F.   21
Aisenstein, B.   28, 96
Aitchison, G.D.   22, 27-30,
  38, 55, 66, 76, 89, 90,
  96, 97, 104, 105, 109, 111,
  112
Alizadeh, A.   103
Allen, R.L.   92
Alpaidze, V.S.   26, 209
Altschaeffl, A.G.   100,
  104, 113
Anderson, M.G.   15, 179,
  235-237
Antonescu, I.P.   48
Arnett, R.R.   153
Arulanandan, K.   97, 103
Atkinson, T.C.   10, 117,
  138, 145, 155, 157, 161
Aubert, G.   18, 228

Baillie, I.C.   7, 26, 38,
  42, 53, 67, 72, 84, 85,
  207, 228
Bally, R.I.   48
Banerjee, A.K.   42, 51,
  214, 228
Barendregt, R.W.   27, 33,
  34, 39, 41, 50, 72, 86,
  88, 151, 195, 206, 214
Barron, C.N.   18, 227
Bayer, A.W.   23, 37, 38,
  44, 217, 218
Beare, J.A.   6, 38
Beattie, J.A.   232
Beckedahl, H.R.   23, 38,
  43, 52, 74, 78, 132, 208
Bell, G.L.   5, 10, 27, 29,
  31, 37, 40, 41, 51, 54,
  55, 66, 70, 72, 88, 217,
  226, 229, 233
Bell, J.P.   15, 73, 82,
  119, 186, 187, 209
Bennett, H.H.   9, 11, 18,
  39, 41, 199
Berg, L.S.   8, 24
Berry, L.   23, 24, 37, 38,
  42, 43, 52, 117, 153, 198
Bertram, G.E.   28
Beven, K.   158, 178, 236

Bishop, W.W.   23, 24, 37, 38,
  224
Black, C.A.   65, 105
Black, R.D.   2, 182
Blench, T.   51
Blong, R.J.   4, 5, 11, 37, 39,
  40, 48-50, 52, 70, 73, 93,
  114, 117 200, 214
Blyth, W.B.   58
Bogutski, A.B.   26, 207
Bond, R.M.   10, 18, 39, 42, 43
Bosazza, V.L.   23, 38
Bowden, D.J.   54, 228
Bower, M.M., see Cruickshank,
  M.M.
Bremer, H.   11, 26, 38, 86
Bretherton, F.P.   237
Brewer, R.   43, 81, 94
Broscoe, A.J.   132
Brown, G.W.   26, 31, 37, 39,
  45, 55, 56, 66, 71, 72, 74,
  78, 80, 87, 88, 239, 240
Brunsden, D.   155
Bryan, K.   1, 17, 35
Bryan, R.B.   5, 27, 39, 52, 73,
  87, 88, 133, 154, 195-197,
  203, 206, 207, 215, 216
Buckham, A.F.   9, 10, 17, 31,
  37, 41, 51, 72, 133, 198,
  199
Büdel, J.   4
Bunting, B.T.   200, 209
Buraczynski, J.   25
Burt, T.P.   179, 235-237
Burton, C.K.   85
Burton, J.R.   48
Butzer, K.   5, 132

Calver, A.   237, 238
Campbell, D.A.   20
Campbell, I.A.   196
Capper, P.L.   28
Carder, D.J.   229
Carmichael, K.M.   48
Carroll, P.H.   9, 11, 12, 18,
  23, 26, 29, 31, 39, 41, 42,
  44, 45, 54, 67, 70, 71, 77,
  78, 86, 88, 89, 95, 106,
  117, 133, 151, 230, 240
Carson, M.A.   4, 5, 94, 237
Casagrande, A.   28
Casinader, R.J.   46
Cassie, W.F.   28
Cedergren, H.   28
Chandler, V.N.   9, 18, 31-33,
  38, 39, 43, 55, 74, 79, 88,
  119, 240

Yong, R.N. 101, 239
Young, A. 5

Zaborski, B. 6, 24, 25, 198,
   206
Zacharias, G. 241

Zaslavsky, D. 71, 97, 102,
   110, 241
Zeitlen, J.G. 97, 110
Zeitlinger, J. 39, 52
Zolnierczyk, A. 25
Zötl, J. 26, 213

SUBJECT INDEX

Channel bank, stream (unit 8), piping, 25, 50, 51, 73,
    107-109, 117-121, 133-136, 186, 212, 213
Channel bed, stream, (unit 9), piping, 51, 107-109, 199, 200
Charcos, 17
Chezy formula, 159
Chiselling, 231
Chloride ions, 100, 217-222
Chlorite, 87, 88
Classification (pipes), 1, 9-15, 19, 94, 95, 182-187,
    234-236
Clay minerals, see also species, 67, 86-92, 97, 102-104
Clay skins, see *argillans, cutans*
Clay soils, 7, 8, 10, 12, 17, 20, 21, 29-31, 44, 46, 53,
    56, 58, 59, 66, 67, 70, 74, 76, 78, 85, 96-106, 110,
    112, 226, 234
Climate (general), 20, 34-41, 72
Climatic change, 41, 58
Collecting areas, 21, 23, 169, 182, 186, 207
Colluvial footslope (unit 6), 49-52, 185
Colebrook-White equation, 127, 129
Compaction, 17, 22, 112
Concentration times (pipe), 170, 172
Conductivity, electrical, 57, 58, 63, 64, 121, 215, 217,
    219
Conductivity, hydraulic, see also *permeability*, 67, 78-86,
    105, 159, 163, 194, 236
Conservation, 20-22, 44, 59, 229-233
Correlation analyses, 115, 116, 119-121, 123-125, 130, 158,
    161, 163, 166, 167, 202, 217
Contour banks, 48, 231
Contour ploughing, 48
Contributing area, see *dynamic contributing area*
Cracks/cracking potential, 12, 13, 15, 21, 25, 29, 31-33,
    40, 41, 44, 46, 53, 55, 63, 71, 75, 78, 79, 82, 86-93,
    234
Creep ratio, 77, 109
Cross-sections (pipe), see also *shape*, 19, 21, 116-130,
    132, 174, 178, 228
Crumb test (Emerson), 239, 242
CSIRO Soil Mechanics Section, see also authors, 22, 27-29,
    96, 239
Cutans, 105
Cyclical salts, 58, 61, 234

Dams, see *earth dams*
Darcy's Law, 107, 112
Darcy-Weisbach equation, 113, 127, 155
Deflocculation, 9, 58, 64, 96-106
Department of Agriculture, Tasmania, see also authors, 229
Desert soils, see also *aridisols*, 226
Desiccation-stress crack piping, 12, 13, 23, 31-33, 35, 40,
    53, 54, 67, 106, 145, 152, 198, 202, 208, 234
Devegetation, 31, 32, 39, 43-46, 53, 230
Dilution experiments, 170, 172
Discharge (pipe), see *pipeflow*
Dispersal Indices, see also *aggregate stability*, 56, 59,
    60-62, 74, 90, 98, 213, 239-242

Dispersive clays, 9, 12, 13, 21, 24, 27-29, 54, 63, 71, 74, 76, 198, 201, 234, 240, 241
Dispersion, 20, 41, 48, 55, 58, 59, 62, 74-77, 86-92, 95-106, 201, 239-242
Doline, 206
Drainpipes, artificial, 43, 47, 84, 123, 136
Draws, 17
Droughts, 31, 33, 34, 37, 39, 40, 45, 55, 82, 149, 173, 234
Drylands, see *arid and semi-arid regions*
Duplex soils, see also soil groups, *podzolic, solodic, solonetz,* 61, 77, 84
Duricrust, 18, 54, 85, 105, 227-229
Dynamic contributing area, 158, 170, 173, 174, 189, 191, 196, 236, 237
Dynamic contributing volume, 184, 185, 194
Dye tracing, 152, 169, 173, 184, 189, 195

Earth dams (piping in), 12, 22, 28, 29, 48, 55, 59, 63, 73, 74, 76, 89-92, 96-113, 231, 239-242
Entisols, 52, 60, 62, 68, 69, 72, 73, 85
Entrainment piping, 12, 105, 107, 208, 210
Erodibility, 32, 44, 47, 48, 54, 66, 67, 71, 74-78, 102, 103, 202-206, 239-242
Erosion cycles (piping), 26, 51, 136, 150, 200, 224-226, 233
Evorsion, 206
Exchangeable sodium percentage (E.S.P.), 19, 29, 32, 33, 54-63, 75, 79, 84, 96-106, 201, 214, 215, 233, 234, 240
Expansion rates (clay), 19, 23, 31, 60, 74, 75, 87, 240
Experiments, see *field experiments, laboratory experiments, modelling*

Ferrallitic soils/ferralsols, 23, 42, 85, 86, 214, 228
Ferricrete, 85, 228
Fersiallitic duricrust, 228
Field experiments, see also *reclamation experiments,* 22, 52, 158-197, 215, 217, 226
Floods, related to piping, 18, 20, 25, 27, 44, 51, 154, 158, 161, 168, 189, 210, 227
Flushes, 138, 172, 179, 183, 184, 189, 210, 223
Fragipan (Btg), 136
Freezing, 25, 41, 54, 208
Friction factor (Darcy Weisbach), 122, 126, 127, 129, 178

Gerlach troughs, 213
Gilgai, 21
Glacial pseudokarst, 8, 35, 150
Gley soils, 115, 133, 136, 192, 210, 213, 214
Great Soil Groups (Australia), 60-62, 66-69
Gullying, see also *arroyos,* 2, 4, 5, 9-13, 15-18, 20, 23, 24, 34, 43, 50-52, 58-60, 70, 72, 75, 77-79, 89, 109, 132, 133, 151-153, 173, 198-206, 210, 217, 224, 225, 228, 230, 231, 233
Gullying, control of, 198, 231
Gypsum/gypsiferous soils, 26, 55, 57, 58, 80, 96, 199, 215
Gypsum treatment, 22, 231

Springs, 1, 19, 24, 28, 45, 95, 116, 179, 182, 209, 213,
   235
Spring sapping, 209
Squirrel-hole gully erosion, 10
Stepped crescents, 18
Streambank piping, see *channel bank*
Stream incision, see *rejuvenation*
Subcutaneous erosion, 9, 11, 19, 20, 44
Subsidence, 11, 12, 50, 88, 105, 106, 134, 145, 152,
   207-211, 234
Subsurface storm flow, see also *throughflow,* 1, 2, 4
Sulphate ions, 55, 100, 221, 222
Swelling capacity, see *expansion rates*
Swelling clays, see also clay species, 33, 34, 40, 55, 63,
   79, 87-91, 98, 99, 101, 107, 113, 234

Temperature, see also *climate,* 32, 35, 37, 40, 67
Texture, see *soil texture*
Thresholds (gullying) 202, 205, 206
    "   (mass movement) 207, 210
    "   (pipeflow) 175-179, 187-188, 237
    "   (piping) 52, 53, 55-58, 64, 66, 67, 70, 75, 79,
   81, 84, 87-113, 240
    "   (soil stability) 76, 94-113, 234
Throughflow, 1, 2, 4, 94, 96, 101, 153-155, 158-161, 174,
   179, 181, 182, 187, 196, 206-208, 220
Tomos, 11
Total dissolved salts (TDS), 65, 90, 92, 98, 99, 223
Transportational mid-slope (unit 5), 46, 49, 51
Triaxial tests, 110
Tunnelling, 9*ff*
Tunnel-gully erosion, 9, 10, 20, 77, 199, 237
Turbulent flow, 129, 155, 187, 195, 234

Ultisols, see *ferralsols*
Unconsolidated sediments, 7, 8, 24, 34, 47, 52, 55, 70,
   85, 89, 196, 209, 224, 226
Unit weights, 88, 107
U.S. Army Waterways Experiment Station (Vicksburg), 29, 241
U.S. Bureau of Reclamation, 76, 240
U.S.D.A. Soil Conservation Service, see also authors, 10,
   18, 20, 22, 27, 29, 47, 74, 87, 199, 239, 241

Vadose zone piping, 31, 228, 234-236
Van der Waals force, 99-101
Variable permeability-consolidation piping, 12, 13, 31,
   106, 208, 234
Velocities (pipeflow), see *pipeflow*
Vertisols, see also *gilgai,* 12, 84, 85, 234
Void ratio, 108, 109
Volcanic ash, 7, 8

Water Research Foundation of Australia, see also authors,
   28
Water stable aggregates, see *aggregate stability*
Wither Hills Reserve Experiment, 20.